HORIZONS IN DNA RESEARCH

HORIZONS IN DNA RESEARCH, VOLUME 1

Horizons in DNA Research

Horizons in DNA Research. Volume 1
Jason R. Chesterton (Editor)
2010. ISBN: 978-1-60876-968-1

HORIZONS IN DNA RESEARCH

HORIZONS IN DNA RESEARCH, VOLUME 1

JASON R. CHESTERTON
EDITOR

Nova Science Publishers, Inc.
New York

LIBRARY OF CONGRESS CATALOGING-IN-PUBLICATION DATA

Available upon request.
ISBN: 978-1-60876-968-1

Published by Nova Science Publishers, Inc. † New York

CONTENTS

PREFACE

This book presents original research results on the leading edge of DNA research. Each article has been carefully selected in an attempt to present substantial research results across a broad spectrum.

Chapter 1 - Carbon nanotubes are of considerable interest in the development of biosensors for detection of DNA hybridization and damage, for affinity biosensors based on DNA/RNA aptamers or antibodies and for detection pathogens. Due to their unique properties, the carbon nanotubes can serve as an efficient support for immobilisation of nucleic acids and proteins. This chapter consists of an introduction into the structural and physical properties of carbon nanotubes and their applications in biosensing. The focus is on development biosensors for detection of DNA hybridization, single point mutations and polymorphisms as well as on affinity biosensors based on proteins and on DNA/RNA aptamers. Special focus on DNA aptamers is due to the unique properties of these new biopolymers that demonstrate substantial advantage over antibodies. While the sensitivity of aptamer-based biosensors is similar or even higher in comparison with that of antibodies, the aptamers can be regenerated without lost of affinity. Aptamers are rather flexible, which allow the applying of simple molecular engineering methods to construct aptamers with improved sensitivity. Immobilisation of aptamers on carbon nanotubes further improve the biosensor properties, which is reflected in better detection limits for various analytes in comparison with biosensors based on conventional immobilisation methods.

Chapter 2 - DNA analysis using advanced molecular biological tools is universally used to unlock identifying information contained in crime scene stains. These advances have allowed scientists to provide critical information to solve crimes and allow the criminal justice community to convict the guilty and exonerate the innocent. Scientific research continues to develop new automated technologies and methods to yield more information from limited samples allowing scientists to optimize time and effort. Although the changes realized over the last 20 years have been remarkable, only small changes have occurred in the identification of the tissue source of a biological stain. Considerable research has been conducted to offer an alternative to these classical identification methods through the use of mRNA analysis. The application of mRNA into the forensic laboratory has not yet taken hold, but offers a new approach that could unequivocally identify difficult to characterize stains and allow the use of molecular biological tools for the identification of stains found at a crime scene. The nature of the research in this chapter is to identify mRNA transcripts that will definitively identify the tissue of origin, determine if such transcripts survive the typical environmental insults that forensic samples may encounter, and develop rapid assays using small amounts of sample. The stability of RNA was evaluated over time using real-time TaqMan®-based PCR assays. Once RNA was shown to be stable in samples aged up to 4 years, 2-3 tissue-specific transcripts were identified for a variety of stains. In addition to specificity, the sensitivity of the assays was also determined using different sized stains. Once candidates were shown to be tissue-specific, the Plexor® One-Step qRT-PCR System was used to develop multiplex assays. This system allows the multiplexing of multiple RNAs in one assay, thus reducing the amount of sample needed and time of analysis. Through collaboration with Promega, a quick, one-tube seminal fluid-saliva screening assay was designed, since analysis of these fluids is a major task for all forensic laboratories. Additionally, a RNA/DNA co-isolation technique was developed which effectively extracts both nucleic acids, thus removing the requirement for separate extractions and allows DNA profiling from the exact stain in which RNA testing was performed. In conjunction with Promega, a likely outcome is a real-time PCR screening assay that determines with certainty whether a stain contains seminal fluid and/or saliva, an invaluable tool to the forensic community.

Chapter 3 - Malignant cells exhibit metabolic alterations affecting growth, lack to respond to physiological stimuli and blockade of differentiation and apoptosis. Selective growth advantage on cells acquired by transformation influences biosynthetic enzymes and alters the internal pools of nucleotides. During the course of studying the induction of erythroid differentiation of murine erythroleukemia (MEL) cells, we observed that agents that inhibit the active methylation cycle, (N^6-methyladenosine, neplanocin A, 3-deazaneplanocin A and cycloleucine), blocked initiation of commitment, erased “memory” response to previous exposure to chemical inducers, altered the expression profile of specific genes and abrogated hemoglobin production in inducer-treated MEL cells presumably by interrupting a central process required for terminal erythroid maturation. In an effort to delineate the mechanism(s) by which such methylation inhibitors abrogate the developmental program of MEL cells, we have shown that exposure to these methylation inhibitors led to cytoplasmic accumulation of relatively small discrete RNA transcripts lacking poly(A) tail. These RNAs were hybridized with ^{32}P-labeled probes derived from 3'-end flanking sequences of the mouse β^{major} globin gene. Structural and functional analysis of this DNA region have revealed a B1 repeat element as well as consensus sequences recognized by transcription factors GATA-1, AP1/NF-E2 and erythroid Krüppel-like factor (EKLF). Successful cloning, characterization and identification of genomic DNA sequence encoding these RNA species would shed some light on how do the active methylation cycle inhibitors activate their biosynthesis and subsequently whether these RNA molecules contribute to the blockade of differentiation of MEL cells by hypomethylating agents.

Chapter 4 - Mitochondria, the powerhouse of the cell, play a critical role in multiple metabolic processes as well as in the apoptotic pathways. Many lines of evidence suggest that mitochondria and its DNA (mtDNA) have a central role in ageing-related neurodegenerative diseases. Moreover, there is a long history of investigations on mitochondria aimed at identifying genetic markers that would relate to aging or neurodegenerative diseases. In this chapter, we highlight some of the major neurodegenerative disorders and discuss the role of the mtDNA variations and haplogroups in the pathogenetic cascade leading to these diseases.

Chapter 5 - In terms of forensic-specific applications of DNA technology, few have had a more profound effect on the field than the launching of databases around the world. The expansion of this phenomenon has not only required an appropriate level of scientific sophistication, but also unprecedented legislative, political and financial backing. In practical terms DNA database implementation and operation has brought massive increases in case submissions and a demonstrable change in case and evidence profiles. Forensic DNA databases also remain high profile and retain some element of controversy, typically in relation to socio-legal issues. Despite this impressive impact, it is important to acknowledge that forensic DNA database applications are still young, still only partially developed and understood and still requiring of our attention, monitoring and effort to ensure an ongoing positive contribution.

Fresh complexities are emerging in the management and development of DNA databases. Some of these problems have been examined in detail in this chapter, and, new solutions that allow cross-comparison of database performance are discussed. These methods should assist, not only in identifying present strengths and weaknesses but also in enhancing future database performance and effectiveness. Effective and creative database management has to be seen as a key priority for the forensic community. Achieving it will require a concerted effort of oversight that scientists themselves must be prepared to participate in as it is crucial to maximising the benefits of their work.

Forensic science has traditionally focussed on the production of evidentiary outcomes – retrospectively, and with a view to confirming or refuting an already-held belief. DNA databases also offer, however, the potential for proactive forensic contributions to investigations. Delivering these requires functional and philosophical adjustments to the process by which forensic outcomes are produced. This chapter also outlines specific steps that need to be taken to progress beyond the traditional operational framework to one that is also conducive to effective policing and crime resolution.

Chapter 6 - Obtaining reliable animal models of human OXPHOS (oxidative phosphorylation) diseases caused by accumulation of mutant mtDNA in cells of certain tissues might be a noticeable step towards understanding their pathogenesis. Despite some attempts, so far no successful approach to the targeted delivery of mutant mtDNA to certain tissues/organs has been elaborated in experiments on animals. We have elaborated a model that allows studying some features of foreign mtDNA distribution at different stages of embryonic development and in organs of transmitochondrial mice developed from zygotes injected with human mitochondria. All pre-implantation embryos and some neonate animals were carrying copies of human mitochondrial genome along with murine mtDNA, which is an indication of artificial heteroplasmy. This feature persisted in the progeny of transmitochondrial animals obtained from injected zygotes i.e. among the mice in generations F_1-F_3 at least 14% were human mtDNA-carriers inheriting foreign genome along the maternal lineage. Only some transmitochondrial embryos had human mtDNA in all the blastomeres. Likewise it was found only in some organs of transmitochondrial mice. Statistical elaboration and Monte Carlo simulation taking into account Markovian process showed, firstly, that human mtDNA was unevenly distributed in zygotic cytoplasm within the time between the injection of human mitochondria and the first cleavage; secondly, that the number of mtDNA segregation units in the entire amount of mitochondria injected (ca. $5 \cdot 10^2$) is about ten. Analysis of human mtDNA distribution among the organs of transmitochondrial animals showed that skeletal muscles, liver and gonads are the targets frequently hit.

Chapter 7 - Trace DNA has become a large part of the average forensic laboratories' workload. Remarkably low DNA amounts (<100pg) have been successfully analysed to obtain profiles from a wide range of sample types. Touched objects constitute the most common source of trace DNA, but any type of biological material present in low amounts may be considered as trace, including minute blood deposits, saliva residue on partially consumed food, or even epithelial cells from the interior surface of condoms. In addition to supplementing existing analysis techniques in serious crime cases, trace DNA can allow investigation of volume crime cases such as burglary or vehicle theft, where previously DNA evidence was not considered usable. However, despite the widespread use of trace DNA, at present there are very few specific validated methods. This has lead to controversy in the use of trace DNA, and particularly the low copy number amplification technique. It has been established that the use of existing methodology (developed for high-copy number samples) leads to significant levels of artefacts with trace DNA, including allele drop-out and drop-in, stutter, and allelic/locus imbalance. To minimize these, there are numerous modifications that can be made to existed methods to increase the success of trace DNA analysis. These include reduced extraction volumes, increased cycle number, reduced PCR volume, and increased injection time for capillary electrophoresis. In addition, new research shows that the introduction of techniques such as whole genome amplification, molecular crowding, and post-PCR purification can significantly increase success rates with trace DNA. It is clear that each step of the analysis procedure (including collection, extraction, amplification and fragment detection) can, and should, be optimized with regards to trace DNA. However, the use of increasingly sensitive trace DNA analysis techniques must bring an increasing awareness of the potential for contamination, both within the laboratory and at crime scenes. In particular, this has implications for the analysis of trace DNA from cold cases, which were not collected or stored with highly sensitive DNA detection techniques in mind. Although it is certain that trace DNA will continue to be used within forensic biology, there may need to be modifications to a wide range of practices to ensure accuracy and reliability.

Chapter 8 - Mitochondrial DNA has been used for the purposes of forensic identification since the mid 90s. Since then dedicated laboratories have been set-up and guidelines for the analysis and interpretation of mitochondrial DNA have been established. The most common method of analysis has been to amplify and sequence the mitochondrial DNA to generate a mitochondrial profile. This method has been applied to species identification used on forensically important animal species, and human identification whereby the mitochondrial DNA is used for forensic identification of low copy number DNA and degraded remains. The use of mitochondrial DNA does not allow discrimination to an individual like nuclear DNA as it can only identify maternally related individuals. However this has the potential for the exclusion of an individual in forensic identification.

Chapter 9 - The production of DNA-wrapped single-wall carbon nanotubes (SWNTs) in aqueous solution has led to a bid advance for separation or isolation of SWNTs. The understanding of the effect of DNA-wrapping on physical properties of SWNTs is important for basic understanding of DNA-SWNT interaction as well as engineering applications of DNA-wrapped SWNTs. Raman scattering of SWNTs are significantly influenced by DNA wrapping. The Raman spectra of DNA-wrapped SWNTs in air strongly depend on relative humidity. In addition, the DNA-wrapping mechanism for smaller diameter SWNTs is different from that for larger ones. Such properties of DNA-wrapped SWNTs can provide further applications in air. In this chapter, first, the production of DNA-wrapped SWNTs is described. Next, the Raman spectral features of DNA-wrapped SWNTs in aqueous solution and air are reviewed. Finally, the morphologies of DNA-wrapped SWNTs are discussed.

Chapter 10 - The knowledge of the haplogroups of the analyzed subjects allows to prevent errors in the analysis of the association between mtDNA variations and some pathologies like diabetes. In this paper a cohort of type 2 diabetic Italian patients with known mitochondrial genetic background (haplogroup) were investigated for the presence in the blood of the 3243A>G mutation and other variants of close to $tRNA^{Leu(UUR)}$ gene (16S/ND1 region) of mitochondrial DNA (mtDNA). The 3243A>G mutation was present in 0.4% of analyzed patients in agreement with the reported frequencies of such mutation in other European groups. This result suggests a limited role for the 3243A>G mutation in the pathogenesis of diabetes also in the Italian population. In the 16S/ND1 region 15 base changes were found. The knowledge of the haplogroup of the analyzed individuals allowed to distinguish between haplogroup-specific polymorphisms and variants potentially associated to diabetes.

Chapter 11 - Mitochondria is the major supplier of intracellular ROS (Radical oxygen species). Since the proposal of free radical theory by Harman in 1956, there has been a long dispute about the impact of mitochondria on longevity. Recently it was reported that several mouse models of mitochondria mutant display the phenotype of premature ageing or age-related disease. Moreover ectopic expression of mitochondria-targeted radical scavenger extends organismal life span in mice

These suggest that mitochondria dysfunction intriguingly correlates with ageing and senescence. However, opposed or different ideas on the role of mitochondria on longevity are also proposed in the research of yeast genetics. Here we review the recent advance on molecular link between mitochondria and senescence. We propose that high efficient mitochondria model might explain clearly the importance of mitochondria on ageing and senescence.

In: Horizons in DNA Research, Volume 1
Editor: Jason R. Chesterton, pp. 1-48
ISBN: 978-1-60876-968-1

Chapter 1

CARBON NANOTUBES IN NUCLEIC ACIDS AND AFFINITY BIOSENSORS

Tibor Hianik[1*] and Gennady Evtugyn[2]

[1]Department of Nuclear Physics and Biophysics, Comenius University, Mlynska dolina F1, 842 48 Bratislava, Slovakia
[2]Analytical Chemistry Department, Kazan State University, 18 Kremlevskaya Street, Kazan, 420008, Russian Federation

ABSTRACT

Carbon nanotubes are of considerable interest in the development of biosensors for detection of DNA hybridization and damage, for affinity biosensors based on DNA/RNA aptamers or antibodies and for detection pathogens. Due to their unique properties, the carbon nanotubes can serve as an efficient support for immobilisation of nucleic acids and proteins. This chapter consists of an introduction into the structural and physical properties of carbon nanotubes and their applications in biosensing. The focus is on development biosensors for detection of DNA hybridization, single point mutations and polymorphisms as well as on affinity biosensors based on proteins and on DNA/RNA aptamers. Special focus on DNA aptamers is due to the unique properties of these new biopolymers that demonstrate substantial advantage over antibodies. While the sensitivity of aptamer-based biosensors is similar or even higher in comparison with that of antibodies, the aptamers can be regenerated without lost of affinity. Aptamers are rather flexible, which allow the applying of simple molecular engineering methods to construct aptamers with improved sensitivity. Immobilisation of aptamers on carbon nanotubes further improve the biosensor properties, which is reflected in better detection limits for various analytes in comparison with biosensors based on conventional immobilisation methods.

* *Corresponding author: e-mail: tibor.hianik@fmph.uniba.sk, Tel: +421-2-60295683; Fax: +421-2-65426774;

Keywords: Carbon nanotubes; DNA; RNA; Aptamers; Immunoglobulins; Field effect transistors; Biosensors.

1. INTRODUCTION

Development of biosensors for label-free detection of various compounds, such as proteins, toxins, pathogens and other molecules, has been growing thanks to recent achievements in nanotechnology. The highly sensitive and selective biosensors are urgently needed for fast and low cost diagnosis in medicine, for detection of food-borne pathogens in the food industry, for detection of environmental pollutants, and for fast control of the quality of the water. The biosensor consists of two basic parts: the sensing layer and the transducer. The sensing layer is composed of receptors such as proteins (enzymes and antibodies), nucleic acids, calixarenes, or even whole cells and tissues. The analyte primarily interacts with the sensing layer, usually in a complex liquid. The role of the sensing layer is specific recognition of the analyte. This interaction is typically of a chemical nature and results in the formation of the analyte-receptor complex, which then initiates further chemical or conformational changes. For example, complex antigen-antibody resulted in increase of thickness or mass of the sensing layer as well as in changes of its electrical properties that is reflected by changes in impedance characteristics. The formation of the complex of analyte with enzyme resulted in degradation of the analyte into the product that may have different properties in comparison with the analyte. For example, this product could have certain redox properties, which allow its detection amperometrically. The transducer transforms the chemical or complexation signal to the physical value that can be measured by standard measuring techniques and analyzed using a computer.

It has been shown that the immobilisation of receptors to the transducer surface is one of the crucial steps in the development of biosensors. Various materials have been examined for this purpose, including carbon paste, carbon roods, glassy carbon, noble metals (gold, silver, platinum), various composite materials including conducting polymers doped by various redox probes, proteins or polymer layers, nanoparticles and in recent years also carbon nanotubes (CNTs). Thanks to unique structural, electrical and mechanical properties of CNTs, they have been widely used in preparation of various biosensors. Among biosensors, those based on gene detection, detection of nucleic acid damage and polymorphism and those based on DNA/RNA aptamers and antibodies are of special interest.

This chapter focuses on an introduction into the physical and structural properties of carbon nanotubes and their application in developed DNA biosensors. Special focus is given to the aptamer-based biosensors. Aptamers are single-stranded DNA or RNA with high specificity to various ligands. Their specificity is comparable and in certain cases even higher than those of antibodies. Recent achievements in improvement of the sensitivity and selectivity of aptamers-based biosensors (aptasensors) are connected with using CNTs as a support for aptamer immobilisation. Current progress in this field is presented in this chapter together with comparison of the sensitivity of aptasensors and antibody-based immunosensors.

2. Carbon Nanotubes, Structure and Properties

Carbon nanotubes (CNTs) have been discovered by Iijima about two decades ago [Iijima 1991]. Since this time, the experimental work with these novel materials has grown substantially [Allen et al. 2007]. The structure of CNTs can be described as a graphene sheet rolled up into a nanoscale tube, alone or with additional tubes. Depending on the number of layers, single-wall and multi-wall carbon nanotubes are defined (SWCNTs and MWCNTs, respectively). Both their ends are normally capped with fullerene like semi-spheres, mainly with pentagon defects. The conditions of CNTs synthesis, mainly the nature of catalysts, temperature and carbon source, determine the number of graphene layers and formation of SWCNTs or MWCNTs.

Three approaches are generally used for CNTs synthesis, i.e., arc discharge, laser ablation, and chemical vapor deposition (CVD) [Dai 2001; Merkoci et al. 2005]. In arc discharge, carbon atoms are evaporated by helium plasma from solid-state precursors at about 3000–3500°C. MWCNTs are obtained by controlling the pressure of inert gas and the arcing current. One advantage of the arc-discharge method is that it does not require metal nanoparticle catalysts whose persistent and residual presence within the MWCNTs can alter behavior of MWCNTs resulting in its misleading interpretation. SWCNTs are synthesized in arc discharge in the presence of transient metals like catalysts.

In laser ablation, intense laser pulses are used to ablate a carbon target placed in a tube-furnace heated to 1200°C and blown with inert gas passed through the growth chamber. Both methods provide gram quantities of nearly perfect CNTs.

Only a few defects, e.g., pentagons or heptagons, are observed on the sidewalls of the nanotubes obtained by arc discharge and laser ablation. As-synthesized CNTs contain many impurities, including amorphous carbon, graphitic particles, and metal catalysts. The purification protocols include combinations of filtration, sonication and oxidation. However, these processes also damage the structural integrity of the CNTs by creating defects on the graphene shells [Monthioux et al. 2001].

In CVD, hydrocarbon gases (preferably ethylene or acetylene) or carbon monoxide are heated and passed through the tube furnace with iron, nickel or cobalt catalysts on alumina support at 550–700°C. Hydrocarbon molecules are dissociated and carbon atoms precipitated onto metal particles to form tubular carbon solids in sp^2 structure. As an example, commercially available HiPCO nanotubes are obtained with high pressure CVD utilizing CO as carbon source and Ni catalyst.

Transmission electron microscopy study of MWCNTs synthesized at different reactor temperatures showed that the nanotube outer diameter directly correlated to the catalyst particle size [Sinnott et al. 1999]. Aligned MWCNTs can be obtained on an iron catalyst placed in mesoporous silica or on a glass substrate at plasma assistance at 700 °C [Li et al. 1996; Fan et al. 1999].

CVD became the preference in production of CNTs in the past decade. Variation of the substrate and deposition conditions offers opportunities for directed regulation of the size and geometry of CNTs. Thus, highly aligned suspended SWCNTs were grown under electric fields on the order of about 1 V/μm. The induced dipole led to a large aligning torque directing CNTs parallel to the electric field. The electric field alignment effect is more stable against thermal fluctuations at the growth temperature [Zhang et al. 2001].

CVD synthesis provides high purity of the product, but significantly lower temperatures comparing with those of arc or laser production methods, which tend to produce CNTs with less defined graphene structure. The post-treatment including heating CNTs in an inert atmosphere to 1800–2600 °C is recommended to remove many of the structural defects and residual iron catalyst [Andrews et al. 2001].

The diameters of SWCNTs and MWCNTs vary from 0.4 to 3 nm (1.4 nm is most stable) and from 2 to 100 nm, respectively [Kim et al. 2007]. In MWCNTs, the graphene layers are separated by a distance of about 0.34 nm [Cohen 2001]. Lengths up to several micrometers are possible by controlling CNTs growth. Oxidative treatment frequently used for chemical modification and dispersion of CNTs results in shortening CNTs and decreasing the number of graphene layers [Liu et al. 1998; Gooding et al. 2003]. SWCNTs show the surface area up to 1600 m^2 g^{-1} [Cinke et al. 2002]. As-produced MWCNTs give the surface area between 200 and 400 m^2 g^{-1}. Chemical treatment with KOH increases this value to about 1000 m^2 g^{-1} [Frackowiak et al. 2002]. It was shown, that the CNTs could be fractured and shortened by ball milling [Gao et al. 2000; Pierard et al. 2001], mechanical cutting [Stepanek et al. 2000] and chemical etching [Ajayan et al. 1993; Liu et al. 1998].

Variety of the features of CNTs mainly considered in sensory devices, i.e. electronic conductivity, electrocatalytic activity and electron emission, is due to their specific structure and influence of defects and impurities. The latter one can be sufficiently altered by post-treatment of CNTs after their synthesis [Charlier 2002].

Comparing with infinite flat graphene sheet, π-electron clouds in SWCNTs exhibit significant curvature that results in partial σ-π hybridization [Blasé et al. 1994]. The C–C bond length is elongated by the curvature from 1.41 Å (graphene) depending on the infinite radius of curvature. The value of 1.44 Å (fullerene) is an upper limit from general consideration.

Among others, curvature leads to the dependence of electric conductivity, absorbance and reactivity of CNTs on the diameter, number of layers and their chirality. Even for the tubes of the same diameter, the plurality of the structure takes place depending on the convolution of a roll. Quantitatively, it can be expressed by the chirality angle, θ, as a function of a pair of integers (n, m), $\theta = \tan^{-1}[3^{1/2}m/(m+2n)]$. The pair (n, m) ascribes the number of lattice units in graphene sheet with sp^2 hybridized carbon atoms that represents repeating part of a SWCNT [Dresselhaus and Eklund 2000]. The n and m integers are the characteristic pair values for each kind of a SWCNT. Depending on them, SWCNTs show metallic (zero bandgap, $n = m$), semi-metallic (bandgap ca. 10 m eV, $n–m = 3k$, where k is an integer) or semi-conducting features (0.6 eV and above, $n - m \neq 3k$). The chirality angle θ determines also stable conformations commonly used for description of SWCNTs geometry, e.g. zigzag ($\theta = 0^o$) or armchair ($\theta = 30^o$) (see Figure 1 as an example).

Great attention is paid to developing methods for selective synthesis of uniform CNTs with selected size, electronic properties and chirality [Bandow et al. 1998; Bower, et al. 1998; Yudasaka et al. 1998; Katauta et al. 1999]. Nevertheless, available samples of CNTs show diversity of the features with rather random distribution in chirality [Odom et al. 1998].

The as-grown semiconducting SWCNTs can be considered as hole-doped p-type field effect transistors (FETs) with hole depletion and diminished conductance (from typically 100 kΩ to 1 MΩ) by positive gate voltages. Molecular oxygen adsorbed on the CNTs is responsible for hole doping [Collins et al. 2000; Chen et al. 2001a]. Other doping molecules as well as charge transfer can highly affect their electrical properties. Metallic SWCNTs are less variable in behavior and less sensitive toward intercalating of low-molecular compounds.

The electron-hole and exitonic interactions weaken the heterogeneity of SWCNTs items so that different chirality and modality of SWCNTs becomes significant only for tube diameter of about 1-2 nm. This might be important for the development of nanoscaled electronic devices like CNTs based transistors [Keren et al. 2003; Allen et al. 2007]. Pentagonal defects of the CNTs sidewalls produce regions with increased charge density, either in planar or fullerene-like caps. There are evidences about the connections of topological defects with electroactivity [Britto et al. 1999; Banks et al. 2005] and electric conductivity of CNTs. The defects followed by dramatic changes in the CNTs parameters can be caused by electron emission [Huang et al. 2004], currents applied for conductivity measurements [Bonard et al. 2003] or by mechanical tip in atomic force microscopy (AFM). Additionally, strain and bends of SWCNTs can reversibly change their conductivity by two orders of magnitude [Dai 2002].

MWCNTs exhibit even higher diversity of features against SWCNTs that originated both from structure and defects significantly altered by specific treatment, e.g. oxidation or functionalization of terminal groups. Within general structure, several morphological variations exist, e.g. "hollow tube", “bamboo-like”, and "herringbone". The tubes remain either opened and the graphite sheets are placed parallel to the MWCNT axis (hollow tube) or offset at a slight angle to the axis ("herringbone"). Another morphology, "bamboo-like", has a similar internal structure to the herringbone MWCNTs except that the tubes are periodically closed off into compartments, similar to the structure of the bamboo plant. The MWCNTs ends have similar properties to the edge-plane face of a graphite crystal; the tube walls can be described as "basal-plane" [Banks et al. 2005; Banks and Compton 2006]. Vacuum annealing at temperatures above 500°C can again close the open ends.

The edge-plane-like sites on the surface of the MWCNTs are decorated with oxygen-containing functional groups, i.e. hydroxyl/phenol, quinonyl and carboxyl groups are almost always present [Masheter et al. 2007; Thorogood et al. 2007; Holloway et al. 2008]. Under certain conditions, redox catalysis by these groups can enhance the rate of electron transfer. But, commonly it is considered negligible in comparison with that through the fullerene like tips.

Although many of the structure defects provide extended possibilities for the following functionalization, various specific techniques for implementation of reactive groups were elaborated to simplify the inclusion of CNTs in the assembly of sensor devices. They can be subdivided into two major groups, i.e. tips opening and sidewall functionalization. The fullerene-like caps of CNTs can be rather easily oxidized in the mixtures of mineral acids or on air at high temperatures to form carboxylic functions which can be then involved in the following functionalization by treating with appropriate condensing agents. Covalent modification of CNTs involves direct fluorination, addition of organic radicals, fluorine displacement in fluorinated nanotubes, the reactions of etherification, amidation, Schiff base formation with terminal carboxylic groups and alkylation with metal organic derivatives (see [Daniel et al. 2007] and references cited therein). Particular examples of covalent modification of CNTs are given below in the part 3.1. considering appropriate DNA sensors

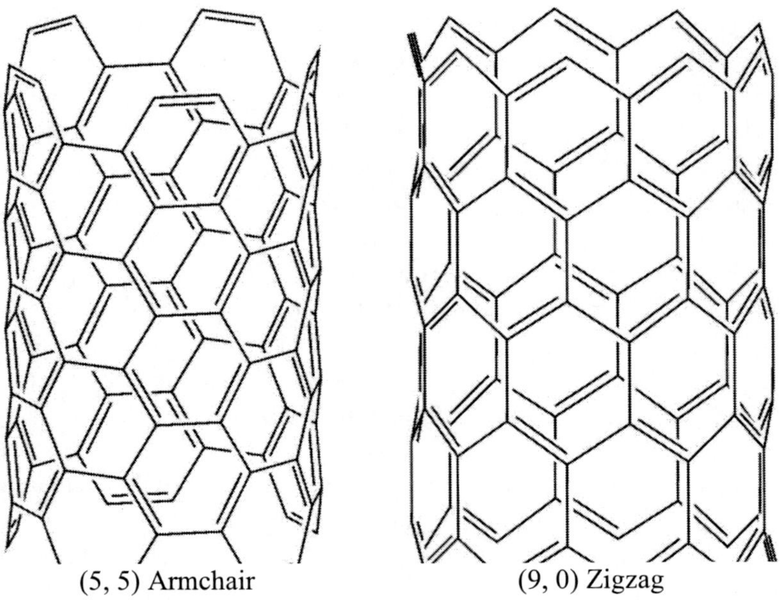

Figure 1. The structure of SWCNTs in two different conformations. The numbers are integers (n,m), respectively.

Oxidation is a most common start for further functionalization. Upon sonication in strong acids, oxidation of the CNTs is initiated from the sidewall defects on the bundle surfaces and the tips, where the reactivity is higher. The reaction then propagates throughout the bundles. This defect-assisted mechanism is supported by the structure of the etched tubes and the overall weight change.

Non-covalent sidewall functionalization is mainly based on weak interactions (hydrogen bonds, π-π stacking and electrostatic forces etc.). Physicosorption was observed for some proteins interacting non-covalently with sidewalls of CNTs [Daniel et al. 2007]. Also, inner volume of CNTs can be filled with some species ranging in size from proteins to small molecules [Kohli and Martin 2005; Taherpour 2009].

Self-assembling of CNTs can be considered as a way of their functionalization because the aggregates obtained show promising features in wiring biomolecules and their targeted adsorption on appropriate transducer interface. SWCNTs form self-assembled layers, which are obtained by a method similar to Langmuir-Blodgett deposition. In this technique, aqueous dispersion of pre-treated SWCNTs is first prepared and then a glass plate is immersed vertically into the suspension. The CNTs are deposited on the substrate along the air/liquid/substrate triple line. The films obtained are smooth and have stepwise sharp edges straight at the substrate/film interface. Electron microscopy studies revealed the SWCNTs bundles are ordered with longitudinal axes lying on the substrate surface and along the air/water/ substrate triple line direction [Shimoda et al. 2002].

In electrophoretic deposition, a dilute dispersion of SWCNTs in organic solvent, e.g. DMF or alcohol, contained a "charger", $MgCl_2$, is used. In direct current electric field, the SWCNTs migrate toward the substrate. The deposition rate depends on the concentration of the SWCNT suspension and the charger and the current. The same techniques can be used to deposit CNTs onto more complicated structures with conducting surfaces [Gao et al. 2001].

MWCNTs can be either grown by CVD directly onto the transducer interface to form ordered layers and arrays of aligned tubes [Wang et al. 2004] or abrasively attached to glassy carbon or basal plane pyrographite electrodes by mechanical treatment [Salimi et al. 2004].

CNTs are characterized by a limited number of physical methods. Among them, scanning tunneling microscopy and transmission electron microscopy can only be used for characterization of CNTs at individual level. Raman spectroscopy is mainly used for detecting chirality, degree of p- and n-doping and sidewall defects as well as interaction of CNTs with DNA molecules [Dresselhaus et al. 2002; Wang 2005]. Photoluminescence excitation and near-IR absorption spectra [O'Connell et al. 2002] provide mapping metallic and semi-metallic SWCNT units. Transmission electron microscopy, electron diffraction and scanning probe microscopy play important role in determining CNT length, diameter of nanotube bundles, and features of aggregate organization along with CNT association with biological and nanostructured materials.

3. Application of Carbon Nanotubes in Biosensors

Due to unique structural, electrical and mechanical properties the CNTs are widely used in development of various biosensors. This was summarized in several reviews [Katz and Willner 2004; Lin et al. 2004; Wang 2005; Wildgoore et al. 2006; He et al. 2006; Wanekaya et al. 2006; Balasubramanian et al. 2006; Allen et al. 2007; Wang and Lin 2008; Lin et al. 2009, Ahammad et al. 2009]. In this chapter we will focus on review of the works dealing with development DNA biosensors, DNA/RNA aptamers based biosensors and immunosensors. The other types of CNTs based biosensors, such as enzyme biosensors [Wang et al. 2005; Rivas et al. 2007; Allen et al. 2007; Wang and Lin 2008] and biosensors for detection pathogens [Yang et al. 2008] have been reviewed recently.

3.1. DNA Biosensors

There are several approaches in development of DNA sensors with the CNTs implementation (see summary in the Table 1). The references and short description of the results obtained with the DNA-CNT sensors are given below.

Interaction of SWCNTs with DNA molecules leads to molecular complex called as wrapped or partially wrapped CNTs. The complexes are used for sequence specific binding RNA and single-stranded DNA (ssDNA), dispersing CNTs and their fractioning by size, diameter and charge with electrophoresis [Vetcher et al. 2006], anion exchange chromatography [Zheng et al. 2003a] and some related techniques. The spontaneous association of ssDNA and CNTs is achieved due to multi-point interaction of the components by electrostatic and/or hydrophobic interactions depending on the charge of CNTs and ionic content of the media. The use of the DNA wrapped CNTs includes, but not limited by, the detection of hybridization events, electron mediation in electrochemical sensors, among them those utilizing enzymatic reactions [Wang et al. 2004a], and the investigation of DNA structure and polymorphism.

Next approach is based on preliminary treatment of the transducer with CNTs, which serve as specific modifiers to establish the immobilization and orientation of DNA molecules required for the signal generation. Both high efficiency of CNTs as electron mediator and high capacity toward biopolymers coupled with a huge increase of the specific surface are utilized in this case. Although this way is often considered as non-covalent immobilization of CNTs it should be mentioned that actually about all of the CNTs are preliminary cleaned from amorphous carbon, fullerene like structures and catalyst residues by mineral acids and oxidizing reagents so that actually CNTs obtain negative charge due to introduction of oxygenated functions (e.g. carboxyl groups).

Table 1. DNA-CNT sensors and their applications (references are given below in the text).

CNTs	DNA-CNT interaction	DNA sensor assembly	Applications	Detection mode
SWCNTs	Wrapping or non-specific adsorption of DNA onto the sidewalls	Surface deposition of DNA wrapping CNTs; Suspension features monitoring.	Nanoeletronics; Investigation of DNA-SWCNT structural specificity binding; Electron transfer mediation.	FETs, nanotransistors; Voltammetry techniques (DPV, stripping chronopotentiometry, EIS)
	Covalent binding to support followed by DNA deposition	Self-assembling layers with thiolated or carboxylated CNTs / ssDNAs; SWCNTs-CNTs binding via aminated and carboxylated terminal groups	Hybridization detection; Mediation of electron transfer to redox active species (intercalators, enzymes).	Direct DNA electrochemistry; Light scattering; Fluorescence quenching.
MWCNTs	Non-covalent	Layer-by-layer adsorptive or electrostatic immobilization; Entrapment in the polymer films; Entrapment into the carbon paste.	Hybridization detection based on redox intercalator signal or electron transfer resistance changes; Detection of DNA binding proteins.	DPV, CV, EIS, QCM
	Covalent	Layer-by-layer immobilization involving thiolated or aminated functional groups; DNA-CNT covalent binding by amidation	Hybridization detection based on redox intercalator signal or electron transfer resistance changes; Mediation of electron transfer to redox active species.	DPV, CV, EIS, QCM

Abbreviations: CV – cyclic voltammetry; DPV – Differential pulse voltammetry; EIS – electrochemical impedance spectroscopy; FET – field effect transistor, QCM – quartz crystal microbalance.

Chemical modification of the CNT structure followed by covalent attachment of DNA and deposition of the DNA-CNT hybrid onto the transducer surface has become increasing attention for biosensor development. The necessity of special treatment and binding protocols followed by separation of the aggregates is a main difference of this approach compared with DNA wrapping CNTs. On the one hand, covalent modification extends the possibilities for the selection of the configuration, size and stoichiometry of DNA-CNT aggregates. This is particularly important when site directed modification via terminal functional groups of DNAs and self-assembly of the ordered layers with preferable orientation of appropriate units is demanded. On the other hand, covalent binding can affect the structure and affinity of DNA molecules and hence the performance of the biosensors against that expected from the behavior of DNA in native environment.

3.1.1. Biosensors based on non-covalent interactions of DNA with CNTs

SWCNTs-DNA adducts. The use of SWCNTs-DNA complexes for biosensor development was initiated by investigations of Zheng group [2003a, 2003b]. They showed the possibility for the dispersion and following separation of CNTs due to their wrapping with DNAs. Being comparable in mass, ssDNAs and SWCNTs form under sonication in aqueous solutions the complexes that differ in charge, size and structure. DNA molecule is quite flexible in bond torsion to wrap SWCNT to form low-energy conformations and nucleotide base - CNT stacking interactions. For uncharged CNTs, sugar-phosphate groups are exposed to water. Chemically pre-oxidized CNTs are charged negatively due to formation of terminal carboxylate groups. They are connected to a DNA molecule by bridging metals ions. In addition to wrapping in right or left turn direction, adsorption with linearity extended DNA structure is described. The DNA-CNT aggregation is sequence specific and depends on the alteration of nucleotides in the primary sequence and the length of the oligonucleotide. Thus, the application of salmon genomic DNA resulted in selective enrichment of (6,5) SWCNTs with the yield up to 86% from CNTs mixture obtained on Co-Mo catalyst [Kim et al. 2008]. Also, preferable wrapping onto A-T alternating sequences was mentioned. The interaction of hydrophilic sugar phosphate groups facing outside the DNA-CNT complex with aqueous environment is strong enough to compete with the tendency of CNTs to cling and form ropes. Base-CNT interactions dominate in aqueous solutions with low ionic strength. The DNA entropy of the DNA backbone and electrostatic interactions between backbone charges contribute to the complex stabilization [Manohar et al. 2007]. DNAs were found more efficient in CNTs dispersing against other charged polymers [Steuerman et al. 2002].

CNTs partly or fully wrapped with DNA were used for assembling various ordered structures, which can be used as platform for nanodevices or DNA sensors with extended stability and sensitivity toward DNA binding species. Thus, charged nanostrips with the length up to 20 nm can be deposited from the DNA-SWCNTs dispersion onto a solid surface. Nanostrips were discovered by AFM or can be re-eluted back from the interface into aqueous phase by treatment with salt solution. Also, hydrogen binding of guanosine residues decorating sidewalls of SWCNT provides their association in the bundles or strips.

The combination of DNA hybridization and partial wrapping of CNTs leads to 3D structures of aligned SWCNTs bridged [Chen et al. 2007]. For this, complementary ssDNAs are separately involved in the complex with SWCNTs. Then, upon mixing, association of initial complexes takes place due to recognition of complementary DNA strands. Among SWCNTs, Au nanoparticles can be used together with thiolated DNAs to synthesize dendrimer-like 3D structures

DNA wrapped SWCNTs can be implemented into the sensing layer of biosensors using electrostatic adsorption onto self-assembled layers or carbon materials. Both deposition and release of the complexes from the interface can be driven by polarization of the electrode [Zheng et al. 2009]. The desorption potentials (+0.90 and -1.40 V) were determined for bare Au electrode and the electrode covered with self-assembled monolayer of octadecanethiol, respectively. As it was shown by EIS and CV, ssDNA-SWCNT complex restores the electron transfer between Au blocked by alkanethiol and redox species. The interaction of ssDNA and SWCNTs in the complex was proved by UV-Vis and FTIR spectrometry.

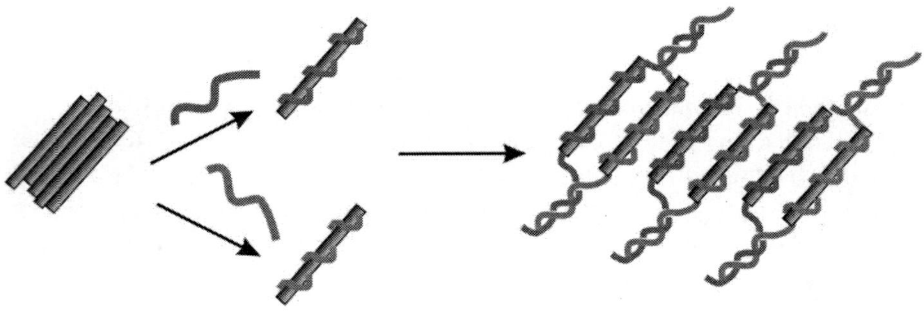

Figure 2. The scheme of formation 3D structures of aligned SWCNTs by combination of DNA hybridization and partial wrapping. Partially adopted from Chen et al. [2007] with permission of the American Chemical Society.

HiPCO SWCNT-DNA adducts were used as specific optic labels for detection of chromosomal DNA of human fibroblast based on the changes in the Raman spectra after Southern blotting procedure [Hwang et al. 2006]. SWCNTs exhibit strong Raman signals as well as fluorescence emissions in the near infrared region. Such signals do not blink or photobleach under prolonged excitation, which is an advantage in optical nano-biomarker applications. 5'-GTG TGT GTG TGT GTG TGT GTA AAA- **AGC CCC GCC TTG ATG AAG**-3' was used as DNA probe. The part marked with bold is complementary to the specific part of human adenosine deiminase encoding gene. Partial wrapping is achieved by weaker interaction of AAAA sequence of the probe.

The adsorption of DNA wrapped SWCNTs onto alkanethiol monolayers at Au electrodes were investigated by reflection absorption FTIR, X-ray photoelectron spectroscopy, and Raman spectroscopy [Zangmeister et al. 2007]. DNA-SWCNTs preferentially adsorb to positively charged amine-terminated monolayers and to bare Au in comparison with hydrophobic methyl-terminated or negatively charged carboxylic acid-terminated monolayers.

DNA partly wrapped SWCNTs were obtained and separated from free nanotubes and DNA by differential centrifugation [Li et al. 2007]. They complexes isolated form uniform porous films on glassy carbon. The micromorphology of the films and their porosity are explained by the balance between electrostatic repulsion of charged parts of the DNA and SWCNTs and high attraction energy due to van der Waals interactions along naked part of SWCNTs. The comparison of electrochemical behavior of two differently charged markers, i.e. $[Fe(CN)_6]^{3-/4-}$ and $Co(bpy)^{2+/3+}$, showed that the DNA partly wrapped SWCNTs are able to facilitate the electron transfer toward positively charged redox active species.

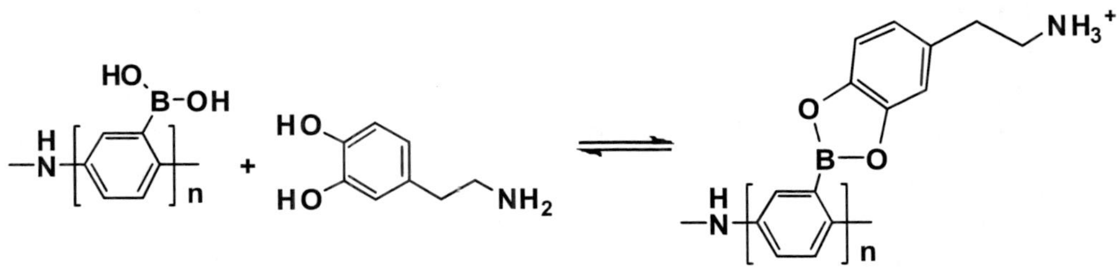

Figure 3. The scheme of formation of the complex between poly(anilineboric acid) and dopamine.

ssDNA-SWCNT complexes were deposited onto poly(anilineboronic acid) by electropolymerization of 2-aminophenylboronic acid monomer onto Au electrode performed in the presence of ssDNA-SWCNT. The biosensor was applied for the detection of dopamine that chemically hinders electrochemical activity of the polymer by reversible binding with boronic acid group (Figure 3). The current decay recorded at pH 7.4 was proportional to the dopamine concentration down to 1 nmol L^{-1} [Ma et al. 2006, 2008]. In comparison with the same polymer obtained in the absence of ssDNA-SWCNTs, the detectable dopamine concentrations were decreased by four orders of magnitude. The influence of the composite was related to its templating effect on polymerization procedure as well as extended electron mediation. The effect of SWCNT and DNA was found to be much milder than that of their adducts. The parameters of electron transfer from aniline boronic acid in strong acids were determined and related to appropriate changes in the absorbance of oligomers depositing on the electrode [Ma et al. 2008].

The electron transfer between SWCNT-DNA adducts and redox indicators are promoted by inserting the components in the ionic liquid [Zhang et al. 2009]. Nanoparticles of cerium dioxide were dispersed in hydrophobic 1-butyl-3-methylimidazolium hexafluorophosphate. The efficiency of electron transfer in the layer of ionic liquid, CeO_2, SWCNTs and ssDNA probe related to phosphoenolpyruvate carboxylase gene 5'-AAA ACT TGT GGT AGT TGG AGC TGA TGG CGT AGG CAA GAG TGC CC-3' was investigated by EIS in the presence of $[Fe(CN)_6]^{3-/4-}$ as the redox probe. Under optimal conditions, the dynamic range for detecting of target DNA was 1.0×10^{-12} - 1.0×10^{-7} mol L^{-1} and LOD was 2.3×10^{-13} mol L^{-1}). The hybridization resulted in consecutive increase of the electron transfer resistance. The DNA sensor was used for detection of transgenic rape tissues.

DNA-SWCNT adducts were obtained by dispersing HiPCO nanotubes in sodium cholate followed by mixing with ssDNA probe and electrophoretic removal of free components and cholate counter ion. Hybridization of the DNA strands with their complements (cDNA 5'-GCC TAC GAG GAA TTC CAT AGC T-3') adsorbed on the SWCNT surface was transduced by a hypsochromic shift in the near-infrared fluorescence of the SWCNT. The (6,5) nanotube fluorescence (λ_{max} = 994 nm) was monitored with a laser excitation of λ_{exc} = 785 nm. The hybridization of DNA onto the sidewalls of SWCNT was confirmed by FRET between fluorescently labeled DNA strands using 3'-terminated fluorescein derivative.

DNA wrapped SWCNTs were used also as fluorescent probe for glucose biosensing. As was shown by AFM, glucose oxidase physically adsorbed onto DNA wrapped SWCNT retains its globular structure and ability to promote glucose oxidation [Karachevtsev et al. 2007]. The redox oxidation of glucose was mediated by ferricyanide anion. Its addition resulted in sharp decay of CNTs emission. Luminescence was excited with diode pumped solid-state green laser (λ_{exc} = 532 nm)

MWCNTs-DNA adducts. Contrary to SWCNTs, MWCNTs do not form DNA adducts with specific pre-organized structure. More commonly, layer-by-layer deposition of MWCNTs and DNA onto transducer is used. In many cases, no special investigations proving or denying interaction of MWCNTs and DNA in such layers were conducted. Electrocatalytic activity of CNTs is used for enhancement of the performance of DNA sensors utilizing MWCNTs as heterogeneous electron mediators.

First, the effect of MWCNTs on the oxidation of DNA itself should be mentioned. The use of MWCNTs deposited from dispersions in organic solvents and mineral acids onto carbon materials, Pt and Au electrodes provide amplification of the oxidation of guanine residues in the DNA sequence. Due to steric limitation of direct electron transfer, oxidation of guanine is commonly hindered and requires either DNA denaturation or the use of specific modification of the electrode surface. MWCNTs provide both higher density of DNA packing onto conducting surface and faster electron transfer yielding increased currents on voltammograms. The signal is well pronounced for ssDNAs and can be recorded using cyclic (CV) or differential pulse voltammetry (DPV) and stripping chronopotentiometry commonly used in direct DNA electrochemistry investigations.

The 11-fold increase in the guanine oxidation current was observed with glassy carbon electrode modified with MWCNTs in comparison with bare electrode [Wang et al. 2003a]. This phenomenon was used for the detection of ssDNA from calf thymus. The oxidation peaks were recorded by stripping chronopotentiometry with preliminary accumulation of ssDNA on the electrode polarized at 1.7 V. The addition of Cu^{2+} ions and acidic pre-treatment of ssDNA increased the signals. The protocol was successfully applied for the detection of hybridization event. For this purpose, DNA probe with guanine residues substituted by inosine fragments was immobilized onto the MWCNTs covered electrode by avidin-biotin binding. The hybridization with target sequence 5'-A AAA GTG TTT TTC ATA AAC CCA TTA TCC AGG ACT GTT TAT AGC TGT TGG AAG GAC TAG GTC-3' related to the BRCA1 breast cancer gene resulted in appearance of guanine oxidation peaks. The accumulation of target oligonucleotides onto magnetic beads provides their quantification in the range from 50 to 250 mg L^{-1}. The LOD was 2 nmol L^{-1}.

High efficiency of oxidation of nucleotides was proved by the electrode covered with ionic liquid film contained MWCNTs [Sun et al. 2009]. Guanosine can be oxidized at 1.098 V vs. SCE. The anodic peak current is proportional to the analyte concentration from 1×10^{-7} to 4×10^{-5} mol L^{-1}. The rate constant of electron transfer determined from CV experiments was found to be 2.94×10^{-4} s^{-1}. The LOD obtained (7.8×10^{-8} mol L^{-1}) was lower than those based on fullerene and Au nanoparticles as electron mediators. Only SWCNTs gave lower LOD value of about 2×10^{-8} mol L^{-1}

Oxidation of calf thymus DNA onto glassy carbon covered with MWCNTs deposited from DMF dispersion was investigated by Luo et al. [2007] using CV method. Two peaks related to guanine and adenine oxidation were observed at first scan at 0.7 and 0.81 V vs. SCE. In the next cycles, additional pair of reversible peaks appeared which was referred to the products of initial guanine oxidation. Electrode reaction proposed is presented on Figure 4.

Preliminary activation of glassy carbon electrode by its partial oxidation in the mixture of sulfuric and nitric acids followed by MWCNTs deposition was suggested to amplify the signal of guanine oxidation [Zhang et al. 2007]. The detection limit of 7.5 nmol L^{-1} for guanine and 150 ng mL^{-1} for acid denatured DNA were obtained.

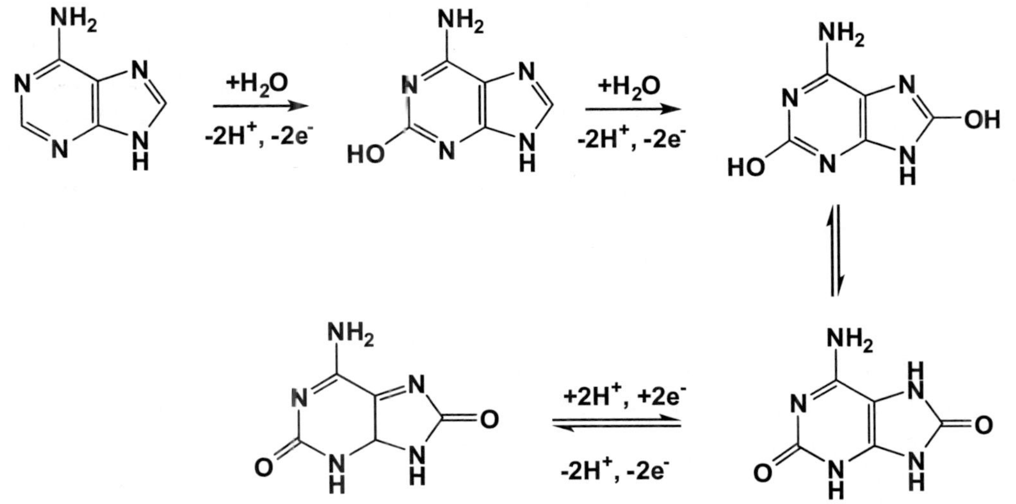

Figure 4. The scheme of adenine oxidation at glassy carbon electrode covered with MWCNTs and calf thymus DNA.

Pencil graphite electrode was modified by pre-oxidized MWCNTs and DNA probe with terminal NH_2 groups [Erdem et al. 2006]. Raman spectra taken at a laser excitation wavelength of 632 nm gave evidences of DNA covalent binding, probably via carboxylate groups of MWCNTs. 70 % increase in the guanine current oxidation recorded by DPV was obtained against unmodified electrode.

The use of screen printed carbon electrode modified with MWCNTs made it possible to shift the potential of guanine and adenine oxidation to less anodic values [Ye and Yu 2005]. The appropriate signals were recorded onto MWCNT/ssDNA modified electrode at 0.595 V and 0.9 V vs. SCE, respectively. With the accumulation time of 5 min, the protocol proposed provides detection of calf thymus ssDNA from 17.0 to 345 μg mL^{-1} with a LOD of 2.0 μg mL^{-1} and yeast tRNA from 8.2 μg mL^{-1} to 4.1 mg mL^{-1}. AC impedance was employed to characterize the surface of modified electrodes.

High efficiency of the DNA oxidation remains with MWCNTs included into the carbon paste. The electrostatic accumulation of target oligo- and poly-nucleotides made it possible to detect down to 2.0 μg L^{-1} of oligonucleotide 5'-ATG TGG AAA ATC TCT AGC AGT-3' and 170 μg L^{-1} of calf thymus dsDNA. The oxidation efficiency increased with the electrode polarization potential and accumulation time. Stripping chronopotentiometry and DC voltammetry was used for guanine oxidation monitoring. The inclusion of CNTs into the electrode body resulted in 29- and 61-fold increase of the response of conventional carbon paste electrode [Pedano and Rivas 2004].

MWCNTs based carbon paste electrode was used for sensitive detection of hybridization event [Qi et al. 2007]. ssDNA probe 5'-ATA GAC TGG CCG TCG TTT TAC AAC GTC GTG GT-3' was immobilized onto the surface of carbon paste electrode by entrapment into electropolymerized polypyrrole film. Oxidation of redox intercalator, ethydium bromide, was recorded by DPV at 0.68 V. The oxidation current linearly depended from the logarithm of target DNA concentration in the range from 1.0×10^{-10} to 1.0×10^{-8} mol L^{-1} (LOD 8.5×10^{-11} mol L^{-1}). The selectivity of the response was checked by completely mismatched and four-point mismatched oligonucleotides.

The combination of MWCNTs and polypyrrole was also applied in impedimetric DNA sensor [Cai et al. 2003a]. Pre-oxidized MWCNTs were deposited on glassy carbon electrode from DMF suspension and covered by electropolymerized polypyrrole. The oligonucleotide sequence specific for colitoxin 5'-GAG CGG CGC AAC ATT TCA GGT CGA-3' was electrostatically adsorbed onto positively charged polymer. Changes in the interfacial charge transfer resistance measured by EIS were referred to the enhanced influence of hybridization on the doping activity of the dsDNA toward polypyrrole. The decay of the logarithm of impedance (-log Z) linearly depended on the amount of complementary oligonucleotide ranging from 1.0×10^{-5} to 3.0×10^{-8} mol L^{-1}.

Similar detection system was proposed for the detection of PAT (phosphinothricin acetyltransferase) gene fragment selected from transgenic plants 5'-GCC ACA AAC ACC ACA AGA GT-3' and the product of PCR amplification of NOS gene (terminator of nopaline synthase gene) 5'-ATC GTT CAA ACA TTT GGC A-3' extracted from transgenic bean [Jiang et al. 2008]. SWCNTs were mixed into the carbon paste and then the electrode was covered with poly-L-lysine obtained by repeating scanning of the potential. DNA probe was electrostatically accumulated onto the polymer layer. Contrary to polypyrrole, polylysine is electrochemically inactive and does not change the conductivity while surface reaction with charged species takes place. Thus the hybridization of DNA probe and target oligonucleotide onto polylysine resulted in decrease of the charge transfer resistance measured by EIS in the presence of redox indicator, $[Fe(CN)_6]^{3-/4-}$ due to diminishing the diffusion rate of the marker. The biosensor provided quantification of the PAT gene fragment in the range from 1.0×10^{-12} to 1.0×10^{-7} mol L^{-1} with the estimated LOD of 3.1×10^{-13} mol L^{-1}.

The ability of MWCNTs and polyaniline nanofibers to spontaneous association was used for the signal amplification of DNA sensor toward the same analytes (PAT and NOS gene fragments) [Yang et al. 2009]. The implementation of the DNA probes together with chitosan into the carbon paste resulted in enormous increase of efficiency of electron transfer and sensitivity of the DNA sensor in detection of hybridization event. DNA probe was accumulated in open circuit mode onto the MWCNTs/polyaniline/carbon paste electrode and the signal related to the oxidation of redox intercalator, Methylene Blue (MB), was recorded by DPV. The association of MWCNTs with polyaniline nanofibers was proved by EIS measurements as well as its contribution toward accelerated electron transfer from $[Fe(CN)_6]^{3-/4-}$ indicator. The dispersion of MWCNTs with polyaniline was also investigated by SEM. Most effective binding corresponded to 1:2 ratio (PANI:MWCNTs, m/m).

One of the common approaches frequently used in electrochemical DNA sensors is the recording of the hybridization event or DNA damage by redox intercalators. They exhibit electrochemical activity, which changes after inclusion of these indicators in dsDNA helix. Intercalation takes place only after the DNA probe hybridization with target oligonucleotide sequence. Some of intercalators are able to promote oxidation of guanine residues in the DNA sequence. The structures of the redox intercalators mentioned in this chapter are presented in Figure 5. MWCNTs were physically entrapped into the composite film of ZnO nanoparticles dispersed in chitosan [Zhang et al. 2008]. Electrostatic attraction of CNTs to positively charged ZnO surface yields in uniform coating with well-reproduced characteristics. The ability of the coating to bind ssDNA probe related to PAT gene was proved by recording the MB signal increased in the presence of ssDNA due to specific binding of the dye to guanine residues of oligonucleotide. Hybridization with complementary target sequence decreased the signal of intercalator. The shift of the cathodic peak current linearly depended on the logarithmic value of the target concentration ranging from 1.0×10^{-11} to 1.0×10^{-6} mol L^{-1}, with the LOD 2.8×10^{-12} mol L^{-1}.

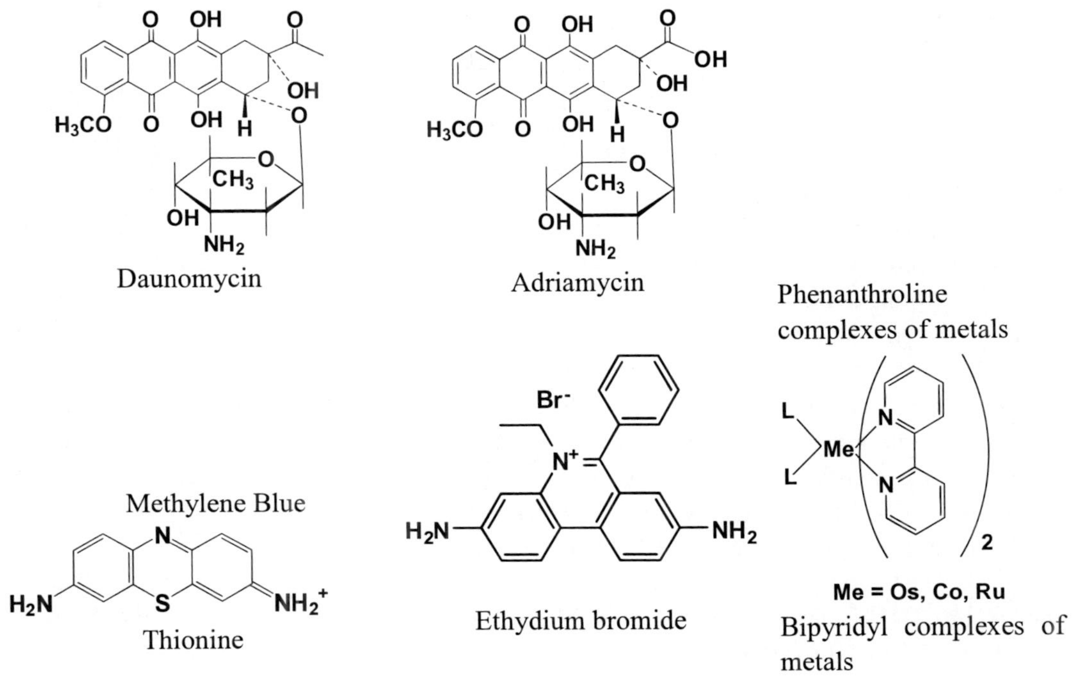

Figure 5. Redox intercalators used in DNA-CNT sensors.

Similar construction of DNA sensor with ZrO_2 particles instead of ZnO and daunomycin as redox indicator was used for detecting specific DNA sequence related to colorectal cancer [Yang et al. 2007a]. The immobilization of DNA probe was due to interaction of ZrO_2 with phosphate groups of the oligonucleotide. The range of target oligonucleotide concentration from 1.49×10^{-10} to 9.32×10^{-8} mol L^{-1} and the LOD of 7.5×10^{-11} mol L^{-1} were obtained.

ZrO_2 aided immobilization of ssDNA probe related to PAT gene was applied for layer-by-layer assembling of biorecognition layer [Yang et al, 2007b]. The deposition of MWCNTs pre-oxidized by nitric acid was performed by electropolymerization of 2,6-pyridinedicarboxylic acid. After that, ZrO_2 layer was chemically deposited onto MWCNT-polymer film and ssDNA probe 5′-GCC ACA AAC ACC ACA AGA GT-3′ immobilized. The efficiency of electron transfer through the surface film was proved by CV and EIS measurements. The hybridization of complementary oligonucleotide sequence increased the resistance of surface charge transfer. DNA sensor makes it possible to quantify 1.0×10^{-11}–1.0×10^{-6} mol L^{-1} target oligonucleotide (LOD 1.38×10^{-12} mol L^{-1}).

Among bridging metal cations, CNTs and DNA can be "glued" to each other with small positively charged organic compounds which can simultaneously participate in the biorecognition or electron transfer. Thus, thionine facilitates interaction of MWCNTs with calf thymus DNA [Gong et al. 2008]. Dispersion of MWCNTs in DMF was first mixed with thionine and the adduct formed was isolated by centrifugation. After that, DNA was electrostatically adsorbed on CNT-thionine adduct and spread onto glassy carbon. The formation of the complex was proved by FTIR and UV spectra. The implementation of thionine in adduct fully suppresses its redox activity observed for CNT-thionine complex and thionine solution. However, the addition of diffusionally free redox intercalator, i.e. $Co(bpy)_3^{2+}$ makes it possible to monitor the DNA onto the electrode surface as well as its interaction with CNT-thionine. The surface layer was quite stable and showed reproducible response to the intercalator used. Amino groups of thionine adsorbed onto CNTs can be used for their further immobilization and development of biosensing structures based on proteins and DNA [Li et al. 2004].

In a similar manner, MB was added to the MWCNTs dispersion in DMF. After coating, composite provided well-defined impedimetric and potentiometric response toward native DNA from chicken erythrocytes and anti-DNA antibodies. As was shown by EQCM and coulometry, the addition of MB to the MWCNTs increases adsorptive capacity of the layer toward native DNA in comparison with that of polymeric MB on bare glassy carbon. The phenothiazine dye involved in the CNTs layer loses its susceptibility to oxidative polymerization and shows quite reversible response toward DNA and DNA-anti-DNA antibody interactions due to effect of the surface charge changes on electrochemical conversion of the indicator. It should be stressed that MB being preliminary adsorbed onto CNTs does not show intercalating behavior and acts like heterogeneous electron mediator. Recently Galandova et al. [2009] reported DNA based MWCNTs electrochemical sensor for detecting DNA damage induced by quinazoline. MWCNTs were dispersed in polyethylenimine (PEI) and covered with the calf thymus dsDNA layer. DNA damage has been characterized by several electrochemical methods including square wave voltammetry of the $[Ru(bpy)_3]^{2+}$ redox indicator and mediator of the guanine moiety oxidation, CV and EIS in the presence of the $[Fe(CN)_6]^{3-/4-}$ indicator in solution. It has been pointed that due to high electroconductivity and large surface area of MWCNT and positive charge of PEI, the MWCNT-PEI composite is an advantageous platform for the DNA immobilization by the polyelectrolyte complexation and its voltammetric and impedimetric detection.

MWCNT-DNA adducts were used for immobilization of cytochrome *c* which was electrostatically accumulated onto dsDNA and participated in electron exchange with the electrode surface [Shie et al. 2008]. Hydrophobic interactions of the protein with naked parts of MWCNTs scaffold are stipulated as a requirement for effective binding of cytochrome *c*. Also, increasing efficiency of electron transfer was reported for electrodes modified with MWCNTs and DNA deposited separately [Wang et al. 2002].

3.1.2. DNA biosensors with covalently modified CNTs

MWCNTs based DNA sensors. Chemical modification of CNTs is a promising tool to improve the stability of biosensing layers and sensitivity of target analyte detection. One of the common purposes of the modification consists of the stabilization of CNTs layer onto the transducer surface and/or the assembling of ordered layers with directed positioning of CNTs to reach better transduction of biorecognition signal and to miniaturize a sensor device. General scheme of MWCNTs immobilization followed by covalent attachment of oligonucleotides is presented on the Figure 6.

Self-assembling of thiolated CNTs and oligonucleotides onto bare Au electrodes or via bridging Au nanoparticles becomes popular for DNA sensors design. Also, direct growth of aligned CNTs and their random abrasive adsorption are described.

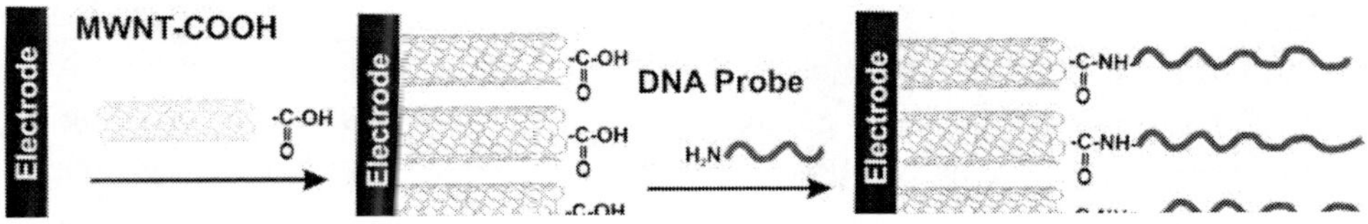

Figure 6. The scheme of general strategy of covalent DNA-CNT modification of electrodes.

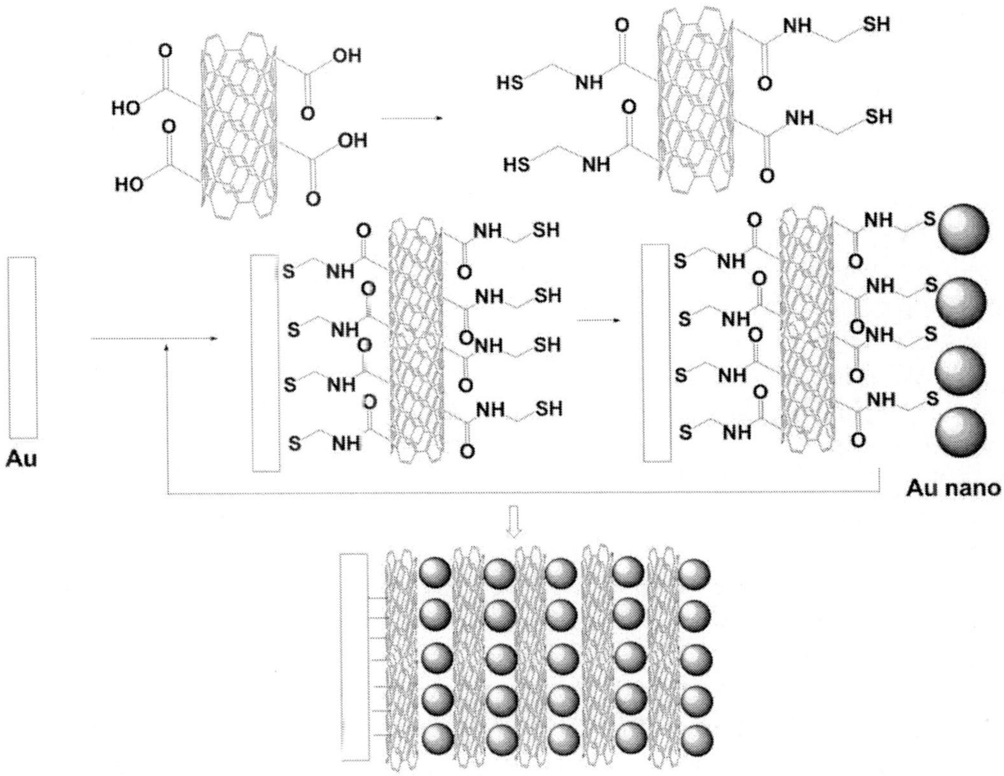

Figure 7. The scheme of preparation of thickness-tunable multiplayer film of MWCNTS and gold nanoparticles (see also [Zhang et al. 2009a]).

Layer-by-layer deposition has been developed to reach high density of recognition elements. MWCNTs were modified by -C(O)NH-$(CH_2)_2$-SH terminal groups and attached to the Au electrode [Zhang et al. 2009]. The modification of the MWCNTs by carbonylation was proved by FTIR spectra. Consecutive treatment of the electrode with modified MWCNTs and Au nanoparticles gives thickness-tunable multilayer films investigated by SEM, EIS, DPV and CV (Figure 7). Thiolated DNA probes 5'-HS-TAG TCG TCA ATC CTC CGC TT-3' were attached to outer layer of Au nanoparticles. The biosensor was used for quantitative detection of complementary oligonucleotides. For that, the oxidation current of doxorubicin electrochemically active intercalator, was measured by DPV. The detection limit of 6.2 pmol L^{-1} was obtained with quite good distinction from mismatched sequences. Also, dsDNA was quantified in the range from 5.0×10^{-10} to 1.0×10^{-11} mol L^{-1} with a detection limit of 6.2 pmol L^{-1} after its 40 min accumulation in open-circuit mode.

Other metals nanoparticles are also used in the CNTs layers to extend their electroactivity. The suspension of MWCNTs with Pt nanoparticles was obtained by chemical reduction of K_2PtCl_6 in the presence of MWCNTs upon sonication in the presence of Nafion as stabilizer [Zhu et al. 2005]. The glassy carbon electrodes were soaked with the mixture obtained and left to dry. Aminated oligonucleotides were immobilized via terminal carboxylic groups of MWCNTs by carbodiimide coupling. The electrochemical detection of hybridization event was based on the accumulation of redox intercalator, daunomycin, in the dsDNA fragments onto the MWCNTs-Pt layer. The appropriate oxidation peak recorded at +0.45 V by DPV changed linearly with logarithm of the target concentration in the range from 2.25×10^{-7} to 2.25×10^{-11} mol L^{-1} with the LOD of 1.0×10^{-11} mol L^{-1}. Similar DNA sensor based on pre-oxidized MWCNTs deposited onto glassy carbon without Pt nanoparticles and covalently bonded to aminated oligonucleotides via terminal carboxylic groups showed much lower sensitivity of complementary oligonucleotide detection. The dynamic range of concentration was from 5×10^{-8} to 2×10^{-10} mol L^{-1} with the LOD of 1.0×10^{-10} mol L^{-1} [Cai et al. 2003b].

Silver nanoparticles were electrochemically deposited onto the glassy carbon electrode modified consecutively with carboxylated MWCNTs and poly-3(3-pyridyl)acrylic acid [Zhang et al. 2009b,c]. Thiolated ssDNA probe SH-$(CH_2)_6$-5'-AAG CGG AGG ATT GAC GAC TA-3' was coupled to Ag nanoparticles by overnight accumulation in open circuit mode. The hybridization was recorded by the signal of redox intercalator, adriamycin that increased proportionally to the logarithm of the concentration of complementary sequence in the range of its concentration from 9.0×10^{-12} to 9.0×10^{-9} mol L^{-1} with a LOD of 3.2×10^{-12} mol L^{-1}. Additional improvement of the sensitivity was reached by implementing Ag nanoparticles taken thiolated DNA probes into self-assembled layer on Au electrode [Gao et al. 2008]. Due to multiplication of reporters number the LOD of 10 fmol L^{-1} was reached for 12-mer target oligonucleotide. Without MWCNTs, the DNA sensor with Ag nanoparticles makes it possible to detect only 1×10^{-9} mol L^{-1} of the target sequence.

Pd nanoparticles were obtained by chemical reduction of Na_2PdCl_4 with $NaBH_4$ and isolated from excess of reagents by centrifugation [Chang et al. 2008]. Glassy carbon electrode was treated with suspension of MWCNTs and Pd nanoparticles in aqueous Nafion solution. Aminated ssDNA probe 5'-NH_2-GAG CGG CGC AAC ATT TCA GGT CGA-3' was covalently attached to terminal carboxyl groups of MWCNTs by carbodiimide binding. Electrochemical detection of hybridization was performed using MB in accordance with its cathodic peak current recorded by DPV. Estimated LOD for complementary target was found to be 1.2×10^{-13} mol L^{-1}.

Advantages of self-ordered MWCNTs layers over random films in DNA sensing were proved with Au electrodes covered with appropriate nanotubes without any polymeric support [Wang et al. 2004]. Aligned MWCNTs were grown on Au surface by CVD techniques with Ni catalyst and then covalently modified with aminated DNA probe via carbodiimine binding. The formation of terminal carboxyl groups involved in covalent binding was proved by FTIR spectra. The efficiency of hybridization was recorded by accumulation of redox intercalator, MB. Its DPV signal progressively decreased with the amounts of complementary oligonucleotides in solution. The effect of hybridization was found to be much higher in the case of ordered MWCNTs layers.

Newly synthesized manganese complex of rutin (Figure 8) was used for detection of salmon sperm DNA and ssDNA probe 5'-TCG ACC TGA AAT GTT GCG CCG CTC -3'.

The indicator showed excellent electrochemical reversibility and sensitivity of the response toward dsDNA and hybridization event. In the latter case, aminated ssDNA probe was immobilized to carboxylated MWCNTs by carbodiimide binding.

In comparison with direct immobilization of ssDNA probe onto glassy carbon, the use of MWCNTs resulted in higher capacity of the surface layer toward DNA probe and higher stability of the response. Target oligonucleotides were determined in the range from 1.60×10^{-9} mol L^{-1} to 4.80×10^{-8} mol L^{-1} (LOD 3.8×10^{-11} mol L^{-1}).

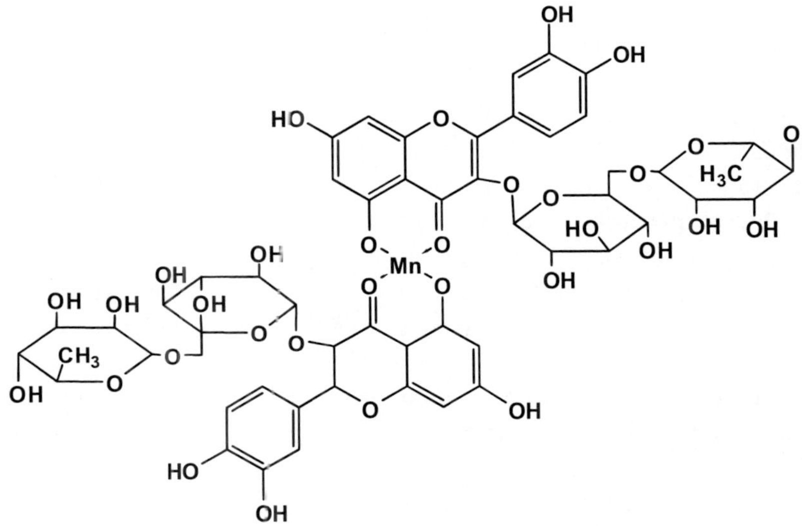

Figure 8. The structure of the magnesium complex of rutin.

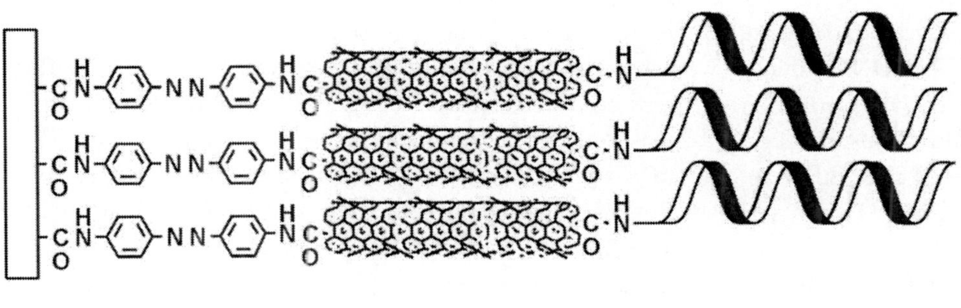

Figure 9. The scheme of covalent attachment of MWCNTs to the glassy carbon surface pretreated by 4,4'- diaminobenzene.

The use of long linkers separating CNTs from the electrode is limited by requirements of electron transfer on the sensor interface. Nevertheless, some immobilization schemes allow the electric wiring of CNTs layer. As example, MWCNTs were covalently attached to the pre-activated glassy carbon by 4,4'-diaminoazobenzene [Li et al. 2008] (Figure 9). The linker can be cathodically reduced to diphenylhydrazide derivative. The peak current recorded by DPV decreased after hybridization probably due to effect of dsDNA onto the charge and hydrogen ion transfer within the reaction layer. The DNA sensor proposed makes it possible to detect short oligonucleotides related to hepatitis B virus between 7.94×10^{-8} and 1.58×10^{-6} mol L^{-1}, with the LOD 1.1×10^{-8} mol L^{-1}.

Reversed order of component immobilization was proposed by [Yang et al. 2007c]. Thiolated ssDNA probe 1 was immobilized onto Au electrode. Second DNA sequence (probe 2) was covalently attached to the terminal groups of CNTs by carbodiimide binding. SWCNTs were treated with ferrocene, which was non-covalently accumulated onto sidewalls by π-stacking interactions. Target DNA hybridized with both DNA-probes so that ferrocene units are attached to the electrode sequence by dsDNA linker. The reaction scheme shown on Figure 10 is similar to sandwich type immunoassay. The current of hydrogen peroxide oxidation has been mediated by the indicator and allowing determination of the ferrocene concentration.

Similar approach based on CNTs as carriers for multiplying the number of indicators attached to DNA probe and hence amplifying the signal was realized with alkaline phosphatase [Wang et al. 2004 a,b], CdSe [Cui et al. 2008] and CdS nanodots [Wang et al. 2003b]. Fluorescence of CdSe quantum dots attached to DNA-CNTs hybrid in supernatant was measured as a signal of hybridization event. The detection limit of 0.2 pM DNA was reached. Besides sandwich-like analysis, hybridization products associated with MWCNTs carrying alkaline phosphatase were isolated using magnetic separation [Wang et al. 2004b]. Cd nanodots were modified by DNA probe by avidine-streptavidine binding. Cd ions were determined after dissolution of nanodots in nitric acid by stripping voltammetry.

SWCNTs based DNA sensors and nanosensors. SWCNTs were covalently bonded to thiolated ssDNAs via caboxylic groups placed both on terminal sites and sidewalls [Baker et al. 2002]. The oxidation of SWCNTs used for their functionalization resulted in 40% loss of mass followed by significant decrease in the amounts of catalyst impurities (Ni and Yt). The immobilization scheme involved consecutive treatment of the oxidized SWCNTs with thionyl chloride, ethylenediamine and succinyl derivative. After that, thiolated ssDNA reacted with maleimine groups to form covalent adducts. The nature of surface oxidized groups as well as the structure of the adduct were proved by X-ray photoelectron spectroscopy and SEM. The SWCNT-DNA adducts were used for sandwich type hybridization assay with fluorescein labeled DNA probe. After reaction, the suspension was washed over polycarbonate filter and resulting absorbance was measured in the standard 96 well microtiter plate. The results showed clear differentiation of complementary and mismatched DNA sequences.

SWCNTs are also successfully applied as templates and semiconductor elements of nanoelectronic devices. They both serve as templates for directed self-assembling of such nanotransistors or recognition elements for detection of DNA-analyte interaction.

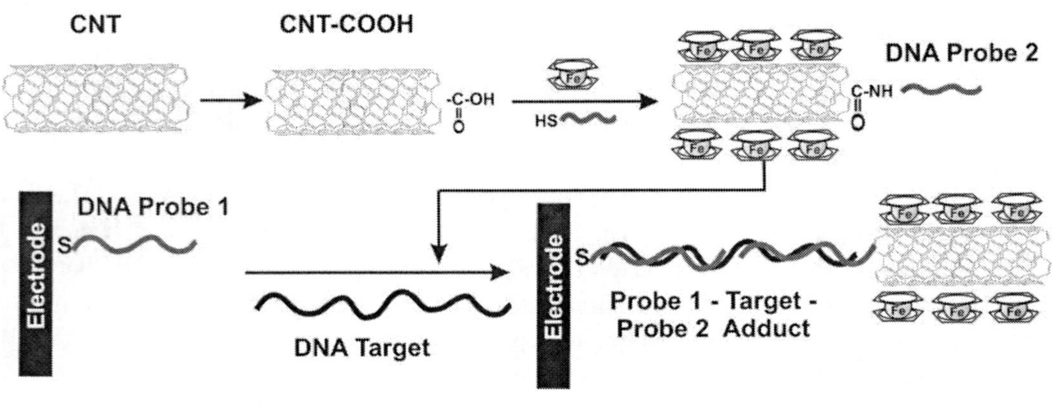

Figure 10. The scheme of reversed order of immobilisation of DNA sensor components. Thiolated ssDNA probe 1 was immobilized onto Au electrode. Second DNA sequence (probe 2) was covalently attached to the terminal groups of CNTs by carbodiimide binding. SWCNTS were treated with ferrocene, which was non-covalently accumulated onto sidewalls by π-stacking interactions. Target DNA hybridized with both DNA-probes so that ferrocene units are attached to the electrode sequence by dsDNA linker. Adapted from [Yang et al. 2007c] with permission of Springer.

The production of nano filed effect transistor (nanoFET) with a single semiconducting SWCNT using SWCNTs and DNA scaffold has been described [Keren et al. 2003]. The general scheme of the template process is illustrated on Figure 11. The assembling was guided by the information encoded in the DNA sequences. First, short ssDNA probe was synthesized to localize further nanoFET positioning. Then, dsDNA - protein A filament was obtained by consecutive polymerization of recombinant protein A from *Esherichia coli* onto dsDNA scaffold (about 48 500 base pairs). The nucleoprotein formed corresponds to the 500 base long DNA fragment complementary to the ssDNA probe. After that, SWCNTs modified with streptavidin was attached to the localized fragment of the dsDNA sequence via sandwich immunoassay protocol with anti-recA and biotinylated antimouse antibodies. Incubation of the filament in $AgNO_3$ solution resulted in chemical deposition of Ag clusters. Electrodes plating of Au onto Ag clusters resulted in formation of nanowires separated with a single SWCNT where metallization is inhibited by oligopeptide layer. The structure of nanoFET was proved by AFM and its usability as transistor by electric characteristics detection.

Another way for creating single SWCNT nanoelectronic device is based on e-beam lithography [Roy et al. 2008]. Electrical contacts to individual SWCNTs were made by consecutive deposition of 10 nm Ti and 50 nm of Au on a silicon waver. A pair of nanoelectrodes was prepared by creating a controlled nanogap in the middle of a single SWCNT by focused ion-beam etching. An 80 base pair long DNA fragment encoding a portion of the H5N1 gene of the avian Influenza A virus was covalently attached in-between carboxylated ends of SWCNTs by covalent binding via amidation. The size of ssDNA probe was chosen on constraint of ion beam etching technology (<25 nm was not reproducible). In hybridization of a target DNA, the current measured between nanoelectrodes increased from about 1 pA to 25-40 pA. Similar construction of transducer based on Si waver – Ti – Au sandwich structure was used in combination with double-walled CNTs as nanoscale resistor for measuring hybridization of 5′-AGG TCG CCG CCC-$(CH_2)_3$-NH_2 covalently attached in the gap. Relative decrease in the drain current linearly depended on the logarithm of target DNA concentration from 80 to 800 nmol L^{-1}.

Field effect transistors (FET) are well suited for integration of sensing layer into the electronic circuit of transducer. In CNTs based FET (NTFET) the SWCNTs are placed between conducting source and drain gold micro electrodes, respectively, at the surface of SiO_2 (Figure 12). The CNTs are aligned with their ends perpendicularly to the electrodes, so the current flows in along the tube axis. The nanotubes are modified by receptors, for example DNA or antibodies. The molecules of analyte when adsorbing to this sensing layer modulates the conductivity of nanotubes, which is reflected in changes of source-drain current, I_{SD}. Usually the current I_{SD} is measured as a function of gate voltage, V_G, or kinetics changes of the current are measured at certain gate voltage (see for example [Star et al. 2004]). NTFET based on SWCNT decorated with ssDNA (Figure 12) was used for detecting vapors of organic compounds able to change the conductivity of CNT-DNA composite [Staii et al. 2005; Johnson et al. 2006]. The ssDNA probe 5'-GAG TCT GTG GAG GAG GTA

GTC-3' was placed onto SWCNT in the FET nanogate. Based on AFM, the average diameter of SWCNT increased from 5.4 to 7.2 nm indicating the formation of DNA-SWCNT adducts. The application of ssDNA caused the threshold value of V_G for measurable conduction to decrease by 3-4 V.

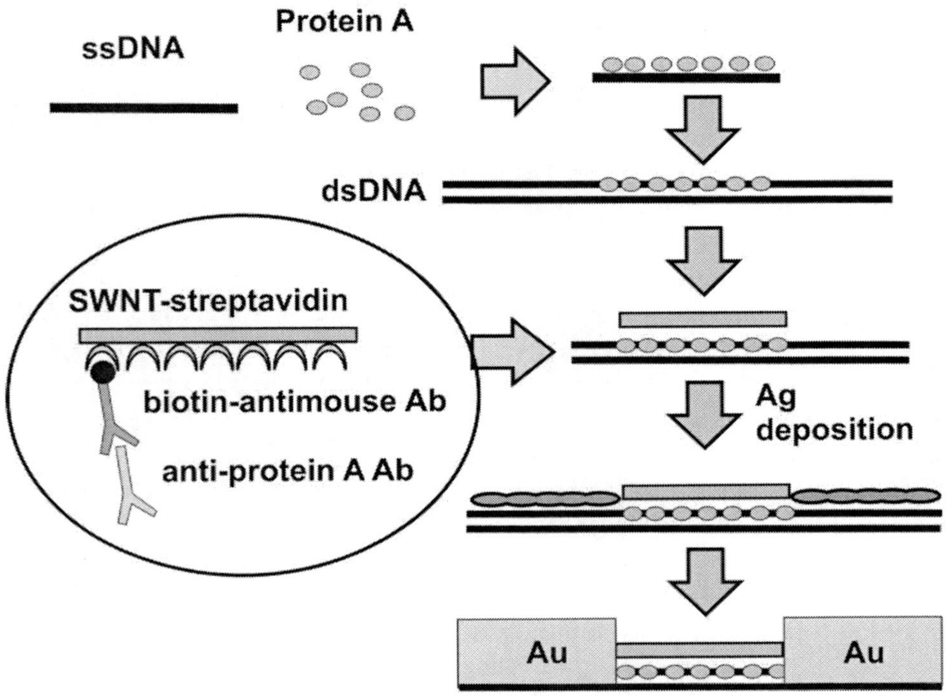

Figure 11. The scheme of preparation sensing layers in NanoFET based on a SWCNT. Adopted from [Keren et al. 2003] with permission of AAAS.

Odor compounds, e.g. propionic acid, methanol, trimethylamine, 2,6-dinitrotoluene and O,O-dimethyl-methylphosphonate (model of chemical nerve agents like sarin) were found to be able to decrease source-drain current, I_{SD}, by 10-20%. In the presence of ssDNA, the response increased by 10-14%. Charge transfer due to acid-base interaction of organic species with residual water and their influence on the conformation of DNA are considered as probably reasons for the nanoFET response. Taking into account the variety of SWCNTs conformations and DNA sequences, nanoscale olfactory sensor array can be constructed for discrimination and quantification of the vapor of organic compounds.

Similar FET with SWCNTs in the gate and Au or Cr contacts were used for detection of hybridization [Gui et al. 2007]. The drain current and barrier energy correlated with DNA immobilization and hybridization. DNA sensing mechanism involves changes in charge density near electrode – SWCNT junction. The photoresist capping study demonstrated that the contribution of electrode-SWCNT junctions (~32% change in I_d) to DNA sensing was up to six times higher than that of the SWCNT channel (~6% change in I_d).

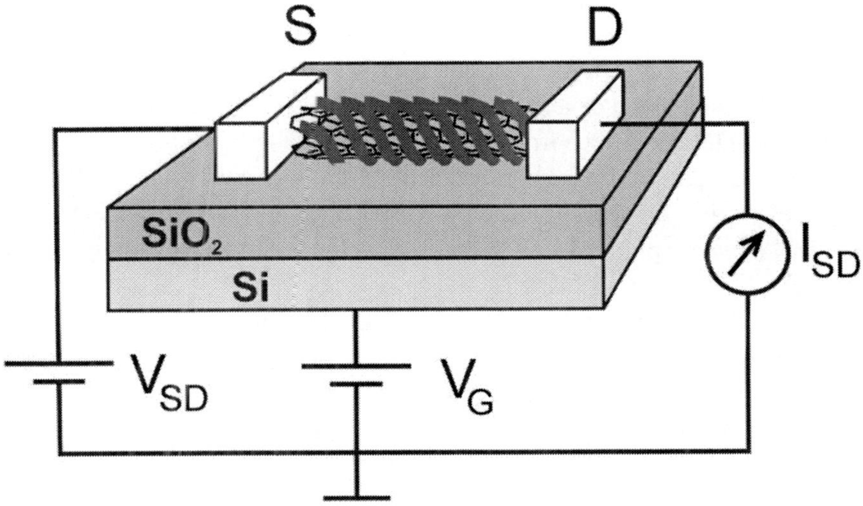

Figure 12. The scheme of field effect transistor based on single SWCNT decorated with ssDNA probe. The CNTs are aligned perpendicularly with their axis to source (S) and drain (D) gold electrodes. V_{SD} – source-drain voltage, V_G – gate voltage, I_{SD} – source-drain current.

Microscale interdigitated CNT network FETs (distance between source and drain about 10 μm) were used to detect hybridization of short target ssDNA 5'-CCT AAT AAC AAT-3' by measuring drain current [Star et al. 2006]. The localization and hybridization of ssDNA probe and target were proved by fluorescence microscopy with fluorescence labeled oligonucleotides. The possibility to specifically recognize target DNA sequences, including single-nucleotide polymorphism discrimination in the HFE gene, responsible for hereditary hemochromatosis was shown.

3.2. DNA/RNA aptamer and immunosensor-based on CNTs

3.2.1. DNA/RNA aptamers

Biosensors based on DNA or RNA aptamers (aptasensors) represent new type of the sensor that utilize unique properties of artificial receptors – aptamers. Aptasensors are of considerable interest due to their application in detection practically unlimited kind of compounds. Aptamers poses also high potential in medical therapy. Aptamers are single stranded DNA or RNA molecules with high specificity to various ligands. Their specificity is comparable and in certain cases even higher than those of antibodies. In contrast to antibodies, aptamers are prepared by in vitro selection procedure developed simultaneously in early 1990s by L. Gold and A. Ellington laboratories [Ellington and Szostak 1990; Tuerk and Gold 1990]. Advantage of aptasensors in comparison with those based on antibodies consists in the possibility to develop aptamers with high specificity to large variety of compounds by means of chemical synthesis without using experimental animals. Aptamers can be developed even against toxins or low molecular weight compounds. Aptamers can be selective to different parts of target molecule. Once the aptamers sequence is identified, this can be synthesize with high purity, reproducibility and relatively low cost, Aptamers can be easily chemically modified by various chemical tags including fluorescence probes, quenchers, electrochemical indicators, nanoparticles or enzymes. This modification allows to immobilize aptamers to various solid supports, provides their high stability, including stability of RNA against endonucleases, and to utilize various methods of detection. Therefore, aptamers can be advantageously used in the development of affinity biosensors. In contrast with antibodies, which at certain conditions can be irreversibly denaturated, aptamers-based biosensors can be regenerated without loss of integrity and selectivity [Hianik and Wang 2009]. Aptamers are also highly perspective for therapy. The antivascular endothelial growth factor aptamers – Macugen - was already approved by US authorities for the treatment of age related macular degeneration [Gold 2006].

The method of selection of aptamers was discovered independently by three groups of investigators. Robertson and Joyce [1990] described the method of selection of RNA with improved enzymatic activity to cleave DNA. Tuerk and Gold [1990] patented the process of selection of DNA ligands as a target for T4 RNA polymerase. This method is known as SELEX (systematic evolution of ligands by exponential enrichment). Ellington and Szostak [1990] reported method of in vitro selection of RNA that specifically binds organic dyes. In this paper also the term ''*aptamer*'' was introduced (from the Latin *aptus*, meaning "to fit" and Greek *meros*, meaning "the part"). The identification of aptamers is based on a combinatorial approach. Specific oligonucleotides are isolated from complex libraries of synthetic nucleic acids. For this purpose, random sequence DNA libraries are obtained by automated DNA synthesis. The size of a randomized region can vary from 30 to 60 nucleotides, flanked on both sides with a specific unique DNA sequence for polymerase chain reaction (PCR) amplification. The theoretical diversity of individual oligonucleotides in these random DNA libraries is rather large. For example, in the case of oligonucleotides composed of 40 bases, it is $4^{40}=1.2\times10^{24}$. In practice, however, a considerably smaller library of approximately 10^{13}-10^{15} molecules is used [Jayasena 1999]. The selection consists of DNA binding with immobilized ligands, such as proteins or low molecular compounds. The stability of complexes is characterized by apparent dissociation constant, K_D. For aptamer–protein complexes, K_D varies within the 1–100 nmol L^{-1} range, which is similar to antibody–antigen complexes. Unbound DNA/RNA molecules are eluted from the column, while bound aptamers are isolated from the complex and then amplified by PCR. This cycle is repeated several times (6 to 10) and as a result the DNA or RNA sequence of a high affinity to target ligand is obtained. The SELEX technique is discussed in detail in a number of papers and reviews (see, for instance Carothers and Szostak [2006]).

Although the first SELEX-related patent was filed in 1989 [Ellington and Szostak 1990], the potential of aptamer-based biosensors has not been fully realized due to the problems with aptamer stability during immobilization and signal transduction. Several problems related to the practical application of aptamers are still under study, for example how the immobilization of aptamers to the supported films and their microenvironment will affect the aptamer structure and aptamer–ligand interactions. Problems are connected with the application of aptamers in the complex biological systems, where interferences with other molecules could take place. So far, mostly radio labeled aptamers were used for quantification of protein kinase [Conrad and Ellington 1996] or in vivo detection of clots [Dougan et al. 1997]. However, to be widely employed in clinical practice, aptamers must be detected via a non-radioisotope method with a comparable sensitivity. For example aptamers can be covalently linked to an enzyme [Osborne et al. 1997], or fluorescently labeled aptamers can be exploited [McCauley et al. 2003]. Moreover, the most reliable and cost-effective way would be exploitation of the direct physical methods that do not require labeling of aptamers by additional chemical ligands. This highly promising direction is currently of considerable focus (see [Hianik and Wang 2009] for recent review). It is also highly advantageous to explore the possibility of immobilization of aptamers onto novel materials, such as carbon nanotubes [So et al. 2005; Allen et al. 2007; Wang 2007; Zhu et al. 2008; Yoon et al. 2008; Evtugyn et al. 2008; Hianik et al. 2008, 2009; Maehashi et al. 2009; Ishikawa et al. 2009] or dendrimers [Snejdarkova et al. 2008], and using nanoparticles [Wang 2007, Pavlov et al. 2004] and quantum dots [Hansen et al. 2006] in detection of the ligand–aptamer interactions.

Aptamers are rather flexible. Using simple molecular engineering methods it is possible to substantially improve their binding properties and to design aptamers that transduce the recognition event into a readily detectable electronic signal through binding-induced conformational changes.

3.2.2. Structure and molecular engineering of aptamers

Currently, the SELEX is a highly automatized procedure and only few days are necessary for development of aptamers for certain ligands. This is much shorter in comparison with the selection of antibodies, where usually several months are required. Due to effectivity of the SELEX, the library of aptamers against various ligands has become wider. On the other hand, the primary procedure does not result in all cases in aptamers with desired affinity. Therefore, optimization of aptamer structure is required. This optimization is performed through biased library generation [Jayasena 1999]. As a result, it is possible to select aptamers with sensitivity to small modification of the ligand. Aptamers can even distinguish the chirality of molecules and their secondary structure. In principle, there is no restriction in the type of target for which the aptamer can be selected. To date, aptamers with affinity to various ligands were synthesized, including metal ions, organic dyes, drugs, amino acids, co-factors, antibiotics and nucleic acid structures (see [Hianik and Wang 2009; Sassolas et al. 2009; Cheng et al. 2009] and references herein). Special interest was focused on the development of aptamers for the detection of proteins including enzymes, antibodies, prions, growth factors, gene regulatory factors, cell adhesion molecules and lectins. Also, aptamers for viral particles and pathogenic bacteria were developed (see [Torres-Chavolla and Alocilja 2009] for review). The aptamers were initially used as therapeutic agents. For example, aptamer that selectively binds thrombin was developed with the purpose of application as an anticoagulant [Bock et al. 1992]. Only about decade ago the aptamers have been used as recognition elements in biosensing (see [Scheller et al. 2001]). To date, the best investigated aptamers are those for thrombin. Thrombin is a multifunctional serine protease that plays an important role in procoagulant and anticoagulant functions. Thrombin converts soluble fibrinogen to insoluble fibrin that forms the fibrin gel, which is responsible either for a physiological plug or for pathological thrombus [Holland et al. 2000]. This process is catalyzed by positively charged fibrinogen binding site at the thrombin molecule. The more positive heparin-binding site is responsible for anticoagulant function of thrombin. These binding sites are spatially separated and localized at opposite poles of thrombin molecule [Tasset et al. 1997]. Initially, Bock et al. [1992] developed a 15-mer DNA aptamer, which selectively binds the fibrinogen site of the thrombin (Figure 13A). This aptamer inhibits the thrombin-catalyzed fibrin clot formation. The dissociation constant for this structure determined by Bock et al. [1992] was K_D = 200 nmol L^{-1}. In this aptamer the intramolecular G-quadruplex plays the crucial role. According to NMR studies [Schultze et al. 1994] and X-ray crystallography [Padmanabhan 1993], the eight guanine residues form G tetrads that are connected at one end by TT loop (T3–T4, Figure 13A) and at the other end by the TGT loop (T7G8T9, Figure 13A). The non-canonical Watson–Crick pair T4–T13 is important for stabilization of the aptamer in conformation that provides binding of T3–T4 side of quadruplex to fibrinogen binding site of the thrombin. Substitution of T4 to adenine (A) resulted in loss of binding affinity. The structure of this aptamer is stabilized by K^+ ions. Later, Tasset et al. [1997] developed 38-mer aptamer that selectively recognizes heparin binding site with considerably higher affinity, K_D = 0.5 nmol L^{-1}. The G-quadruplex also represents important part of this aptamer (Figure 13B).

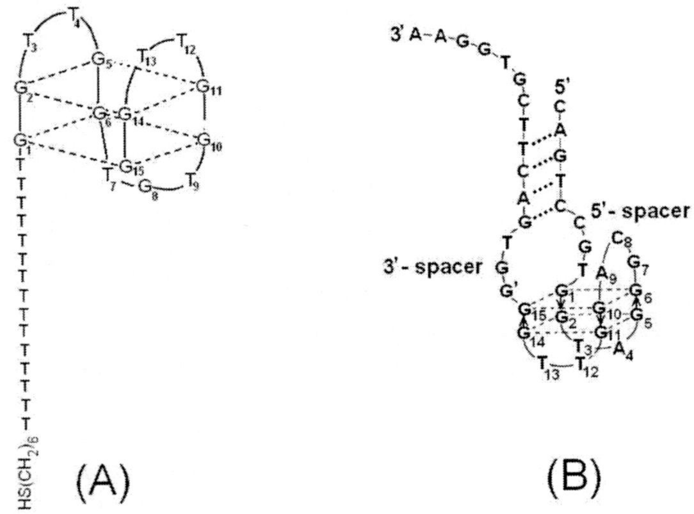

Figure 13. Structures of guanine quadruplex binding motifs for fibrinogen (A) and heparin (B) binding sites of DNA aptamers sensitive to human thrombin. Aptamer A contains $(dT)_{15}$ chain connected to hydrocarbon chain terminated by thiol group that serves for chemisorption of aptamers to a gold surface. According to Hianik and Wang [2009] with permission of Wiley-VCH.

This aptamer had certain similarities with that developed by Bock et al. [1992]. The substantial difference consisted in replacement of T4 with A4 in a quadruplex loop. However, the quadruplex structure remained stable due to extension of aptamer, in which four Watson–Crick pairs stabilized the aptamer conformation. It is interesting that for this aptamer K^+ ions had no influence on the binding properties. Special investigations by means of photocrosslinking of 5-iodo-2′-deoxyuridine (5-IdU) performed in this work showed that exclusively one nucleotide T12 is responsible for binding to thrombin through Phe245 in heparin binding site. However, other nucleotides also play crucial role in providing high affinity to this aptamer. The adenine at position 4 of the quadruplex is important for determination of binding site of this DNA aptamer. The first guanine (G′) of the 3′ spacer contributes to high affinity of the aptamer to heparin binding site of thrombin (see Tasset et al. [1997] for structure of the aptamer–thrombin complex). Our recent studies using circular dichroism spectroscopy (CD) suggest that G-quadruplex for this aptamers is also stable even without stabilizing sequence, however K^+ ions are necessary for stability [Ponikova et al.

2008]. Apart from DNA aptamers, RNA aptamers that selectively bind thrombin were also developed. The RNA aptamer developed by Kubik et al. [1994] was selective to the heparin-binding site of thrombin. The K_D for this RNA aptamer was 2–5 nmol L^{-1}, which was 10 time lower than that for DNA aptamer of high affinity to heparin binding site, i.e., DNA aptamer revealed higher affinity to thrombin. Later, White et al. [2001] developed a RNA aptamer specific for the fibrinogen-binding site of thrombin. The binding affinity of this aptamer to thrombin on the surface was determined by Gronewold et al. [2005] using filter binding assay, surface plasmon resonance (SPR) and acoustic methods. The K_D values determined by these methods were 294, 113 and 181 nmol L^{-1}, respectively.

Two-stacked guanine tetramers in the active binding site of the DNA aptamers against thrombin were typical also for other aptamers subsequently developed, such as those for insulin [Yoshida et al. 2009] and RNA aptamers (see for example aptamer against cellular prions [Weiss et al. 1997; Rhie et al. 2003]). Apart from aptamers that maintain the G-quadruplex structure of binding site, also so-called aptamer beacons were developed. These aptamers contain special complementary spacers at both 3′ and 5′ terminals. In absence of target (for example thrombin), these terminals form double helix that stabilizes the molecular beacon structure [Tan et al. 2004]. However, in presence of target, the aptamer is transformed into a typical three-dimensional structure contained G-quadruplex, allowing specific interaction with the ligand. The advantage of aptamer beacon consists in possibility of sensitive detection of ligands, using, for example, fluorescence method. One terminal of aptamer is modified by fluorescence label, while the other terminal contains quencher. Thus, without the target the quencher and probe are in close proximity and the intensity of fluorescence is low. The addition of target resulted in aptamers conformational changes that move fluorescence probe away from the quencher. As a result the fluorescence intensity increases [Hamaguchi et al. 2001]. However this method requires modification of aptamers both by fluorescence probe and by quencher. In addition the non-specific interactions may cause false positive fluorescence signal. Recently it has been shown, that CNTs can effectively quench the fluorescence. This peculiarity has been used in development of aptamer-based bioassay using conventional aptamers modified by fluorescence probe only at one end. This was possible also due to the fact that single stranded DNA aptamers non-covalently interact with CNTs. Thus, addition of the target resulted in folding of aptamers into 3D structure and desorption from the surface of CNTs. This approach has been used recently for targeted release of photosensitizer for controllable single oxygen generation [Zhu et al. 2008]. The molecular beacon approach has been used also for electrochemical detection of various analytes. One end of the aptamer was modified by an electrochemical indicator, such as MB or ferrocene. Thus, conformational changes of aptamers resulted in changes of the distance of the indicator from electrode surface and as a result changes of the redox current took place (see [Hianik and Wang 2009] for review). It is interesting that recently developed DNA aptamers against cellular prions [Takemura et al. 2006] or against a tumor marker protein, human vascular endothelial growth factor (VEGF) [Hasegawa et al. 2008a] contain the loop parts that are important for binding the target protein.

Except the molecular beacons, another design of aptamers has been proposed recently. Di Giusto and King [2004] reported multivalent circular anticoagulant aptamers containing four binding motifs to heparin binding site of thrombin. These aptamers were more stable and exhibited 2-3-fold higher anticoagulant activity in comparison with canonical single stranded aptamers. Fredrikson et al. [2002] proposed the method of fabrication of aptamer dimers using the ligation of oligonucleotides. This substantially improved sensitivity of aptamer based assay of homodimeric proteins. Increased inhibitory activity against HIV-1 reverse transcriptase has been shown at presence of bimodular DNA aptamers [Michalowski et al. 2008]. In aptamer-based assay the affinity of aptamers to the target can be increased using aptamer dimers. Hasegawa et al. [2008b] designed an aptamer dimer for thrombin and $VEGF_{165}$. The 15-mer thrombin-binding aptamer sensitive to heparin binding site has been connected via thymine linker with a 29-mer thrombin-binding aptamer sensitive to heparin binding site of the thrombin. The dissociation constant, K_D, of this aptamer was 0.1 of that for monomers. They also connected two identical aptamers against $VEGF_{165}$, which is a homodimeric protein. Similarly like for anti-thrombin aptamer, the dimeric anti-VEGF aptamer had a much lower K_D value than that of the monomer. This study showed, that the dimerization of aptamers improves the affinities of aptamers for target proteins. In our recent papers we proposed another simple design for aptamers engineering based on DNA hybridization [Hianik et al. 2008, 2009a]. Two thrombin binding aptamers were connected by hybridization of their complementary supporting parts. Thus, so-called aptabody was formed that contained two binding sites for thrombin (Figure 14).

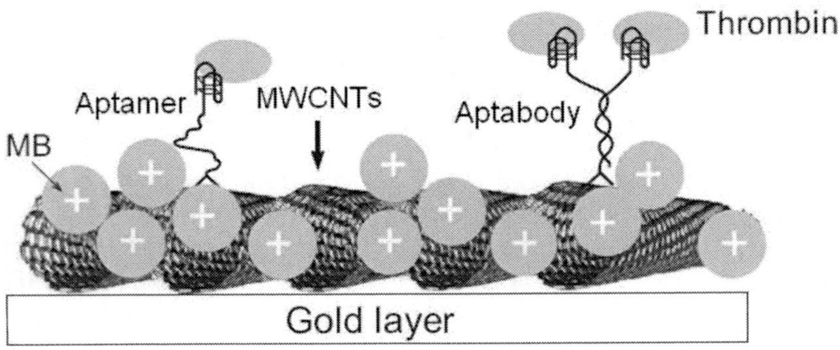

Figure 14. Schematic representation of the immobilization of aptamer and aptabody at layer formed by electropolymerisation of a mixture of MWCNTs with MB According to Ref. [Hianik et al. 2008] with permission of Bentham Science Publishers Ltd.

The aptasensor prepared by immobilisation of these aptabodies onto the surface of MWCNTs revealed approx. 3 times higher sensitivity in comparison with traditional aptamers [Hianik et al. 2008]. Similar approach has been used most recently for design of aptamers against platelet-derived growth factor BB (PDGF-BB) [Zhang et al. 2009].

3.2.3. Immobilisation of aptamers on a solid support

The crucial step in biosensor fabrication is the immobilisation of aptamers to a solid support. Control of chemistry and surface coverage is essential for enssuring high reactivity, orientation, accessibility, and stability of the surface-bound aptamer, as well as for minimizing non-specific binding or adsorption events. In principle these methods are similar to those applied previously for the immobilization of single- or double- stranded DNA in genosensors or DNA biosensors for detection of DNA damage (see Pividori et al. [2000] for review). The methods of immobilization based on physical adsorption of DNA by means of electrostatic interactions are in general not suitable due to low stability caused by aptamers desorption from the surface. However, adsorption of aptamers onto MWCNTs and their composites with electrochemical indicators, such as MB or Methylene Green (MG) results in stable structures, especially when the immobilisation process is accompanied by cycling of the voltage [Evtugyn et al. 2008]. Simple and effective method is based on chemisorption of thiol-labeled aptamers onto a gold surface [Pavlov et al. 2004; Hansen et al. 2006] or on strong affinity of biotin to avidin, streptavidin or neutravidin. In latter case, one end of DNA or RNA aptamer is modified by biotin. The solid support is covered by avidin, streptavidin or neutravidin [Liss et al. 2002; Cavic and Thompson 2002; Centi et al. 2007; Ostatna et al. 2008]. Avidin is usually chemically linked to the organic layer formed by 3,3′-dithiopropionic acid di(N-succinimidylester) (DSP) [Liss et al. 2002]. Neutravidin is very convenient because it can be directly chemisorbed on a gold and does not require additional chemical modification of the surface [Cavic and Thompson 2002; Ostatna et al. 2008]. Mixed two-component alkanethiol self-assembly monolayers of recognition and shielding components are very attractive for achieving the desired balance between high loading, minimal non-specific interactions and preferred orientation. Such mixed co-assembly monolayers have been widely used in DNA hybridization sensors and are being employed for the design of electrochemical aptasensors [Radi et al. 2006; Hansen et al. 2006]. A recent study by Plaxco's group examines the influence of the structure of such alkanethiol monolayers on the performance of electrochemical aptasensors [White 2008]. Poly(amidoamine) dendrimers (PAMAM) were also used for aptamer immobilization. PAMAM dendrimers are globular macromolecules with amino terminal groups. Thus, the biotinylated aptamers can be immobilised on a dendrimer surface preliminary modified by avidin (attachment of avidin has been performed by its croslinking to the dendrimer amino groups using glutaraldehyde). The advantage of dendrimers is their high stability and relatively large surface in comparison with flat electrode [Snejdarkova et al. 2008]. So et al. [2005] reported immobilization of thrombin aptamer onto the surface of CNTs. The nanotubes were pretreated with carbodiimidazole-activated Tween 20 (CDI-Tween). 3'- end of thrombin aptamer was modified by $-NH_2$ groups, which allowed covalent binding of aptamer to CDI-Tween. Application of conducting SWCNTs allowed fabricating FET- based biosensor for detection of thrombin [So et al. 2005] and immunoglobulin E (IgE) [Maehashi et al. 2007, 2009]. Except SWCNTs also MWCNTs were used. The advantage of MWCNTs layers is their larger surface. This allows increasing the amount of the aptamers at the surface. The aptasensor based on MWCNTs showed excellent properties, especially in combination with electrochemical quartz crystal microbalance detection [Evtugyn et al. 2008] for ligand-aptamer [Hianik et al. 2008, 2009b] and ligand-antibody interactions [Hianik et al. 2009b]. For electrochemical detection of the analyte-aptamer interaction it is also advantageous to immobilize aptamers onto composite films with desired properties, that provide good charge transfer between the sensing layer and the support. For example, Kang et al. [2008] used for this purpose nanogold-chitosan composite film. The film was formed at the surface of glassy carbon electrode. The composite layer was deposited electrochemically. The surface was then modified by glutaraldehyde for subsequent crosslinking with NH_2 terminated receptor. In this work the sandwich assay was used (see below) and instead of aptamers first the antibody was immobilised at the surface. However, this method of immobilisation can be used also for attachment of NH_2-terminated aptamers. This work demonstrated substantial improvement of the detection of thrombin by DPV using MB as an indicator. Another example of optimal selection of the support was presented by Rodriguez et al. [2005] who used streptavidin-polymer-coated indium-tin oxide electrode (ITO) for immobilizing a DNA aptamer sensitive to lysozyme and for detecting this protein using electrochemical impedance spectroscopy. When selecting the method of immobilization, it is important to provide sufficient conformational freedom to the aptamer. Particularly this can be achieved also by using sufficiently long spacer. For example, Liss et al. [2002] showed that substantial improvement in the binding properties of anti-IgE aptamer could be reached by extension of the aptamer length. Aptamer can be immobilized also on glass slides, on a Si surface using UV radiation [Potyrailo et al. 1998] or on a supported self assembled monolayers [Gronewold et al. 2005].

3.2.4. Carbon nanotubes in immunosensors and aptasensors

Affinity biosensors are of great promise for applications in medical diagnosis, detection of environmental pollutants, toxins and pathogens in food. As a bioreceptors in affinity biosensors so far mostly antibodies were used. However, growing area in this field is connected with application of DNA/RNA aptamers. The substantial advantage of aptamers over antibodies is possibility of their regeneration after binding of the analyte, while regeneration of antibodies is rarely possible and usually caused in their damage. Carbon nanotubes are suitable support for immobilisation both antibodies and aptamers. CNTs can enhance the electroactivity of biopolymers. Historically the CNTs as a support in affinity biosensors served first for immobilisation of antibodies. Therefore in this section we will first describe the works related to development protein based affinity biosensors and then we will focus on aptasensors based on CNTs. The achievements in development affinity biosensors have been reviewed recently by Allen et al. [2007], Wang and Lin [2008] and Wang and Lu [2009]. However, this field of research is so rapidly developing that even since latest review additional information appeared in literature.

Affinity biosensors based on proteins and antibodies immobilised on CNTs

Large variety of proteins binds to CNTs via non-specific adsorption. This adsorption is due to at least two reasons. 1. affinity of amino groups of protein aminoacids to CNTs [Bradley et al. 2004]. 2. intermolecular interactions between aromatic amino acids (i.e. histidine, tryptophan or tyrosine) with CNTs. Nonspecific interaction can be facilitated also by coadsorption of surfactant Triton and poly(ethyleneglycol) (PEG) [Shim et al. 2002] The presence of PEG at CNTs surface can also eliminate nonspecific binding of analyte. Some proteins, such as streptavidin crystallize at the surface of CNTs resulting in formation of protein arrays [Balavoine et al. 1999]. Many reports demonstrated also possibility of chemical modifications of CNTs for covalent binding of proteins (see for instance [Chen et al. 2001b]). For this purpose the CNTs are oxidized to have free carboxyl groups. These groups react with aminogroups at proteins [Chen et al. 2001b]. However, oxidation of CNTs affects negatively their mechanical and electrical properties. Noncovalent modification of CNTs is more favorable in respect of improved solubility of CNTs in water and in preserving mechanical and electrical properties of carbon nanotubes.

NTFET are very suitable for detection of affinity interactions. Star et al. [2003] developed NTFET biosensor for detection of streptavidin. The SWCNTs has been coated with a mixture of poly(ethylene imine) (PEI) and PEG. PEI provided amino groups for covalent attachment of biotin-N-hydroxy-succinimidylester to CNTs and PEG prevented CNTs surface from nonspecific adsorption of proteins. The sensor response has been studied by measurement of the dependence of source-drain current on the gate voltage. This dependence is presented on Figure 15. As it can be see from this Figure significant changes of the current took place upon binding of streptavidin. The mechanism of current changes was explained by effect of electron doping of CNTs channel (formed between source and drain, see Figure 12 for NTFET scheme) upon binding of charged streptavidin molecules.

Substantial sensitivity of NTFET immunosensor has been demonstrated by Li et al. [2005]. They used SWCNTs for development biosensor for detection prostate-specific antigen (PSA). The antibodies against PSA (PSA-ABs) were anchored to SWCNTs surface using π-π stacking method with a pyrene moiety {Figure 16). The interaction of PSA with PSA-ABs was explained by charge-transfer mechanism similar to those proposed by Star et al. [2003]. The LOD of PSA was 14 pmol L^{-1}. Improved LOD of PSA (7 pmol L^{-1}) was reported recently by Okuno et al. [2007]. In this work PSA-ABs were covalently linked to SWCNTs. The detection of PSA was performed by DPV. They found that the signal derived from the oxidation of tyrosine (Tyr) and tryptophan (Trp) residues in PSA increased with the interaction between PSA and PSA-ABs. Extremely low LOD of PSA (0.1 pmol L^{-1}) was reported by Yu et al. [2006]. They exploited SWCNTs for PSA-ABs immobilisation.

High sensitivity of detection was achieved by multichannel secondary antibody-nanotube conjugate with horseradish peroxidase (HRP). The immunochip based on SWNTs has been used also as a platform for HRP labeled sandwiched Enzyme-Linked ImmunoSorbent Assay (ELISA) for high sensitive detection of IgG (LOD: 2.5 pmol L^{-1}) [Lu et al. 2009]. Amperometric detection based on square-wave voltammetry becomes effective also for CNTs-based immunosensor for detection of cholera toxin (CT) [Viswanathan et al. 2006]. In this case Nafion-supported MWCNTs film was prepared on the surface of glassy carbon electrode. The potassium ferrocyanide-encapsulated into liposomes served as specific recognition label. Similar method has been used also for development of disposable electrochemical immunosensor for the detection of carcinoembryonic antigen (CEA) in saliva and serum [Viswanathan et al. 2009].

Monoclonal anti-CEA antibodies (alpha CEA) were covalently immobilized on polyethyleneimine wrapped MWCNTs screen-printed electrode. A sandwich immunoassay was performed with CEA and alpha CEA tagged ferrocene carboxylic acid encapsulated liposomes (alpha CEA-FCL). The square wave voltammetry (SWV) was employed to analyze faradic redox responses of the released ferrocene carboxylic acid from the immunoconjugated liposomes on the electrode surface. The magnitude of the SWV peak current was directly related to the concentration of CEA. The high sensitivity of CEA detection (LOD: 1 $ng.L^{-1}$) has been reported. Nafion-coated MWCNTs immunosensor for detection of alpha-1-fetoprotein (AFP) was reported by Su et al. [2009]. This electrochemical sensor was prepared by dropping of nafion-MWCNTs homogeneous composite onto a surface of glassy carbon electrode. Then stepwisely thionine (Thi) and negative-charged nano-gold particles were chemisorbed onto which anti-AFP antibody was immobilised. The sensor allowed detecting AFP with LOD of 0.26 $\mu g.L^{-1}$. Combination of CNTs and poly (diallyldimethylammonium chloride) (PDDA) allowed to substantially enhancing electrochemiluminescence (ECL) of CdSe quantum dots (QD). This finding was used for construction of immunosensor for the detection of human IgG. The principle of ECL detection of IgG was based on the increment of steric hindrance after immunoreaction, which resulted in the decrease of ECL intensity. The IgG concentration was determined with a LOD of 0.6 ng L^{-1}, which corresponds to 4 fmol L^{-1} [Jie et al. 2009]. Lin et al. [2009] reported CNTs based immunosensor for detection alpha-fetoprotein (AFP) in human serum. The immunosensor was prepared by immobilizing AFP antigen onto the glassy carbon electrode modified by gold nanoparticles and carbon nanotubes doped chitosan (GNP/CNT/Ch) film. The electrochemical properties of GNP/CNT/Ch films were characterized by impedance spectroscopy and cyclic voltammetry.

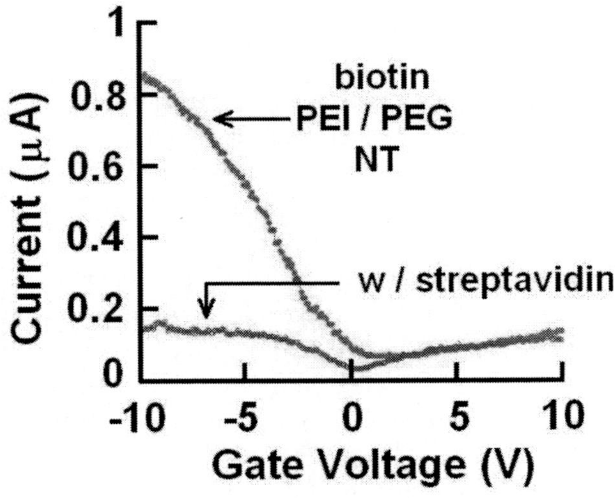

Figure 15. The source-drain current dependence on the gate voltage of the NTFET device based on SWNT functionalized with biotin in the absence and presence of streptavidin. Adopted from [Star et al. 2003] with permission of The American Chemical Society.

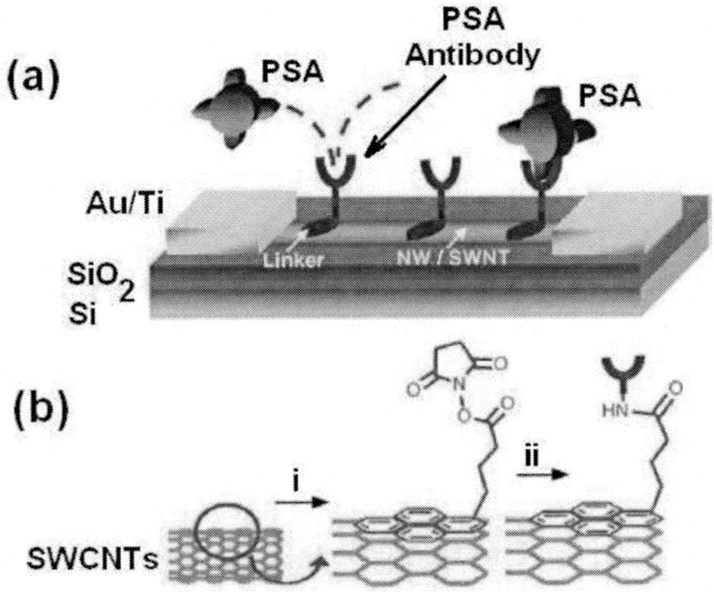

Figure 16. a) Schematic illustration of a nanosensor. Prostate-specific antigen antibodies (PSA-ABs) are anchored to the NW/SWNT surface and function as specific recognition groups for PSA binding. b) Reaction sequence for the modification of the SWNT. i) deposition of 1-pyrenebutanoic acid succinimidyl ester, ii) PSA–AB incubation. Reproduced from [Li et al. 2005] with permission of The American Chemical Society.

They found that GNP/CNT nanohybrid acted as an electron promoter and accelerated the electron transfer. Sample AFP, immobilized AFP, and alkaline phosphatase (ALP)-labeled antibody were incubated together for the determination based on a competitive immunoassay format. After the immunoassay reaction, the bound ALP label on the modified GC led to an amperometric response of 1-naphthyl phosphate (1-NP), which was changed with the different antigen concentrations in solution. Under the optimized experimental conditions, the resulting immunosensor detected AFP with a LOD of 0.6 μg L^{-1}.

NTFET has been used also for detection small molecule – bisphenol A in water [Sanchez-Acevedo et al. 2009]. SWCNTs were functionalized with the estrogen receptor alpha (ER-alpha). This biosensor allowed detecting selectively picomolar concentrations of bisphenol A. Zhang and Yi [2009] reported MWCNTs-gold nanoparticle composite based electrochemical biosensor with immobilised hemoglobin as a biocatalyst for detection nitrite. The biosensor exhibited enhanced electrocatalytic behavior to the reduction of nitrite at -0.10 V vs. SCE in 0.05 M H_2SO_4 solution. The current response of the sensor increases linearly with nitrite concentration from a range from 3.6 μmol L^{-1} to 3.0 mmol L^{-1} with a LOD of 0.96 μmol L^{-1}.

Aptasensors based on CNTs

In section 3.1. we mentioned advantage of CNTs as a support for immobilisation of DNA. 1D – dimension of conducting CNTs exhibit substantial advantage over the supports based on thin films. The advantage of SWCNTs arrays consist in direct transport of charge to the solid support with minimized loss of signal intensity in lateral direction [Yoon et al. 2008]. The next advantage is size compatibility and biocompatibility between biomolecules and CNTs. In a biosensor based on SWCNTs each atom of carbon nanotubes is exposed to environment. Thus, even small changes at the sensor surface substantially affect the electrical properties of CNTs. The electrical properties of CNTs are of great advantage especially in biosensors based on FETs. The preparation of FETs based on CNTs with immobilised DNA aptamers are similar to those of DNA biosensors discussed above. DNA aptamers like nucleic acids can be immobilised on a surface of CNTs noncovalently or covalently (see section 3.1).

In this section we will describe recent results obtained in development CNTs biosensors based on aptamers and in certain cases will compare this with those based on antibodies. Most of these sensors utilized the advantage of FETs. The principles of operation of FET and ion selective FET (ISFET) can be found in several books (see for example Eggins [2002]) and have been briefly explained above. In order to sense the interaction at the FET surface it is important that the analyte interacts with the surface in framework of so-called Debye length, which is approx. 3 nm for 10 mmol L^{-1} ionic concentrations. This rule is in particularly valid for aptamers (dimensions 1-2 nm). Thus the interaction of small molecules or proteins with aptamers at the surface of FET takes place in the layer of the thickness comparable with Debye length and therefore the sensitivity of detection is rather high.

Direct detection of thrombin using SWCNTs-FET was reported by So et al. [2005]. In this work aptamer was covalently immobilized on CNTs modified by Twin 20. Addition of thrombin resulted in drop of the conductance of the device. This has been explained by screening of negative charge of DNA by thrombin that was positively charged under experimental conditions (pH 5.4). Certainly, the isoelectric point of thrombin is rather high: 7-7.6. The lowest LOD of the sensor was 10 nmol L^{-1}. The sensor was selective to thrombin.

This was proved by experiment in which elastase instead of thrombin was used. Elastase is serine protease with similar isoelectric point and molecular weight like thrombin. Addition of elastase resulted in considerable lower changes of the current in comparison with thrombin. Sensor was regenerable by treatment with 6 M guanidine hydrochloride solution that removed bounded thrombin molecules. Another design of FET biosensor for detection thrombin was proposed by Yoon et al. [2008]. In this work the thrombin-sensitive DNA aptamers were immobilised on as surface of carboxylic acid functionalised polypyrrol nanotubes (CPPy). The CPPy nanotubes were covalently immobilised on the microelectrode substrate and the amine-terminated thrombin aptamers were covalently linked to the nanotubes. At presence of thrombin the FET sensor revealed decrease in current flow, which was attributed to the dipole-dipole and dipole-charge interactions between thrombin protein and the aptamers-conjugated polymer chains. The sensor worked effectively also in blood serum. The LOD of this sensor was 50 nmol L^{-1}.

Maehashi et al. [2007, 2009] proposed similar biosensor format with improved sensitivity of detection IgE. As the receptors they used anti-IgE antibodies or anti-IgE aptamers immobilised on a surface of CNT-FET. They showed, that aptamers based biosensor provide better sensitivity of detection of IgE in comparison with antibody based FET with LOD approx. 250 pmol L^{-1}.

Except of the FET technology also the electrochemical quartz crystal microbalance (EQCM) method has been used for development aptasensors based on MWCNTs with immobilised aptamers. EQCM is an acoustic method based on the changes of oscillation frequency of quartz crystal following binding of analyte to the receptors immobilised in a thin layer at the crystal surface. EQCM combine the mass detection with possibility of simultaneous electrochemical analysis of the surface properties. The mass sensitive method known as quartz crystal microbalance was originally developed by Sauerbrey [1959] who also introduced the equation connected the changes of the frequency, Δf, with changes of the mass, Δm: $\Delta f = - 2f_0^2\ \rho_S\ (\mu_q\rho_q)^{-1/2}$ (f_0 is the fundamental frequency, typically 5-10 MHz, ρ_S is the mass density, $\mu_q = 2.95 \times 10^{11}$ dyne cm^{-2} is shear stiffness of quartz, $\rho_q = 2.65$ g cm^{-3} is the quartz density). This equation can be transformed into form: $\Delta f = -2.26 \times 10^{-6} f_0^2\ \Delta m/A$ (A is working area of the electrode sputtered at the quartz crystal surface, typically around 0.2 cm^2). It is known that Sauerbrey equation is strongly valid only for dry crystal in vacuum and that in solution the viscoelastic forces affect the oscillation frequency. However, this method is widely using for development of biosensors operating in liquids. The sensor response should be, however analyzed with special care due to above mentioned viscoelastic contribution [Lucklum et al. 1999; Ellis and Thompson, 2004; Snejdarkova et al. 2008]. Recently we applied EQCM method in the thrombin specific aptasensors based on MWCNTs – MB composite [Evtugyn et al. 2008]. First we optimized the method of immobilisation of aptamers on a surface of MWCNTs. Among various options such as direct non-covalent binding of aptamers to the MWCNTs surface, binding of aptamers to the MWCNTs-MB composite and those contained also avidin, the film composed of MWCNTs-MB composite revealed best sensitivity. Simultaneously with measurements of the changes of the oscillation frequency of quartz crystal, which reflect both mass and viscosity changes at the crystal surface, we performed the cycling of the voltage in the range -0.7 to + 0.6 V (vs. Ag/AgCl reference electrode). This voltage cycling provided better condition for interaction of the thrombin with DNA aptamers. This method has improved detection limit substantially. While conventional QCM method with aptamers immobilisation by avidin-biotin technology resulted in detection thrombin with LOD of 10 nmol L^{-1} [Hianik et al. 2005], the immobilisation of aptamers to the MWCNTs-MB composite layer and simultaneous cycling of the voltage allowed to detect thrombin with LOD 1 nmol L^{-1} for canonical single stranded aptamers sensitive to the fibrinogen binding site at thrombin. 3-fold better LOD has been obtained for aptamer dimers composed of two single stranded aptamers sensitive to fibrinogen binding site at thrombin (see Figure 14). This aptasensor revealed also good selectivity in more complex liquids such those contained human serum albumin (HSA) and in a blood serum [Hianik et al. 2008].

Similar method of preparation MWCNTs based biosensor we have used also for development biosensor for detection cellular prions (PrP^C). In this case we also compared the sensitivity of detection PrP^C by aptamers and antibodies. Both aptamers and antibodies were immobilised at the surface of MWCNTs-MB composite by physical adsorption. The schematic picture of this biosensor is shown on Figure 17a and the sensor response to the PrP^C is shown on Figure 17b both for aptamers and antibodies as a selective receptors [Hianik et al. 2009b]. As it is seen from Figure 17b the changes of the resonant frequency following addition of PrP^C start already at rather low prion concentrations. The binding curves for both aptamers (curve 1) and antibodies (curve 2) are well described by Langmuir isotherm: $-\Delta f = -\Delta f_{max} [K_C C/(1+K_C C)]$, where Δf_{max} is a maximal change of the frequency at high PrP^C concentration, C is the concentration of PrP^C and K_C is the binding or association constant, $K_C=1/K_D$, where K_D is dissociation constant. Free energy of prion binding can be calculated from the standard Gibbs equation $\Delta G = -RT \ln K_C$. The parameters of Δf_{max}, K_C, K_D, ΔG as well as LOD for both biosensors are presented in Table 2. One can see from the Table 2, that the values of K_C, K_D and ΔG are similar for both biosensors and not significantly different. Better LOD for BAR 223 in comparison with PRI 308 may be related to higher affinity of BAR 223 to PrP^C. The maximal changes in frequency were higher for antibody BAR 223 based biosensor in comparison with those for aptamer biosensor and differed significantly according to the Student`s t-test with $p<0.01$. The frequency changes for PRI 308 based biosensor were similar to those for aptamer biosensor. Thus aptasensors and immunosensors based on MWCNTs exhibit similar sensitivity. However, considering possibility of further improvements of aptamers sensitivity by bioengineering, for instance by preparation of aptamer dimers, the improved LOD for these sensors can be potentially obtain.

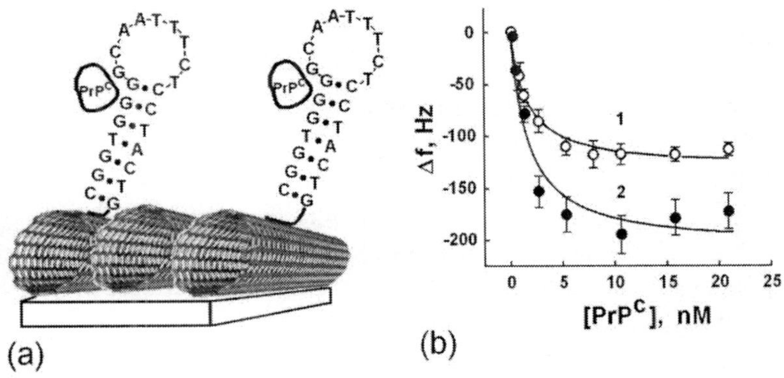

Figure 17. (a) Schematic representation of the sensor surface composed of MWCNTs and immobilised DNA aptamers with bounded PrP^c. (b) The plot of the frequency changes as a function of PrP^C concentration for biosensor based on: 1 – single-stranded aptamer and 2 – BAR 223 antibody immobilized on the surface of MWCNTs. The points are experimental results and the lines are fit according to Langmuir isotherm (error bars: SD, N = 3). Adapted from [Hianik et al. 2009b] with permission of Bentham Science Publishers Ltd.

Table 2 The parameters of Δf_{max}, K_C, K_D, ΔG as well as LOD for biosensors based on aptamer and antibodies BAR 223 and PRI 308 determined from Langmuir isotherms according to Eq. (2) and Gibbs equation $\Delta G = -RT \ln K_C$, where $K_C = 1/K_D$. Results represent mean±S.D. obtained from 3 independent experiments in each series.

Sensing element	Δf_{max}, Hz	K_C, $nmol^{-1}.L$	K_D, $nmol\ L^{-1}$	ΔG, $kJ.mol^{-1}$	LOD, $pmol\ L^{-1}$
Aptamer	-136.7±7.9	0.58±0.12	1.72±0.37	49.2±17.0	50
Antibody BAR 223	-207.0±16.4	0.61±0.20	1.64±0.53	49.3±15.9	20
Antibody PRI 308	-137.7±8.7	0.51±0.13	1.96±0.5	48.9±12.5	48

4. Conclusion

Since their discovery by Iijima [1991], the carbon naotubes have demonstrated high potential as unique nanomaterials with exceptional electrical and mechanical properties. The surface of CNTs can be easily modified by biopolymers both by physical adsorption and by covalent attachment. Such complexes revealed high stability against degradation. The electrical properties of CNTs are highly sensitive to affinity interactions at the surface, to substrate degradations by enzymes and to redox processes. Especially in SWCNTs each atom of carbon nanotubes is exposed to the environment. Thus, even small changes at the sensor surface substantially affect the electrical properties of CNTs. Therefore CNTs can be used not only as an immobilisation support, but also as transducers in biosensors. Moreover, CNTs, can be used as the conducting wires for electron transport. They can be incorporated into the field effect transistors. These CNT-FET integrate the CNTs transduction properties into the electrical circuit. Thus, the CNT-FET modified with bioreceptors can serve as highly sensitive biosensors. CNTs can be doped by nanoparticles and electrochemical indicators, which further extend their use in biosensing with respect to application of various electrochemical techniques for detection of analyte-receptor interactions. The overview of applications of CNTs in the development of DNA and affinity biosensors (based on antibodies and DNA aptamers) presented in this chapter clearly demonstrate the substantial advantage of CNTs over other materials used so far in biosensor development. At the same time, further effort is required to optimize fabrication of CNT-FET and other CNT-based biosensors. Additional exciting routes of CNTs application could be related to molecular and cell biology. CNTs can be used as the probes for direct contacts with the cells to probe their properties [Hood et al. 2004, Allen et al. 2007]. The unique electronic properties of CNTs in combination with biopolymers may accelerate their use in molecular electronics and in biocomputers. Considering the high biocompatibility of CNTs with a living system, we can also expect more extensive use of this nanomaterial in development of targeted drug delivery systems, cancer therapy, and in other medical applications.

ACKNOWLEDGMENTS

The financial support by Slovak Grant Agency (Project No. 1/4016/07), Agency for Promotion Research and Development under the contract No. APVV-0362-07, European Commission under FP6 (DVT-IMP, IST-034256), NoE NeuroPrion (Contract No. FOOD-CT-2004-506579) and by Russian Foundation for Basic Research (grants 08-03-00161 and 09-03-92420 to GE) is greatly acknowledged.

5. REFERENCES

Ahammad, A. J. S., Lee, J. J. & Rahman, M. A. (2009). Electrochemical sensors based on carbon nanotubes. *Sensors*, *9*, 2289-2319.

Ajayan, P. M., Ebbesen, T. W., Ichihashi, T., Iijima, S., Tanigaki, K. & Hiura, H. (1993). Opening carbon nanotubes with oxygen and implications for filling. *Nature*, *362*, 522-525.

Allen, B. L., Kichambare, P. D. & Star, A. (2007). Carbon nanotube field-effect-transistor-based biosensors. *Adv. Mater.*, *19*, 1439-1451.

Andrews, R., Jacques, D., Qian, D. & Dickey, E. C. (2001). Purification and structural annealing of multiwalled carbon nanotubes at graphitization temperatures. *Carbon*, *39*, 1681-1687.

Andrews, R., Jacques, D., Qian, D. & Rantell, T. (2002). Multiwall carbon nanotubes: synthesis and application. *Acc. Chem. Res.*, *35*, 1008-1017.

Baker, S. E., Cai, W., Lasseter, T. L., Weidkamp, K. P. & Hamers, R. J. (2002). Covalently bonded adducts of deoxyribonucleic acid (DNA) oligonucleotides with single-wall carbon nanotubes: synthesis and hybridization. *Nano Lett.*, *2*, 1413-1417.

Balasubramanian, K. & Burghard, M. (2006). Biosensors based on carbon nanotubes. *Anal. Bioanal. Chem.*, *385*, 452-468.

Balavoine, F., Schultz, P., Richard, C., Mallouh, V., Ebbesen, T. W. & Mioskowski, C. (1999). Helical crystallization of proteins on carbon nanotubes: A first step towards the development of new biosensors. *Angew. Chem. Int. Ed.*, *38*, 1912-1915.

Bandow, S., Asaka, S., Saito, Y., Rao, A. M., Grigorian, L., Richter, E. & Ecklund, P. C. (1998). Effect of the growth temperature on the diameter distribution and chirality of single-wall carbon nanotubes. *Phys. Rev. Lett.*, *80*, 3779-3782.

Banks, C. E., Davies, T. J., Wildgoose, G. G. & Compton, R. G. (2005). Electrocatalysis at graphite and carbon nanotube modified electrodes: edge-plane sites and tube ends are the reactive sites. *Chem. Commun.*, 829-841.

Banks, C. E. & Compton, R. G. (2006). New electrodes for old: From carbon nanotubes to edge plane pyrolytic graphite. *Analyst*, *131*, 15-21.

Blase, X., Benedict, L. X., Shirley, E. L. & Louie, S. G. (1994). Hybridization effects and metallicity in small radius carbon nanotubes. *Phys. Rev. Lett.*, *72*, 1878-1881.

Bock, L. C., Griffin, L. C., Latham, J. A., Vermaas, E. H. & Toole, J. J. (1992). Selection of single-stranded DNA molecules that bind and inhibit human thrombin, *Nature*, *355*, 564-566.

Bonard, J. M., Klinke, C., Dean, K. A. & Coll, B. F. (2003). Degradation and failure of carbon nanotube field emitters. *Phys. Rev. B. 67*, 115406.

Bower, C., Suzuki, S., Tanigaki, K. & Zhou, O. (1998). Synthesis and structure of pristine and cesium intercalated single-walled carbon nanotubes. *Appl. Phys., A67*, 47-52.

Bradley, K., Briman, M., Star, A. & Grüner, G. (2004). Charge transfer from adsorbed proteins. *Nano Lett.*, 4, 253-256.

Britto, P. J., Santhanam, K. S. V., Alonso, V., Rubio, A. & Ajayan, P. M. (1999). Improved charge transfer at carbon nanotube electrodes. *Adv. Mater.*, *11*, 154-157.

Cai, H., Xu, Y., He, P. G. & Fang, Y. Z. (2003a). Indicator free DNA hybridization detection by impedance measurement based on the DNA-doped conducting polymer film formed on the carbon nanotube modified electrode. *Electroanalysis*, *15*, 1864-1870.

Cai, H., Cao, X., Jiang, Y., He, P. & Fang, Y. (2003b). Carbon nanotube-enhanced electrochemical DNA biosensor for DNA hybridization detection. *Anal. Bioanal. Chem.*, *375*, 287-293.

Carothers, J. M. & Szostak, J. W. (2006). In vitro selection of functional oligonucleotides and the origins of biochemical activity. In: Klussmann, S. (Ed.) *The Aptamer Handbook,* Weinheim Wiley-VCH, 3-28.

Cavic, B. A. & Thompson, M. (2002). Interfacial nucleic acid chemistry studied by acoustic shear wave propagation. *Anal. Chim. Acta*, *469*, 101-113.

Centi, S., Tombelli, S., Minunni, M. & Mascini, M. (2007). Aptamer-based detection of plasma proteins by an electrochemical assay coupled to magnetic beads. *Anal. Chem.*, *79*, 1466-1473.

Chang, Z., Fan, H., Zhao, K., Chen, M., He, P. & Fang, Y. (2008). Electrochemical DNA biosensors based on palladium nanoparticles combined with carbon nanotubes. *Electroanalysis*, *20*, 131-136.

Charlier, J. C. (2002). Defects in carbon nanotubes. *Acc. Chem. Res.*, *35*, 1063-1069.

Chen, R. J., Franklin, N. R., Kong, J., Cao, J., Tombler, T. W., Zhang, Y. & Dai, H. (2001a). Molecular photodesorption from carbon nanotubes. *Appl. Phys. Lett.*, *79*, 2258-2260.

Chen, J., Dyer, M. J. & Yu, M. F. (2001b). Cyclodextrin-mediated soft cutting of single-walled carbon nanotubes. *J. Am. Chem. Soc.*, *123*, 6201-6202.

Chen, Y., Liu, H., Ye, T., Kim, J. & Mao, C. (2007). DNA-directed assembly of single-wall carbon nanotubes. *J. Am. Chem. Soc., 129*, 8696-8697.

Cheng, A. K. H., Sen, D. & Yu, H. Z. (2009). Design and testing of aptamer-based electrochemical biosensors for proteins and small molecules. *Bioelectrochemistry*, *77*, 1-12.

Cinke, M., Li, J., Chen, B., Cassell, A., Delzeit, L., Han, J. & Meyyappan, M. (2002). Pore structure of raw and purified HiPco single-walled carbon nanotubes. *Chem. Phys. Lett.*, *365*, 69-74.

Collins, P. G., Bradley, K., Ishigami, M. & Zettl, A. (2000). Extreme oxygen sensitivity of electronic properties of carbon nanotubes. *Science*, *287*, 1801-1804.

Cohen, M. L. (2001). Nanotubes, nanoscience, and nanotechnology. *Mater. Sci. Eng. C*, *15*, 1-11.

Conrad, R. & Ellington, A. D. (1996). Detecting immobilized protein kinase C isozymes with RNA aptamers. *Anal. Biochem.*, *242*, 261-265.

Cui, D., Pan, B., Zhang, H., Gao, F., Wu, R., Wang, J., He, R. & Asahi, T. (2008). Self-assembly of quantum dots and carbon nanotubes for ultrasensitive DNA and antigen detection. *Anal. Chem.*, *80*, 7996-8001.

Dai, H. (2001). Nanotube growth and characterization. In: M. S. Dresselhaus, G. Dresselhaus, & Ph. Avouris (Eds.), *Carbon nanotubes. Topics Appl. Phys.*, Springer: Berlin, Heidelberg. 80, 29-53.

Dai, H. (2002). Carbon nanotubes: synthesis, integration, and properties. *Acc. Chem. Res.*, *35*, 1025-1044.

Daniel, S., Rao, T. P., Rao, K. S., Rani, S. U., Naidu, G. R. K., Lee, H. & Kawai, T. (2007). A review of DNA functionalized/grafted carbon nanotubes and their characterization. *Sensors Actuators B*, *122*, 672-682.

Di Giusto, D. A. & King, G. C. (2004). Construction, stability and activity of multivalent circular anticoagulant aptamers, *J. Biol. Chem.*, *279*, 46483-46489.

Dougan, H., Hobbs, J. B., Weitz, J. I. & Lyster, D. M. (1997). Synthesis and radioiodination of a stannyl oligodeoxyribonucleotide. *Nucleic Acids Res.*, *25*, 2897-2901.

Dresselhaus, M. S. & Eklund P. C. (2000). Phonons in carbon nanotubes. *Adv. Phys.*, *49*, 705-814.

Dresselhaus, M. S., Dresselhaus, G., Jorio, A., Souza Filho, A. G., Pimenta, M. A. & Saito, R. (2002). Single nanotube Raman spectroscopy. *Acc. Chem. Res.*, *35*, 1070-1078.

Eggins, B. R. (2002). *Chemical Sensors and Biosensors*, Wiley, Chichester.

Ellis, J. S. & Thompson, M. (2004). Slip and coupling phenomena at the liquid-solid interface. *Phys. Chem. Chem. Phys.*, *6*, 4928-4938.

Ellington, A. D. & Szostak, J. W. (1990). In vitro selection of RNA molecules that bind specific ligands. *Nature*, *346*, 818-822.

Erdem, A., Papakonstantinou, P. & Murphy, H. (2006). Direct DNA hybridization at disposable graphite electrodes modified with carbon nanotubes. *Anal. Chem.*, *78*, 6656-6659.

Evtugyn, G. A., Porfireva, A. V. Ryabova, M. & Hianik, T. (2008). Aptasensor for thrombin based on carbon nanotubes-methylene blue composites, *Electroanalysis*, *20*, 2310-2316.

Fan, S., Chapline, M. G., Franklin, N. R., Tombler, T. W., Cassell, A. M. & Dai, H. (1999) Self-oriented regular arrays of carbon nanotubes and their field emission properties. *Science*, *283*, 512-514.

Frackowiak, E., Delpeux, S., Jurewicz, K., Szostak, K., Cazorla-Amoros, D. & Béguin, F. (2002). Enhanced capacitance of carbon nanotubes through chemical activation. *Chem. Phys. Lett.*, *336*, 35-41.

Fredriksson, S., Gullberg, M., Jarvius, J., Olsson, C., Pietras, K. Gustafsdottir, S. M., Oestman, A. & Landegren, U. (2002). Protein detection using proximity-dependent DNA ligation assays. *Nat. Biotechnol.*, *20*, 473-477.

Galandova, J., Ovadekova, R., Ferancova, A. & Labuda, J. (2009). Disposable DNA biosensor with the carbon nanotubes-polyethyleneimine interface at a screen-printed carbon electrode for tests of DNA layer damage by quinazolines. *Anal. Bioanal. Chem.*, *394*, 855-861.

Gao, B., Bower, C., Lorentze, J., Fleming, L., Kleinhammes, A., Tang, X. P., McNeil, L. E., Wu, Y. & Zhou, O. (2000). Enhanced saturation Li composition in ball-milled single-walled carbon nanotubes. *Chem. Phys. Lett.*, *327*, 69-75.

Gao, B., Yue, G.Z., Qiu, Q., Cheng, Y., Shimoda, H., Fleming, L. & Zhou, O. (2001). Fabrication and electron field emission properties of carbon nanotube films by electrophoretic deposition. *Adv. Mater.*, *13*, 1770-1774.

Gao, M., Qi, H., Gao, Q. & Zhang, C. (2008). Electrochemical detection of DNA hybridization based on the probe labeled with carbon-nanotubes loaded with silver nanoparticles. *Electroanalysis*, *20*, 123-130.

Gold, L. (2006). A personal perspective: Aptamers after 15 years. In: Klussmann S. (Ed.) *The Aptamer Handbook*, Weinheim, Wiley-VCH, 461-469.

Gong, M., Han, T., Cai, C., Lu, T. & Du, J. (2008). Fabrication and characterization of DNA-thionine-carbon nanotube nanocomposites. *J. Electroanal. Chem.*, *623*, 8-14.

Gooding, J. J., Wibowo, R., Liu, J., Yang, W., Losic, D., Orbons, S., Mearns, F. J., Shapter, J. G. & Hibbert, D. B. (2003). Protein electrochemistry using aligned carbon nanotube arrays. *J. Am. Chem. Soc.*, *125*, 9006-9007.

Gronewold, T. M. A., Glass, S., Quantdt, E. & Famulok, M. (2005). Monitoring complex formation in the blood-coagulation cascade using aptamer-coated SAW sensors. *Biosens. Bioelectron.*, *20*, 2044-2052.

Gui, E. L., Li, L. J., Zhang, K., Xu, Y., Dong, X., Ho, X., Lee, P. S., Kasim, J., Shen, Z. X., Rogers, J. A. & Mhaisalkar, S. G. (2007). DNA sensing by field-effect transistors based on networks of carbon nanotubes. *J. Am. Chem. Soc.*, *129*, 14427-14432.

Hamaguchi, N., Ellington, A. & Stanton, M. (2001). Aptamer beacons for the direct detection of proteins. *Anal. Biochem.*, 294, 126-133.

Hansen, J. A., Wang, J., Kawde, A. N., Xiang, Y., Gothelf, K. V. & Collins, G. (2006). Quantum-dot/aptamer-based ultrasensitive multi-analyte electrochemical biosensor. *J. Am. Chem. Soc.* *128*, 2228-2229.

Hasegawa, H., Sode, K. & Ikebukuro, K. (2008a). Selection of DNA aptamers against VEGF(165) using a protein competitor and the aptamer blotting method. *Biotechnol. Lett.*, *30*, 829-834.

Hasegava, H., Tiara, K., Sode, K. & Ikebukuro, K. (2008b). Improvement of aptamer affinity by dimerization, *Sensors*, *8*, 1090-1098.

He, P., Xu, Y. & Fang, Y. (2006). Applications of carbon nano- tubes in electrochemical DNA biosensors. *Microchim. Acta*, *152*, 175-186.

Hianik, T., Ostatna, V., Zajacova, Z., Stoikova, E. & Evtugyn, G. (2005). Detection of aptamer-protein interactions using QCM and electrochemical indicator methods. *Bioorg. Med. Chem. Lett.*, *15*, 291-295.

Hianik, T., Porfireva, A., Grman, I. & Evtugyn, G. (2008). Aptabodies – new type of artificial receptors for detection proteins. *Protein and Peptide Lett.*, *15*, 799-805.

Hianik, T., Grman, I. & Karpisova, I. (2009a). The effect of DNA aptamers configuration on the sensitivity of detection thrombin at surface by acoustic method. *Chem. Commun.* 6303-6305.

Hianik, T., Porfireva, A., Grman, I. & Evtugyn, G. (2009b). EQCM biosensors based on DNA aptamers and antibodies for rapid detection of prions. *Protein and Peptide Lett. 16*, 363-367.

Hianik, T. & Wang, J. (2009). Electrochemical aptasensors - Recent achievements and perspectives. *Electroanalysis*, *21*, 1223-1235.

Holland, C. A., Henry, A. T., Whinna, H. C. & Church, F. C. (2000). Effect of oligodeoxynucleotide thrombin aptamer on thrombin inhibition by heparin cofactor II and antithrombin, *FEBS Lett.*, *484*, 87-91.

Holloway, A. F., Wildgoose, G. G., Compton, R. G., Shao, L. & Green, M. L. H. (2008). The influence of edge-plane defects and oxygen-containing surface groups on the voltammetry of acid-treated, annealed and "super-annealed" multiwalled carbon nanotubes. *J. Solid State Electrochem., 12*, 1337-1348.

Hood, L., Heath, J. R., Phelps, M. E. & Lin, B. Y. (2004). Systems biology and new technologies enable predictive and preventative medicine. *Science*, *306*, 640-643.

Huang, N. Y., She, J. C., Chen, J., Deng, S. Z., Xu, N. S., Bishop, H., Huq, S. E., Wang, L., Zhong, D. Y., Wang, E. G. & Chen, D. M. (2004). Mechanism responsible for initiating carbon nanotube vacuum breakdown. *Phys. Rev. Lett., 93*, 075501.

Hwang, E. S., Cao, C., Hong, S., Jung, H. J., Cha, C. Y., Choi, J. B., Kim, Y. J. & Baik, S. (2006). The DNA hybridization assay using single-walled carbon nanotubes as ultrasensitive, long-term optical labels. *Nanotechnology*, *17*, 3442-3445.

Iijima, S. (1991) Helical microtubules of graphitic carbon. *Nature*, *354*, 56-58.

Ishikawa, F. N., Stauffer, B., Caron, D. A. & Zhou C. (2009). Rapid and label-free cell detection by metal-cluster-decorated carbon nanotube biosensors. *Biosens. Bioelectr.*, *24*, 2967-2972.

Jayasena, S. D. (1999). Aptamers: an emerging class of molecules that rival antibodies in diagnostics. *Clin. Chem.*, *45*, 1628-1650.

Jiang, C., Yang, T., Jiao, K. & Ga, H. (2008). A DNA electrochemical sensor with poly-l-lysine/single-walled carbon nanotubes films and its application for the highly sensitive EIS detection of PAT gene fragment and PCR amplification of NOS gene. *Electrochimica Acta, 53*, 2917-2924.

Jie, G. F., Li, L. L., Chen, C., Xuan, J. & Zhu, J. J. (2009). Enhanced electrochemiluminescence of CdSe quantum dots composited with CNTs and PDDA for sensitive immunoassay. *Biosens. Bioelectr.*, *24*, 3352-3358.

Johnson, A. T. C., Staii, C., Chen, M., Khamis, S., Johnson, R., Klein, M. L. & Gelperin, A. (2006). DNA-decorated carbon nanotubes for chemical sensing. *Semicond. Sci. Technol., 21*, S17-S21.

Kang, Y., Feng, K. J., Chen, J. W., Jiang, J. H., Shen, G. L. & Yu, R. Q. (2008). Electrochemical detection of thrombin by sandwich approach using antibody and aptamers. *Bioelectrochemistry*, *73*, 76-81.

Karachevtsev, V. A., Glamazda, A. Yu., Leontiev, V. S., Lytvyn, O. S. & Dettlaff-Weglikowska, U. (2007). Glucose sensing based on NIR fluorescence of DNA-wrapped single-walled carbon nanotubes. *Chem. Phys. Lett.*, *435*, 104-108.

Katauta, H., Kumazawa, Y., Maniwa, Y., Umezu, I., Suzuki, S., Ohtsuka, Y. & Achiba, Y. (1999). Optical properties of single-wall carbon nanotubes. *Synth. Met.*, *103*, 2555-2558.

Katz, E., Willner, I. (2004). Biomolecule-functionalized carbon nanotubes: Applications in nanobioelectronics. *Chem. Phys. Chem.*, *5*, 1084-1104.

Keren, K., Berman, R. S., Buchstab, E., Sivan, U. & Braun, E. (2003). DNA-Templated carbon nanotube field-effect transistor. *Science*, 302, 1380-1382.

Kim, S. N., Rusling, J. F. & Papadimitrakopoulos, F. (2007). Carbon nanotubes for electronic and electrochemical fetection of biomolecules. *Adv. Mater.*, *19*, 3214-3228.

Kim, S. N., Kuang, Z., Grote, J. G., Farmer, B. L. & Naik, R. R. (2008). Enrichment of (6,5) single wall carbon nanotubes using genomic DNA. *Nano Lett., 8*, 4415-4420.

Kohli, P. & Martin, C. R. (2005). Template-synthesized nanotubes for biotechnology and biomedical applications, *J. Drug Delivery Sci. Technol.,5*, 49-57.

Kubik, M. F., Stephens, A. W., Schneider, D., Marlar, R. A. & Tasset, D. (1994). High-affinity RNA ligands to human α-thrombin. N*ucleic Acids Res*., *22*, 2619-2626.

Li, C., Currelli, M., Lin, H., Lei, B., Ishikawa, F. N., Datar, R., Cote, R., Thompson, M. & Zhou, C. (2005). Complementary detection of prostate-specific antigen using In_2O_3 nanowires and carbon nanotubes. *J. Am. Chem. Soc*., *127*, 12484-12485.

Li, H. H., Zhang, Z. L., Shen, X. C., Zhang, M. X. & Pang, D. W. (2007). Glassy carbon electrodes modified with DNA-partly-wrapped single-walled carbon nanotubes. Electrochem. *Commun*., *9*, 2729-2733.

Li, W. Z., Xie, S. S., Qian, L. X., Chang, B. H., Zou, B. S., Zhou, W. Y., Zhao, R. A. & Wang, G. (1996). Large-scale synthesis of aligned carbon nanotubes. *Science*, *274*, 1701-1703.

Li, X. M., Zhan, Z. M., Ju, H. Q. & Zhang, S. S. (2008). Label-free electrochemical detection of short sequences related to the hepatitis B virus using 4,4′-diaminoazobenzene based on multiwalled carbon nanotube-modified GCE. *Oligonucleotides, 18*, 321-328.

Li, Q., Zhang, J., Yan, H., He, M. & Liu, Z. (2004). Thionine-mediated chemistry of carbon nanotubes. *Carbon, 42*, 287-291.

Lin, Y., Taylor, S., Li, H., Fernando, K. A. S., Qu, L., Wang, W., Gu, L., Zhou, B. & Sun, Y. P. J. (2004). Advances toward bioapplications of carbon nanotubes. *J. Mater. Chem*., *14*, 527-541.

Lin, J. H., He, C. Y., Zhang, L. J. & Zhang, S. S. (2009). Sensitive amperometric immunosensor for alpha-fetoprotein based on carbon nanotube/gold nanoparticle doped chitosan film. *Anal. Biochem*., *384*, 130-135.

Liss, M., Petersen, B., Wolf, H. & Prohaska, E. (2002). An aptamer-based quartz crystal protein biosensor. *Anal. Chem., 74*, 4488-4495.

Liu, J., Rinzler, A., Dai, H., Hafner, J., Bradley, A. R., Boul, P., Lu, A., Iverson, T., Shelimov, A. K., Huffman, C., Rodriguez-Macias, F., Shon, Y., Lee, R., Colbert, D. & Smalley, R. E. (1998). Fullerene pipes. *Science*, *280*, 1253-1256.

Lu, M., Lee, D. J., Xue, W. & Cui, T. H. (2009). Flexible and disposable immunosensors based on layer-by-layer self-assembled carbon nanotubes and biomolecules. *Sens. Act. A*, *150*, 280-285.

Lucklum, R., Behling, C. & Hauptmann, P. (1999). Role of mass accumulation and viscoelastic film properties for the response of acoustic-wave-based chemical sensors. *Anal. Chem., 71*, 2488-2496.

Luo, H. X., Guo, Z. X. & He, N. (2007). Reversible electrochemistry of DNA on multi-walled carbon nanotube modified electrode. *Chinese Chem. Lett., 18*, 861-864.

Ma, Y., Ali, S. R., Dodoo, A. S. & He, H. (2006). Enhanced sensitivity for biosensors: multiple functions of DNA-wrapped single-walled carbon nanotubes in self-doped polyaniline nanocomposites. *J. Phys. Chem. B., 110*, 16359-16365.

Ma, Y., Chiu, P. L., Serrano, A., Ali, S. R., Chen, A. M. & He, X. (2008). The electronic role of DNA-functionalized carbon nanotubes: efficacy for in Situ polymerization of conducting polymer nanocomposites. *J. Am. Chem. Soc., 130*, 7921-7928.

Maehashi, K., Katsura, T., Kernan, K., Takamura, Y., Matsumoto, K. & Tamyia, E. (2007). Label-free protein biosensor based on aptamer-modified carbon nanotube field-effect transistors. *Anal. Chem., 79*, 782-787.

Maehashi, K., Matsumoto, K., Takamura, Y. & Tamiya, E. (2009). Aptamer-based label-free immunosensors using carbon nanotube field-effect transisors. *Electroanalysis*, *21*, 1285-1290.

Manohar, S., Tang, T. & Jagota, A. (2007). Structure of homopolymer DNA-CNT hybrids. *J. Phys. Chem. C.*, *111*, 17835-17845.

Masheter, A. T., Xiao, L., Wildgoose, G. G., Crossley, A., Jones, J. H. & Compton R. G. (2007). Voltammetric and X-ray photoelectron spectroscopic fingerprinting of carboxylic acid groups on the surface of carbon nanotubes via derivatisation with arylnitro labels. *J. Mater. Chem., 17*, 3515-3524.

McCauley, G. T., Hamaguchi, N. & Stanton, M. (2003). Aptamer-based biosensor arrays for detection and quantification of biological molecules. *Anal. Biochem.*, *319*, 244-250.

Merkoci, A., Pumera, M., Llopis, X., Perez, B., del Valle, M. & Alegret, S. (2005). New materials for electrochemical sensing VI: Carbon nanotubes. *Trends Anal. Chem., 24*, 826-839.

Michalowski, D., Chitima-Matsiga, R., Held, D. M. & Burke, D. H. (2008). Novel bimodular DNA aptamers with guanosine quadruplexes inhibit hylogenetically diverse HIV-1 reverse transcriptases, *Nucleic Acids Res., 36*, 7124-7135.

Monthioux, M., Smith, B. W., Burteaux, B., Claye, A., Fischer, J. E. & Luzzi, D. E. (2001). Sensitivity of single-wall carbon nanotubes to chemical processing: an electron microscopy investigation. *Carbon*, *39*, 1251-1272.

O'Connell, M. J., Bachilo, S. M., Huffman, C. B., Moore, V. C., Strano, M. S., Haroz, E. H., Rialon, K. L., Boul, P. J., Noon, W. H., Kittrell, C., Ma, J., Hauge, R. H., Weisman, B. & Smalley, R. E. (2002). Band gap fluorescence from individual single-walled carbon nanotubes. *Science*, *297*, 593-596.

Odom, T. W., Huang, J. L., Kim, P. & Lieber, C. M. (1998). Atomic structure and electronic properties of single-walled carbon nanotubes. *Nature*, *391*, 62-64.

Okuno, J., Maehashi, K., Kerman, K., Takamura, Y., Matsumoto, K. & Tamiya, E. (2007). Label-free immunosensor for prostate-specific antigen based on single-walled carbon nanotube arraymodified microelectrodes. *Biosens. Bioelectron.*, *22*, 2377-2381.

Osborne, S. E., Matsumura, I. & Ellington, A. D. (1997). Aptamers as therapeutic and diagnsotic reagents: problems and prospects. *Curr. Opin. Chem. Biol.*, *1*, 5-9.

Ostatna, V., Vaisocherova, H., Homola, J. & Hianik, T. (2008). Effect of the immobilisation of DNA aptamers on the detection of thrombin by means of surface plasmon resonance. *Anal. Bioanal. Chem.*, *391*, 1861-1869.

Padmanabhan, K., Padmanabhan, K. P., Ferrara, J. D., Sadler, J. E. & Tulinsky, A. (1993). The structure of α-thrombin inhibited by a 15-mer single-stranded DNA aptamer. *J. Biol. Chem., 268,* 17651-17654.

Pavlov, V., Xiao, Y., Shlyahovsky, B. & Willner, I. (2004). Aptamer-functionalized Au nanoparticles for the amplified optical detection of thrombin. *J. Am. Chem. Soc., 126*, 11768-11769.

Pedano, M. L. & Rivas, G. A. (2004). Adsorption and electrooxidation of nucleic acids at carbon nanotubes paste electrodes. *Electrochem. Commun.*, *6*, 10-16.

Pierard, N., Fonseca, A., Konya, Z., Willems, I., Van Tendeloo, G. & Nagy, J. B. (2001). Production of short carbon nanotubes with open tips by ball milling. *Chem. Phys. Lett.*, *335*, 1-8.

Pividori, M. I., Merkoci, A. & Alegret, S. (2000). Electrochemical genosensor design: immobilization of oligonucleotides onto transducer surfaces and detection methods. *Biosens. Bioelectron.*, *15*, 291-303.

Ponikova, S., Antalik, M. & Hianik, T. (2008). A circular dichroism study of the stability of guanine quadruplexes of thrombin DNA aptamers at presence of K^+ and Na^+ ions. *Gen. Physiol. Biophys.*, *27*, 271–277.

Potyrailo, R. A., Conrad, R. C., Ellington, A. D. & Hieftje, G. M. (1998). Adapting selected nucleid acid ligands (aptamers) to biosensors. *Anal. Chem.*, *70*, 3419-3425.

Qi, H., Li, X., Chen, P. & Zhang, C. (2007). Electrochemical detection of DNA hybridization based on polypyrrole/ss-DNA/multi-wall carbon nanotubes paste electrode. *Talanta*, *72*, 1030-1035.

Radi, A. E., Acero Sanchez, J. L., Baldrich, E. & O'Sullivan, C. K. (2006). Reagentless, reusable, ultrasensitive electrochemical molecular beacon aptasensor, *J. Am. Chem. Soc.* *128*, 117-124.

Rhie, A., Kirby, L., Sayer, N., Wellesley, R., Disterer, P., Sylvester, I., Gill, A., Hope, J., James, W. & Tahiri-Alaouit, A. (2003). Characterization of 2l-fluoro-RNA aptamers that bind preferentially to disease-associated conformations of prion protein and inhibit conversion. *J. Biol. Chem.*, *278*, 39697-39705.

Rivas, G. A., Rubianes, M. D., Rodriguez M. C., Ferreyra, N. F., Luque, G. L., Pedano M. L., Miscoria S. A. & Parrado, C. (2007). Carbon nanotubes for electrochemical biosensing. *Talanta*, *74*, 291-307.

Robertson, D. L. & Joyce, G. F. (1990). Selection in vitro of an RNA enzyme that specifically cleaves single-stranded DNA. *Nature*, *344*, 467-468

Rodriguez, M. C., Kawde, A. N. & Wang, J. (2005). Aptamer biosensor for label-free impedance spectroscopy detection of proteins based on recognition-induced switching of the surface charge. *Chem. Commun.*, 4267-4269.

Roy, S., Vedala, H., Roy, A. D., Kim, D., Doud, M., Mathee, K., Shin, H., Shimamoto, N., Prasad, V. & Choi, W. (2008). Direct electrical measurements on single-molecule genomic DNA using single-walled carbon nanotubes. *Nano Lett.*, *8*, 26-30.

Salimi, A., Banks, C. E. & Compton, R. G. (2004). Abrasive immobilization of carbon nanotubes on a basal plane pyrolytic graphite electrode: application to the detection of epinephrine. *Analyst*, *129*, 225-228.

Sanchez-Acevedo, Z. C., Riu, J. & Rius, F. X. (2009). Fast picomolar selective detection of bisphenol A in water using a carbon nanotube field effect transistor functionalized with estrogen receptor-alpha. *Biosens. Bioelectr.*, *24*, 2842-2846.

Sassolas, A., Blum L. J. & Leca-Bouvier, B. D. (2009). Electrochemical aptasensors. *Electroanalysis*, *21*, 1237-1250.

Sauerbrey, G. (1959). The use of oscillator for weighing thin layers and for microweighing, *Z. Phys.*, *155*, 206-210.

Scheller, F. W., Wollenberg, U., Warsinke, A. & Lisdat, F. (2001) Research and development in biosensors. *Curr. Opin. Biotechnol.*, *12*, 15-40.

Schultze, P., Macaya, R. F. & Feigon, J. (1994). Three-dimensional solution structure of the thrombin-binding DNA aptamer d(GGTTGGTGTGGTTGG). *J. Mol. Biol., 235*, 1532-1547.

Shie, J. W., Yogeswaran, U. & Chen, S. M. (2008). Electroanalytical properties of cytochrome c by direct electrochemistry on multi-walled carbon nanotubes incorporated with DNA biocomposite film. *Talanta, 74*, 1659-1669.

Shim, M., Kam, N., Chen, R., Li, Z. & Dai, H. (2002). Functionalization of carbon nanotubes for biocompatibility and biomolecular recognition. *Nano Lett.,* 2, 285-288.

Shimoda, H., Oh, S. J., Geng, H. Z., Walker, R. J., Zhang, X. B., McNeil, L. E. & Zhou, O. (2002). Self-assembly of carbon nanotubes. *Adv. Mater., 14*, 899-901.

Sinnott, S. B., Andrews, R., Qian, D., Rao, A. M., Mao, Z., Dickey, E. C. & Derbyshire, F. (1999). Model for carbon nanotube growth through chemical vapor deposition. *Chem. Phys. Lett., 315*, 25-30.

Snejdarkova, M., Svobodova, L., Polohova, V. & Hianik, T. (2008). The study of surface properties of an IgE-sensitive aptasensor using an acoustic method. *Anal. Bioanal. Chem., 390*, 1087-1091.

So, H. M., Won, K., Kim, Y. H., Kim, B. K., Ryu, B. H., Na, P. S., Kim, H. & Lee, J. O. (2005). Single-walled carbon nanotube biosensors using aptamers as molecular recognition elements. *J. Am. Chem. Soc., 127*, 11906-11907.

Staii, C., Johnson, A. T., Chen, M. & Gelperin, A. (2005). DNA-decorated carbon nanotubes for chemical sensing. *Nano Lett., 5*, 1774-1778.

Star, A., Gabriel, J. C. P., Bradley, K. & Grüner, G. (2003). Electronic detection of specific protein binding using nanotube FET devices. *Nano Lett., 3*, 459-463.

Star, A., Han, T. R., Joshi, V. Gabriel, J. C. P. & Grüner. G. (2004). Nanoelectronic carbon dioxide sensors. *Adv. Mat., 16*, 2049-2052.

Star, A., Tu, E., Niemann, J., Gabriel, J. C. P., Joiner, C. S. & Valcke, C. (2006). Label-free detection of DNA hybridization using carbon nanotube network field-effect transistors. *PNAS, 103*, 921-926.

Stepanek, I., Maurin, G., Bernier, P., Gavillet, J., Loiseau, A., Edwards, R. & Jaschinski, O. (2000). Nano-mechanical cutting and opening of single wall carbon nanotubes. *Chem. Phys. Lett., 331*, 125-131.

Steuerman, D. W., Star, A., Narizzano, R., Choi, H., Ries, R. S., Nicolini, C, Stoddart, J. F. & Heath, J. R. (2002). Interactions between conjugated polymers and single-walled carbon nanotubes. *J. Phys. Chem. B., 106*, 3124-3130.

Su, H. L., Yuan, R. Chai, Y. Q., Zhuo, Y., Hong, C. L., Liu, Z. Y. & Yang, X. (2009). Multilayer structured amperometric immunosensor built by self-assembly of a redox multi-wall carbon nanotube composite. *Electrochim. Acta, 54*, 4149-4154.

Sun, W., Li, Y., Duan, Y. & Jiao, K. (2009). Direct electrochemistry of guanosine on multi-walled carbon nanotubes modified carbon ionic liquid electrode. *Electrochimica Acta, 54*, 4105-4110.

Taherpour, A. A. (2009). Quantitative structural relationship and theoretical study of electrochemical properties of C_{60}@[SWCN(5,5)-armchair-C_nH_{20}] complexes. *Chem. Phys. Lett., 469*, 135-139.

Takemura, K., Wang, P., Vorberg, I., Surewicz, W., Priola, S. A., Kanthasamy, A., Pottathil, R. & Chen, S. G. (2006). Sreevatsan, DNA aptamers that bind to PrP^C and not PrP^{Sc} show sequence and structure specificity. *S. Exp. Biol. Med., 231*, 204-214.

Tan, W., Wang, K. & Drake, T. J. (2004). Molecular beacons. Curr. *Opinion Chem. Biol., 8*, 547-553.

Tasset, M. D., Kubik, F. M. & Steiner, W. (1997). Oligonucleotide inhibitors of human thrombin that bind distinct epitopes, *J. Mol. Biol., 272*, 688-698.

Thorogood, C. A., Wildgoose, G. G., Crossley, A., Jacobs, R. M. J., Jones, J. H. & Compton R.G. (2007). Differentiating between ortho- and para-quinone surface groups on graphite, glassy carbon, and carbon nanotubes using organic and inorganic voltammetric and X-ray photoelectron spectroscopy labels. *Chem. Mater., 19*, 4964-4974.

Torres-Chavolla, E. & Alocilja, E. C. (2009). Aptasensors for detection of microbial and viral pathogens. *Biosens. Bioelectr., 24*, 3175-3182.

Tuerk, C. & Gold, L. (1990). Systematic evolution of ligands by exponential enrichment: RNA ligands to bacteriophage T4 DNA polymerase, *Science*, *249*, 505-510.

Vetcher, A. A., Srinivasan, S., Vetcher, I. A., Abramov, S. M., Kozlov, M., Baughman, R. H. & Levene, S. D. (2006). Fractionation of SWCNT/nucleic acid complexes by agarose gel electrophoresis. *Nanotechnology, 17*, 4263-4269.

Viswanathan, S., Wu, L. C., Huang, M. R. & Ho, J. A. A. (2006). Electrochemical immunosensor for cholera toxin using liposomes and poly(3,4-ethylenedioxythiophene)-coated carbon nanotubes. *Anal. Chem., 78*, 1115-1121.

Viswanathan, S., Rani, C., Anand, A. V. & Ho, J. A. A. (2009). Disposable electrochemical immunosensor for carcinoembryonic antigen using ferrocene liposomes and MWCNT screen-printed electrode. *Biosens. Bioelectr., 24*, 1984-1989.

Wanekaya, A. K., Chen, W., Myung, N. V. & Mulchandani, A. (2006). Nanowire-based electrochemical biosensors. *Electroanalysis, 18*, 533-550.

Wang, G., Xu, J. J. & Chen, H. Y. (2002). Interfacing cytochrome c to electrodes with a DNA - carbon nanotube composite films. E*lectrochem. Commun*., *42* 506-509.

Wang, J., Kawde, A. N. & Musameh, M. (2003a). Carbon-nanotube-modified glassy carbon electrodes for amplified label-free electrochemical detection of DNA hybridization. *Analyst*, *128*, 912-916.

Wang, J., Liu, G., Jan, M. R. & Zhu, Q. (2003b). Electrochemical detection of DNA hybridization based on carbon-nanotubes loaded with CdS tags. *Electrochem. Commun.,* *5*, 1000-1004.

Wang, J., Kawde, A. N. & Jan, M. R. (2004a). Carbon-nanotube-modified electrodes for amplified enzyme-based electrical detection of DNA hybridization. *Biosens. Bioelectron., 20*, 995-1000.

Wang, J., Liu, G. & Jan, M. R. (2004b). Ultrasensitive electrical biosensing of proteins and DNA: carbon-nanotube derived amplification of the recognition and transduction events. *J. Am. Chem. Soc.*, *126*, 3010-3011.

Wang, J. (2005.) Carbon-nanotube based electrochemical biosensors: A review. *Electroanalysis*, *17*, 7-14.

Wang, J. (2007). Nanoparticle-based electrochemical bioassays of proteins. *Electroanalysis, 19*, 769-776.

Wang, J. & Lin, Y. (2008). Functionalized carbon nanotubes and nanofibers for biosensing applications. *Trends in Anal. Chem., 27*, 619-626.

Wang, S. G., Wang, R., Sellin, P. J. & Zhang, Q. (2004). DNA biosensors based on self-assembled carbon nanotubes. *Biochem. Biophys. Research Commun.,* 325, 1433-1437.

Wang, Y., Iqbal, Z. & Malhotra, S. V. (2005). Functionalization of carbon nanotubes with amines and enzymes. *Chem. Phys. Lett., 402*, 96-101.

Wang, Z. & Lu, Yi. (2009). Functional DNA directed assembly of nanomaterials for biosensing. *J. Mat. Chem. 19*, 1788-1798.

Weiss, S., Proske, D., Neumann, M., Groschup, M. H., Kretzschmar, H. A. Famulok, M. & Winnacker, E. L. (1997). RNA aptamers specifically interact with the prion protein PrP. *J. Virol., 71*, 8790-8797.

Wildgoose, G., Banks, C., Leventis, H. & Compton, R. (2006). Chemically modified carbon nanotubes for use in electroanalysis. *Microchim. Acta, 152*, 187-214.

White, R., Rusconi, C., Scantino, E., Wolberg, A., Lawson, J., Hoffman, M. & Sullenger, B. (2001). Generation of species cross-reactive aptamers using toggle "SELEX". *Mol. Ther., 4*, 567-573.

White, R., Phares, N., Lubin. A., Xiao, Y. & Plaxco, K. W. (2008). Optimization of electrochemical aptamer-based sensors via optimization of probe packing density and surface chemistry Langmuir, *24*, 10513-10518.

Yang, Y., Wang, Z., Yang, M., Li, J., Zheng, F., Shen, G. & Yu, R. (2007a). Electrical detection of deoxyribonucleic acid hybridization based on carbon-nanotubes/nano zirconium dioxide/chitosan-modified electrodes. *Anal. Chim. Acta, 584*, 268-274.

Yang, J., Jiao, K. & Yang, T. (2007b). A DNA electrochemical sensor prepared by electrodepositing zirconia on composite films of single-walled carbon nanotubes and poly(2,6-pyridinedicarboxylic acid), and its application to detection of the PAT gene fragment. *Anal. Bioanal. Chem., 389*, 913-921.

Yang, X., Lu, Y., Ma, Y., Liu, Z. Du, F. & Chen, Y. (2007c). DNA electrochemical sensor based on an adduct of single-walled carbon nanotubes and ferrocene. *Biotechnol. Lett., 29*, 1775-1779.

Yang, H., Li, H. & Jiang, X. (2008). Detection of foodborne pathogens using bioconjugated nanomaterials. *Microfuidics and Nanofluidics., 5*, 571-583.

Yang, T., Zhou, N., Zhang, Y., Zhang, W., Jiao, K. & Li, G. (2009). Synergistically improved sensitivity for the detection of specific DNA sequences using polyaniline nanofibers and multi-walled carbon nanotubes composites. *Biosens. Bioelectron., 24*, 2165-2170.

Ye, Y. & Ju, H. (2005). Rapid detection of ssDNA and RNA using multi-walled carbon nanotubes modified screen-printed carbon electrode. *Biosens. Bioelectron., 21*, 735-741.

Yoshida, W., Mochizuki, E., Takase, M. Hasegawa, H., Morita, Y., Yamazaki, H., Sode, K. & Ikebukuro, K. (2009). Selection of DNA aptamers against insulin and construction of an aptameric enzyme subunit for insulin sensing. *Biosens. Bioelectr., 24*, 1116-1120.

Yoon, H., Kim, J. H., Lee, N., Kim, B. G. & Jang, J. (2008). A novel sensor platform based on aptamers-conjugated polzpyrrole nanotubes for label-free electrochemical protein detection. *ChemBioChem., 9*, 634-641.

Yu, X., Munge, B., Patel, V., Jensen, G., Bhirde, A., Gong, J. D., Kim, S. N., Gillespie, J., Gutkind, J. S., Papadimitrakopoulos, F. & Rusling, J. F. (2006). Carbon nanotube amplification strategies for highly sensitive immunodetection of cancer biomarkers. *J. Am. Chem. Soc., 128*, 11199-11205.

Yudasaka, M., Komatsu, T., Ichihashi, T., Achiba, Y. & Iijima, S. (1998). Pressure dependence of the structures of carbonaceous deposits formed by laser ablation on targets composed of carbon, nickel, and cobalt. *J. Phys. Chem. B, 102*, 4892-4896.

Zangmeister, R. A., Maslar, J. E., Opdahl, A. & Tarlov, M. J. (2007). Adsorption behavior of DNA-wrapped carbon nanotubes on self-assembled monolayer surfaces. *Langmuir, 23*, 6252-6256.

Zhang, Y., Chan, A., Cao, J., Wang, Q., Kim, W., Li, Y., Morris, N., Yenilmez, E., Kong, J. & Dai, H. (2001). Electric field-directed growth of aligned single-walled carbon nanotubes. *Appl. Phys. Lett., 79*, 3155-3157.

Zhang, Y., Ma, H., Zhang, K., Zhang, S. & Wang, J. (2009). An improved DNA biosensor built by layer-by-layer covalent attachment of multi-walled carbon nanotubes and gold nanoparticles. *Electrochimica Acta, 54*, 2385-2391.

Zhang, R., Wang, X. & Chen, C. (2007). Electrochemical biosensing platform using carbon nanotube activated glassy carbon electrode. *Electroanalysis, 19*, 1623-1627.

Zhang, W., Yang, T., Huang, D. M. & Jiao, K. (2008). Electrochemical sensing of DNA immobilization and hybridization based on carbon nanotubes/nano zinc oxide/chitosan composite film. *Chinese Chem. Lett., 19*, 589-591.

Zhang, W., Yang, T., Zhuang, X., Guo, Z. & Jiao, K. (2009). An ionic liquid supported CeO_2 nanoshuttles-carbon nanotubes composite as a platform for impedance DNA hybridization sensing. *Biosens. Bioelectron., 24*, 2417-2422.

Zhang, Y., Ma, H., Zhang, K., Zhang, S. & Wang, J. (2009a). An improved DNA biosensor built by layer-by-layer covalent attachment of multi-walled carbon nanotubes and gold nanoparticles. *Electrochimica Acta, 54*, 2385-2391.

Zhang, Y., Zhang, K. & Ma, H. (2009b). Electrochemical DNA biosensor based on silver nanoparticles/poly(3-(3-pyridyl)acrylic acid)/carbon nanotubes modified electrode. *Anal. Biochem., 387*, 13-19.

Zhang, Y., Pang, P., Jiang, J. H., Shen, G. L. & Yu, R. Q. (2009c). A simple and sensitive biosensor based on silver enhancement of aptamer-gold nanoparticle aggregation. *Electroanalysis, 21*, 1316-1320.

Zhang, L. Y. & Yi, M. (2009). Electrochemical nitrite biosensor based on the immobilization of hemoglobin on an electrode modified by multiwall carbon nanotubes and positively charged gold nanoparticle. *Bioproc. Biosyst. Eng., 32*, 485-492.

Zheng, D., Li, X. & Ye J. (2009). Adsorption and release behavior of bare and DNA-wrapped-carbon nanotubes on self-assembled monolayer surface. *Bioelectrochemistry, 74*, 240-245.

Zheng, M., Jagota, A., Strano, M. S., Santos, A. P., Barone, P., Chou, S. G., Diner, B. A., Dresselhaus, M. S., Mclean, R. S., Onoa, G. B., Samsonidze, G. G., Semke, E. D., Usrey, M. & Walls D. J. (2003a). Structure-based carbon nanotube sorting by sequence-dependent DNA assembly. *Science, 302*, 1545-1548.

Zheng, M, Jagota, A., Semke, E. D., Diner, B. A., Mclean, R. S., Lustig, S. R., Richardson, R. E. & Tassi, N. G. (2003b). DNA-assisted dispersion and separation of carbon nanotubes. *Nature Mat., 7*, 338-342.

Zhu, N., Chang, Z., He, P. & Fang, Y. (2005). Electrochemical DNA biosensors based on platinum nanoparticles combined carbon nanotubes. *Anal. Chim. Acta*, *545*, 21-26.

Zhu, Z., Tang, Z., Phillips, J. A., Yang, R., Wang, H. & Tan, W. (2008). Regulation of singlet oxygen generation using single-walled carbon nanotubes. *J. Am. Chem. Soc., 130*, 10856-10857.

In: Horizons in DNA Research, Volume 1
Editor: Jason R. Chesterton, pp. 49-80
ISBN: 978-1-60876-968-1

Chapter 2

BRINGING TISSUE IDENTIFICATION INTO THE 21ST CENTURY: MRNA ANALYSIS AS THE NEXT MOLECULAR BIOLOGY REVOLUTION IN FORENSIC SCIENCE?

Trisha L. Noreault-Conti and Eric Buel*
Vermont Forensic Laboratory, Department of Public Safety,
103 South Main Street, Waterbury, VT 05671, USA

ABSTRACT

DNA analysis using advanced molecular biological tools is universally used to unlock identifying information contained in crime scene stains. These advances have allowed scientists to provide critical information to solve crimes and allow the criminal justice community to convict the guilty and exonerate the innocent. Scientific research continues to develop new automated technologies and methods to yield more information from limited samples allowing scientists to optimize time and effort. Although the changes realized over the last 20 years have been remarkable, only small changes have occurred in the identification of the tissue source of a biological stain. Considerable research has been conducted to offer an alternative to these classical identification methods through the use of mRNA analysis. The application of mRNA into the forensic laboratory has not yet taken hold, but offers a new approach that could unequivocally identify difficult to characterize stains and allow the use of molecular biological tools for the identification of stains found at a crime scene. The nature of the research in this chapter is to identify mRNA transcripts that will definitively identify the tissue of origin, determine if such transcripts survive the typical environmental insults that forensic samples may encounter, and develop rapid assays using small amounts of sample. The stability of RNA was evaluated over time using real-time TaqMan®-based PCR assays. Once RNA was shown to be stable in samples aged up to 4 years, 2-3 tissue-specific transcripts were identified for a variety of stains. In addition to specificity, the sensitivity of the assays was also determined using different sized stains. Once candidates were

* Corresponding author: E-mail: ebuel@dps.state.vt.us

shown to be tissue-specific, the Plexor® One-Step qRT-PCR System was used to develop multiplex assays. This system allows the multiplexing of multiple RNAs in one assay, thus reducing the amount of sample needed and time of analysis. Through collaboration with Promega, a quick, one-tube seminal fluid-saliva screening assay was designed, since analysis of these fluids is a major task for all forensic laboratories. Additionally, a RNA/DNA co-isolation technique was developed which effectively extracts both nucleic acids, thus removing the requirement for separate extractions and allows DNA profiling from the exact stain in which RNA testing was performed. In conjunction with Promega, a likely outcome is a real-time PCR screening assay that determines with certainty whether a stain contains seminal fluid and/or saliva, an invaluable tool to the forensic community.

INTRODUCTION

Twenty years ago forensic scientists offered a statistical interpretation of their biological stain analysis from a homicide or sexual assault case that could represent ten percent of the population. The prosecuting attorney was quite happy with this statistic as it ruled out ninety percent of the population as a possible donor. Alex Jeffreys changed all that in 1985 with his use of multi-locus probe RFLP analysis.

The application of DNA analysis to obtain the type of information that could link a particular individual to a crime scene stain has revolutionized the way forensic scientists analyze biological material. Since the early developments of DNA analysis, much effort has been made to improve the analytical capabilities of the forensic DNA typing laboratory. Methods have been developed to improve the extraction of DNA from a variety of sample types using robotics or other semi-automated procedures. Quantitation of total human DNA extracted was once a labor intensive and semi-quantitative procedure using a slot-blot approach. This procedure was replaced by real-time PCR assays which yield faster, more quantitative results with far less effort. Furthermore, improvements occurred within this technology to allow the determination of the male component in addition to the total human DNA present in the sample.

The biggest change seen since the introduction of forensic DNA analysis is the rapid migration from RFLP analysis to PCR. PCR has revolutionized biotechnology and has allowed the forensic community to offer analyses that were beyond reach only a handful of years ago. The application of PCR has seen great changes throughout the years from a limited number of loci identified on hybridization strips to silver staining of STRs on polyacrylamide gels to the use of multi-loci amplification analyzed semi-automatically on capillary electrophoresis instruments. These sweeping, fundamental changes have forever changed how forensic scientists approach biological stain analysis. Gone are the days where one speaks about absorption elution techniques to determine ABO blood type or where isoelectric focusing is used to sub-type an enzyme polymorphism. Today we have robots that can extract a stain for rapid instrumental analysis. Much has been accomplished to improve the "back-end" analysis of biological evidence.

The examination of biological evidence can be viewed as two separate steps. The "back-end" or the analysis of a particular stain as described above and the "front-end" of the analysis which involves the detection and identification of the tissue type of a potential body fluid stain. Evidence of a biological nature is routinely examined visually for stains which may then undergo presumptive and confirmatory testing. As described above, the analytical procedures to individualize a stain to a person have seen remarkable changes over the last 20 years. However, the front-end of the analysis, the presumptive and confirmatory testing of a biological stain, has not seen similar changes. A review of the progress made in the analysis of biological evidence was recently published by Fourney et al. (2007). This review details the advances seen in the two disciplines and points out the lack of significant progress in the screening of biological evidence. Traditional serological approaches for stain identification often involve a presumptive color test, followed by a confirmatory test that typically employs a specific antibody designed to complex with a known protein; such as hemoglobin for human blood and P30 or prostatic antigen for seminal fluid. Modest improvements have been made in the identification of some body fluids, but many of the tests still rely upon the fundamental antibody-antigen reactions used by the forensic community for years. These technical changes, dominated by the use of simple immunochromatographic tests, are less labor intensive than the classical Ouchterlony or crossover electrophoresis methods and have essentially replaced these older procedures.

Although these newly employed techniques offer a simplified and faster analysis, they still suffer from similar problems encountered with the classical tests. Major shortcomings of the broadly employed front-end methods include consuming portions of a perhaps limited sample, lack of acceptable confirmatory tests, limited capacity for automation and cross-reactivity with other fluids or species. For example, the traditional test for saliva involves detecting the presence of the enzyme amylase. While this enzyme is found in relatively high concentrations in saliva, it is also present at lower levels in other fluids which makes analysis of saliva via conventional methods problematic.

Over the past few years considerable work by many researchers has been conducted to change the very way forensic scientists identify the tissue source of a stain. With the advent of innovative molecular biological techniques, it is plausible to imagine the eventual replacement of the serological testing methods traditionally used to identify questioned stains with tests similar to those used for DNA analysis. As described above, the forensic community has embraced the use of PCR analysis for DNA profiling, making polymorphic enzyme assays used to differentiate individuals relics of the past. An analogous transition from the conventional biochemical approach to a molecular biological approach seems inevitable and could conceivably replace traditional tissue identification techniques. New tests that are tissue specific and designed to be multiplexed in a semi-automated approach would yield rapid results on a minimal amount of sample. Such testing could employ mRNA as the tissue-specific determinant to allow the identification of forensically important fluids/tissues by pinpointing the appropriate tissue-specific mRNAs. While the DNA of all tissues from an individual is essentially identical, the mRNA spectrum made by the different cells in each tissue is very different. Each tissue or cell type makes a unique constellation of mRNAs, some specific for only that tissue or cell type. Some body fluids, such as blood, contain cells as part of their function while other fluids, such as urine, contain cells that have been shed from their tissue of origin. Therefore, analysis of the "RNA profile" in a sample can uniquely identify the fluid or tissue of origin.

The purpose of this RNA-based stain identification research is to develop a simple mRNA extraction and analysis method to allow the quick, unequivocal identification of body fluid stains and tissues (for review, see Bauer, 2007). The proposed assays will be designed so that a single stain will amplify with only the corresponding mRNA(s) of its fluid/tissue type;

a mixture, however, should amplify representing various fluids/tissues. The assays are designed to function more qualitatively than quantitatively, so although they may indicate which fluids/tissues are present, they may not give the exact ratio of each fluid contained in the stain. However, it is of greater concern to know what kind of mixture exists, than to know the exact amount of each fluid present. Furthermore, a major advantage to these assays is that a single test will be used to classify the sample as blood, semen, vaginal secretions, brain, heart, etc. which will drastically reduce the number of identification analyses performed prior to DNA profiling.

When molecular biologists began isolating mRNA for experiments, it was thought that mRNA was very ephemeral and that tissues needed to be processed rapidly in separate rooms with dedicated instruments and often with hazardous chemicals. However, due to the development of new techniques and the recent increase in knowledge concerning mRNA, it has been shown to be relatively stable. Groups have isolated mRNAs from blood, semen and saliva for research and diagnostic purposes. For blood, groups have isolated mRNA from dried blood spots for RT-PCR and restriction or quantitation (Matsubara et al., 1992; Cao and Cousins, 2000; Zhang and McCabe, 1992; Watanapokasin et al., 2000). For semen, a number of groups have developed methods to detect hepatitis C viral mRNA or HIV mRNA in seminal fluid (Bourlet et al., 2002; Dulioust et al., 1998). Researchers have also isolated nuclear mRNAs such as calcium channel subunit mRNAs from the sperm in semen (Goodwin et al, 2000). In fact, mRNAs of a number of genes have been found in human spermatozoa (Miller, 2000; Richter et al., 1999). In addition, beta-hCG mRNA has been isolated from the prostate cells in human ejaculate (Daja et al., 2000). For saliva, mRNA of viruses such as measles has been detected (Jin et al., 2002). In terms of forensic analysis, Bauer et al. (1999) have detected mRNAs specific for epithelial (endometrial) cells in menstrual blood samples. They found that they could isolate mRNA after 6 months of room temperature storage and detect a number of mRNAs species. Bauer et al. (2003) have a subsequent paper where they studied 106 bloodstains stored up to 15 years. They found that mRNA levels as measured by laser-induced fluorescence capillary electrophoresis correlates with the age of the sample and that "mRNA suitable for RT-PCR can be isolated from samples stored up to 15 years". Various groups have demonstrated the stability of RNA in forensic stains by traditional RT-PCR and more recently, RT reactions coupled with real-time PCR (Juusola and Ballantyne, 2003; Alvarez et al., 2004; Bauer et al., 2003; Nussbaumer et al., 2006; Noreault-Conti and Buel, 2007). Furthermore, Setzer et al. (2008) addressed concerns on the stability, and hence recoverability, of RNA in forensic samples (blood, saliva, semen, and vaginal secretions) exposed to a range of environmental conditions from 1 to 547 days. The results demonstrate that RNA can be recovered from biological stains in sufficient quantity and quality for mRNA analysis.

In 2003, Juusola and Ballantyne, using RT-PCR to identify candidate tissue-specific genes for saliva, were able to demonstrate that RNA is stable in biological stains and can be recovered in sufficient quantity and quality for analysis. Also in 2003, Bauer and Patzelt (2003) showed that the detection of protamine gene expression in dried semen stains by RT-PCR provides evidence for the presence of sperm. More recent work from Juusola and Ballantyne (2005) identified tissue-specific genes for the more common forensic casework samples: blood, saliva, semen and vaginal secretions. Furthermore, they developed a multiplex RT-PCR assay composed of eight genes which could be analyzed on a standard capillary electrophoresis platform; a more accurate and more sensitive detection method than agarose gel electrophoresis. Nussbaumer et al. (2006) reported the development of real-time PCR assays for body fluid identification using one gene each for blood, semen, vaginal secretions and saliva genes. The blood and semen gene transcripts were specific, but the vaginal secretion and saliva genes were variably expressed within and between different body fluids. In contrast, Juusola and Ballantyne (2007) chose to develop triplex real-time PCR assays that detect two body fluid-specific genes for blood, saliva, semen and menstrual blood in addition to a housekeeping gene; the idea being that the use of two markers in each assay would improve assay specificity and provide analytical redundancy by taking into account possible biological variation in gene expression levels between individuals. An approach to identify stable tissue-specific mRNA markers using real-time PCR verification of blood and saliva stains was reported by Zubakov et al. (2008); 5 mRNAs specific for saliva and 9 mRNAs specific for whole blood were identified. In a recent publication, Bauer and Patzelt (2008) expanded their work using real-time PCR and menstrual blood markers. In this report, samples were identified as non-menstrual by determining a cut-off value for the control marker GAPDH, below which negative results for the menstrual blood marker MMP-7 allow the conclusion that the sample was indeed not menstrual blood.

Technology in the field of multiplexing gene expression assays is rapidly improving. The Plexor® One-Step qRT-PCR System takes advantage of the specific interaction between two modified nucleotides to achieve quantitative PCR analysis. As part of the Plexor® One-Step qRT-PCR reaction, a reverse transcription step is performed in order to streamline the RNA analysis. It is possible to design assays to quantify multiple targets within the same reaction using primer pairs with a different fluorophore for each target sequence. The limiting factor in assay design is the dye capability of the real-time PCR instrument. However, since DNA quantitation is moving towards a molecular biological approach which utilizes three- or four-dye real-time PCR instruments, most labs could implement this technology without purchasing new equipment.

As the instrumentation necessary to perform mRNA analysis is in place in most labs, the next major obstacle to overcome prior to laboratories investing in RNA technology is the development of a straight-forward extraction procedure. A method that could co-purify RNA and DNA from a single sample with a minimal number of steps would be attractive to those seeking new technologies. A number of methods describing the simultaneous isolation of DNA and RNA have been reported (Alvarez et al., 2004; Bauer and Patzelt, 2003; Chomczynski, 1993). However, most of these have not been optimized to deal with the reduced quantity or quality of samples encountered in forensic casework. Alternatively, the co-isolation reports using forensically relevant samples (Alvarez et al., 2004; Bauer and Patzelt, 2003) require numerous time-consuming steps that would not benefit fast and simple stain identification assays. Making improvements to the extraction process to allow a streamlined approach is among the goals of our research and one which we envision developing with our commercial partner.

Change in a forensic laboratory can and does happen as witnessed by the revolution seen with DNA analysis. Changing any analytical approach requires documented external research detailing suitable methods and techniques and internal validation of published reports. The forensic laboratory must weigh the advantages and potential pitfalls of any new method and efforts required to bring such technology on-line in the face of pressures to maintain heavy caseloads and meet the demands of the criminal justice community. The research community has published considerable work on mRNA and has amassed a substantial body of information for the forensic community. Presently we understand that mRNA is stable in forensic stains and can be extracted in a sufficient quantity to be detected. Many markers have been identified as potential tissue determinants and a number of detection schemes have been described. This research has been published for many years but the general forensic community has not embraced mRNA for routine tissue identification. The analysis of mRNA is a radical change from the present approach, but the predominant reason for laboratory inactivity in this area is the lack of commercial support. The forensic community has become dependent upon commercialized products for forensic applications. This is not due to a lack of interest on the part of the forensic laboratory scientists to bring newly researched technologies on-line, but rather a realization that in the face of mounting casework, the efforts to bring published methods on-line in a laboratory is a very difficult task. A commercially available assay for particular mRNA transcripts could allow this technology to become a standard tool for forensic analysis.

Our laboratory is but one of many active in mRNA research and we have been fortunate to forge a relationship with a commercial organization to study the possibility of developing a product that could be useful for the forensic community. We intend to develop methods that broaden real-time PCR applications and to investigate a new technology that could radically change the analysis of biological stains and tissues. The biochemical approaches currently used in tissue identification are limited in scope and often imply, but not truly identify, the source fluid/tissue. We believe that the positive identification of fluids and tissues can be performed quickly and efficiently, and that tissues not routinely evaluated could be easily assessed by all laboratories so that an equality of testing could be realized across the country. The evaluation of mRNA through real-time PCR will be a technique that can offer a level of confidence and expand our knowledge of the materials we routinely examine. The nature of our research is to identify mRNA transcripts that will definitively identify the tissue of origin, determine if such transcripts survive the typical environmental insults that forensic samples may potentially encounter, and develop rapid assays to assess these molecules using small amounts of sample. The research into new technologies will demonstrate the power of multiplexing for forensic analysis. This chapter describes our efforts in this regard and our hope, in conjunction with our commercial partners, to develop an assay that could be routinely employed in a forensic laboratory.

MATERIALS AND METHODS

Sample Preparation

Blood, seminal fluid, saliva and urine samples were collected by pipetting known amounts of fresh fluid onto swatches of cotton cloth and allowing the spots to dry at room temperature. Vaginal swabs were collected on cotton swabs, whereas menstrual blood was collected on tampons and allowed to dry at room temperature. Each sample was stored in a glassine envelope, which was kept at room temperature in a box designated for the particular fluid at the Vermont Forensic Laboratory. Human tissues (kidney, colon, adipose, skin) were collected from the Surgical Pathology Suite at Fletcher Allen Health Care (Burlington, VT) and frozen at -20°C until use.

Human control RNAs (brain, heart, liver, kidney, intestine) were purchased from Ambion (Austin, TX). Additional human control RNAs (submaxillary gland, salivary gland, skin, blood) were purchased from BioChain (Hayward, CA). These RNAs were supplied as 1 μg/μl and diluted as noted for experiments.

Serology Tests

Acid Phosphatase Presumptive Test (Brentamine Reaction): This presumptive test for seminal fluid involves the detection of the enzyme acid phosphatase (AP). The reaction is a result of the liberation of napthol from sodium α-naphthyl phosphate by the enzyme acid phosphatase and the subsequent formation of a purple azo dye by the coupling of napthol with buffered fast blue B. The visible result of this reaction is the development of a deep magenta-purple color.

Amylase Diffusion: Amylase diffusion is a test to indicate whether amylase is present in a sample or stain at levels indicative of the presence of saliva. Amylase in vaginal swabs, semen stains, or other body fluid stains may also show activity, although usually at lower levels. Amylase breaks down starch into a variety of sugar products, and it is this reaction that allows for the detection of amylase. An agarose gel containing starch is prepared and the samples are placed into wells in the gel. If amylase is present, it will migrate away from the point of introduction and break down the starch into sugars. An iodine solution added to the plate will stain blue any areas still containing starch and any areas that were broken down into sugars by the present of amylase will remain clear.

RNA/DNA Extraction

Organic DNA Extraction: The organic extraction method (Forensic Science Research Training Center, 1989) as modified in Akane et al. (1993) was used in this chapter. Samples were lysed in Stain Extraction Buffer with Proteinase K overnight at 56°C. Phenol:chloroform:isoamyl alcohol (25:24:1, v/v) from Invitrogen (Carlsbad, CA) was added to the lysed samples. Samples were mixed into a milky emulsion and centrifuged to separate the DNA aqueous phase from the organic solvent. Tris-EDTA (TE) was added to a Microcon® 100 Concentrator (Millipore Corporation, Bedford, MA) to wet the membrane. The aqueous phase was added to the Microcon® 100 and centrifuged. A volume of TE was added to the Microcon® 100 and centrifuged. The filtrate cup was removed from the concentrator and placed in a new retentate cup. TE was added to the concentrator and the concentrator was inverted onto the retentate cup to allow for release of the DNA from the concentrator during centrifugation.

Tris-Buffered Phenol Method (Stain ID extraction): This protocol is optimized for the simultaneous extraction of both DNA and RNA from a sample. To the sample, RNagents denaturation solution (Promega, Madison, WI), 1X PBS and 20 mg/ml proteinase K is added. The sample is vortexed and incubated for 10 minutes at 56°C. The sample is transferred to a spin basket and centrifuged at max speed for 2 minutes. To the retentate, 2M sodium acetate is added and mixed followed by the addition and mixing of 1M Tris (pH 9.0). A volume of phenol:chloroform:isoamyl alcohol is added and the sample vortexed and centrifuged at max speed for 5 minutes. The aqueous phase is removed into a new tube and chloroform:isoamyl alcohol is added and vortexed. The sample is centrifuged at max speed for 5 minutes and the aqueous phase is removed into a new 1.5 mL tube. Glycogen (5 μg/μl) and isopropanol is added and vortexed. The sample is centrifuged at max speed for 10 minutes and following removal of the supernatant, the sample is washed with ice-cold 95% ethanol. Centrifuge at max speed for 5 minutes and remove supernatant and wash the pellet with 75% ethanol. Centrifuge at max speed for 5 minutes and remove supernatant and allow pellet to air dry for 1 minute. The pellet is resuspended in TE^{-4} and stored at -20°C.

TRIzol® Extraction: Samples were combined with TRIzol® Reagent (Invitrogen) and incubated at 56°C for 20 minutes. To the samples, chloroform was added and the tube was shaken vigorously for 15 seconds. The sample was incubated at 56°C for 10-15 minutes and vortexed every few minutes. The tube was centrifuged for 15 minutes at 11,000 x g causing the sample to separate into a lower, red phenol-chloroform phase, an interphase, and a colorless upper aqueous phase. The aqueous phase containing the RNA was removed into a new tube and combined with isopropanol. The sample was incubated at room temperature for 10 minutes and centrifuged for 15 minutes at 11,000 x g. The supernatant was removed and the pellet was washed with 75% ethanol and then centrifuged for 5 minutes at 7,500 x g. The supernatant was removed and the pellet allowed to air dry prior to resuspension in sterile water. Samples were stored at -20°C.

Reverse Transcription

Reverse transcription was performed on isolated RNA using the MessageSensor™ RT Kit (Ambion) with minor modifications. RNA extracts were combined with random decamer primers (Ambion) and nuclease-free water. The reaction mixture was heated at 80°C for 3 minutes. To the sample, 10X RT buffer, dNTPs, RNase inhibitor, and reverse transcriptase were added. The sample was mixed gently and incubated at 43°C for 1 hour followed by 92°C for 10 minutes. cDNAs were stored at -20°C.

Real-Time PCR

TaqMan: Singleplex assays using real-time PCR on cDNAs were performed using Assays-on-Demand™ Gene Expression Products (Applied Biosystems, Foster City, CA). These are a comprehensive collection of pre-designed and tested primer and probe sets that allow researchers to perform quantitative gene expression studies on any human gene. They are designed against GenBank transcripts, transcripts from the Mammalian Gene Collection, and human Celera transcripts. Each assay is built on 5' nuclease chemistry and consists of two unlabeled PCR primers and a FAM™ dye-labeled TaqMan® minor groove binder (MGB) probe. The components are formulated into a single 20X mix and designed to run under universal conditions for reverse transcription and PCR.

For the real-time PCR reactions, TaqMan® Universal PCR Master Mix, No AmpErase UNG (Applied Biosystems), cDNA, Assay-on-Demand™ primer/probe mix and nuclease-free water were combined. Cycling was performed in a RotorGene 3000 or 6000 (Corbett Research, Sydney, Australia) according to the following temperature profile: hold at 95°C for 10 minutes, 50 cycles of 95°C for 15 seconds and 60°C for 60 seconds. Data were analyzed using RotorGene software from Corbett Research.

Plexor®: The Plexor® technology (Promega) takes advantage of the specific interaction between two modified nucleotides to achieve quantitative PCR analysis. One of the primers contains a modified nucleotide (iso-dC) linked to a fluorescent label at the 5' end. The second PCR primer is unlabeled. The reaction mix includes deoxynucelotides and iso-dGTP modified with the quencher dabcyl. Dabcyl-iso-dGTP is incorporated opposite the iso-dC residue in the primer. The incorporation of the dabcyl-iso-dGTP at this position results in quenching of the fluorescent dye on the complementary strand and a reduction in fluorescence, which allows quantitation during amplification. Therefore, with the Plexor® technology, accumulation of product is accompanied by a decrease in fluorescence. As a quality check, the Plexor® System allows you to measure the melting temperature of the PCR products as homogeneous products create a well defined melting curve.

A single Plexor® reaction can contain multiple Plexor® primer sets; each primer pair is specific to a different target sequence, and labeled with different fluorophores. The dabcyl-iso-dGTP in the Plexor® master mix will quench the fluorescence of all the dyes present in the reaction. Plexor® primers were designed by Promega using the Plexor® Primer Design Software (Promega).

For the Plexor® Stain ID reactions, 2X Master Mix, RNasin® RNase Inhibitor, ImProm-II™ Reverse Transcriptase, Primer Mix and water were combined with unknown sample RNA. Cycling was performed in a RotorGene 6000 according to the following temperature profile: hold at 45°C for 5 minutes, hold at 95°C for 2 minutes, 38 cycles of 95°C for 5 seconds and 60°C for 35 seconds. This program was followed by a dissociation protocol of 95°C for 5 seconds, 50°C-95°C (2.5 acquisitions per °C), and 50°C for 30 seconds. Data were analyzed using Plexor® Analysis Software from Promega.

Quantitation of DNA Extracts

The Plexor® HY System is a real-time PCR assay specifically designed to determine the concentration of total human DNA and male human DNA simultaneously in one reaction. The kit contains an internal PCR control (IPC) to test for false-negative results that may occur in the presence of PCR inhibitors, and a melt curve function to confirm the correct product was amplified. Plexor® HY is a sensitive multiplex kit, with routine detection of approximately 6.4 pg total DNA. PCR amplification setup is performed at room temperature and is compatible with automated platforms. After PCR, a melt analysis can be performed to provide an internal control for the final assay design or to expedite troubleshooting.

THE RESULTS

Determining the half-life of mRNA in different types of stains

In order for a candidate to be implemented into a stain identification assay, it needs to be detectable for years following deposition of the sample. The amount of RNA present was determined by real-time PCR TaqMan® assays using HBB (blood), PRM2 (seminal fluid) or the housekeeping gene B2M. Applied Biosystems has created TaqMan® primer/probe sets for many human genes (Assay-on-Demand™ products). These have been carefully designed and thoroughly tested for specificity. Most of these sets are mRNA specific (cross exon-exon boundaries and have no cross-reaction with pseudogenes) although some do react with genomic DNA. Since these sets are available, optimized and cost only $150 each, we decided to use these pre-designed sets for our initial studies rather than expend the time and resources necessary to design our own. Table 1 lists the Assay-on-Demand™ numbers for each of the genes of interest. The mRNA specific assays (probe spans an exon/exon junction) end in "_m1". Those ending in "_g1" are not guaranteed to be mRNA specific. We checked the results when using two of the _g1 probes (HBB and PSA) by amplifying with DNA. Figure 1 shows the result of amplification of cDNA and genomic DNA (256 ng in a 10 ul reaction) with five TaqMan® assays; only the cDNA was amplified.

One of the earliest experiments we performed was to determine if amplifiable RNA could be detected from an aged blood sample. From preliminary studies, we had 20 µl and 40 µl blood stains on cotton cloth which had been stored at room temperature in the dark for 13 and 26 months, respectively. Following amplification with the TaqMan® HBB assay, it was apparent that this candidate gene was very stable over time (Figure 2). Similar results were seen for detection of the candidate gene PRM2 in seminal fluid. Twenty µl seminal fluid stains aged for 16 or 24 months were easily detected using this assay (Figure 3).

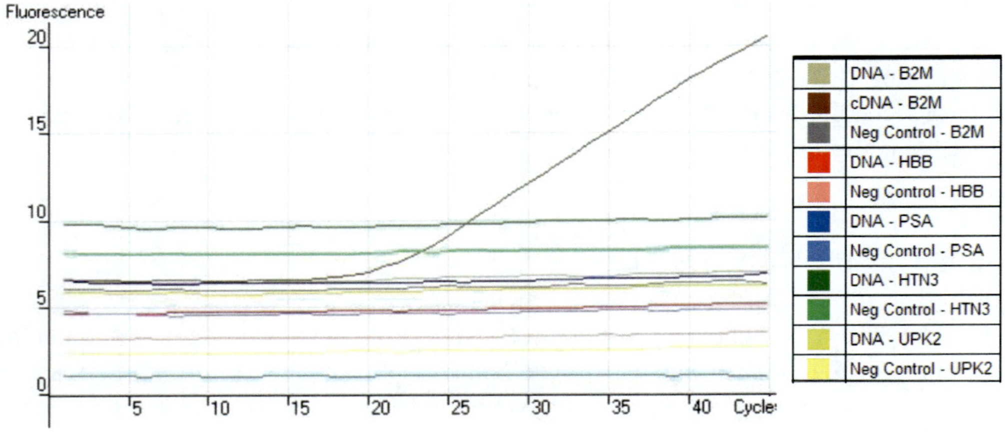

Figure 1. DNA and cDNA amplified with five cDNA-specific assays. After room temperature storage for 18 hours, a 20 µl blood stain was extracted using the TRIzol® reagent. Six µl of RNA was reverse transcribed using the MessageSensor™ RT Kit. Two µl of the resulting cDNA was used with the B2M Assays-on-Demand™ Gene Expression Product. Water as a negative control or control DNA (Promega; 256 ng in a 10 ul reaction) was used with the B2M, HBB, PSA, HTN3 and UPK2 Assays-on-Demand™ Gene Expression Products. Real-time PCR reactions were performed on a Corbett Rotor-Gene 6000 and data analyzed using the Rotor-Gene Software.

Table 1. Assays-on-Demand™ Gene Expression products used in the project.

TISSUE	Assays on Demand (ABI)		
Blood	hemoglobin, beta [HBB] (Hs00747223_g1)	CD3G antigen, gamma polypeptide [CD3G] (Hs00173941_m1)	spectrin, beta, erythrocytic [SPTB] (Hs00165820_m1)
Semen	semenogelin I [SEMG1] (Hs00268141_m1)	kallikrein 3 [PSA] (Hs00426859_g1)	acid phosphatase, prostate [ACPP] (Hs00173475_m1)
	cysteine-rich secretory protein 1[CRISP1] (Hs00538261_m1)	microseminoprotein, beta- [MSMB] (Hs00159303_m1)	transglutaminase 4 [TGM4] (Hs00162710_m1)
Sperm	protamine 2 [PRM2] (Hs00172513_m1)		
Saliva	statherin [STATH] (Hs00162389_m1)	histatin 3 [HTN3] (Hs00264790_m1)	

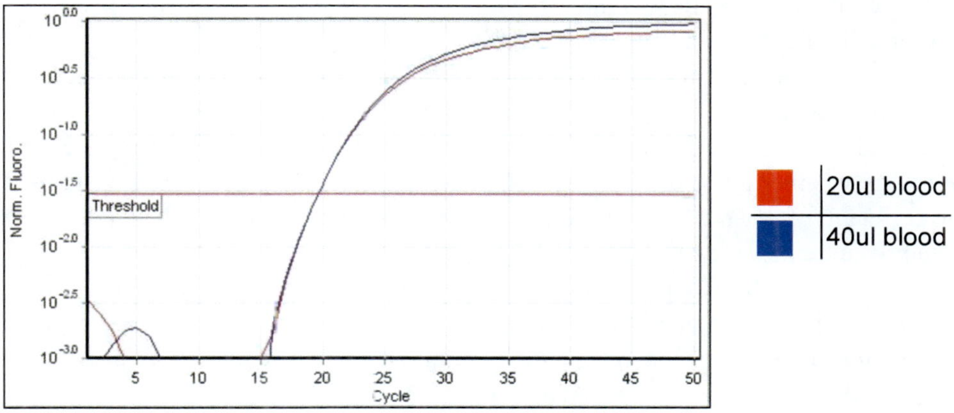

Figure 2. Amplification of aged blood with the HBB TaqMan® assay. After room temperature storage for 13 or 26 months, dried blood stains (20 and 40 µl, respectively) were extracted using the TRIzol® reagent. Six µl of RNA was reverse transcribed using the MessageSensor™ RT Kit. Two µl of the resulting cDNA was used with the HBB Assays-on-Demand™ Gene Expression Product. Real-time PCR reactions were performed on a Corbett Rotor-Gene 6000 and data analyzed using the Rotor-Gene Software.

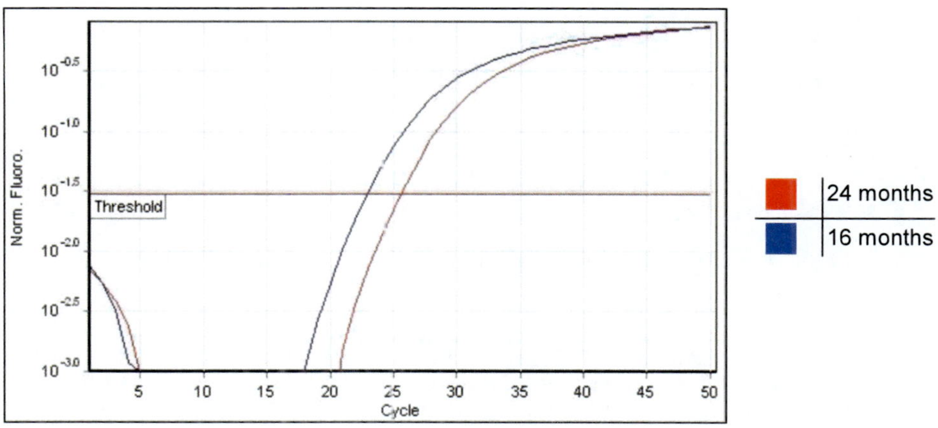

Figure 3. Amplification of aged semen with the PRM2 TaqMan® assay. After room temperature storage for 16 or 24 months, dried semen stains (20 µl) were extracted using the TRIzol® reagent. Six µl of RNA was reverse transcribed using the MessageSensor™ RT Kit. Two µl of the resulting cDNA was used with the PRM2 Assays-on-Demand™ Gene Expression Product. Real-time PCR reactions were performed on a Corbett Rotor-Gene 6000 and data analyzed using the Rotor-Gene Software.

Now that we had determined the HBB and PRM2 genes were amplifiable in samples aged for two years, we performed time course studies to look at the stabilities over time. For blood, we used two sample volumes, 1 μl and 10 μl that had been aging at room temperature in the dark for 3 days to 510 days (1 μl) or 864 days (10 μl). As depicted in Table 2, there is a slight decrease in expression over time, but samples aged well over a year (or more) maintained a high HBB expression level.

In a similar experiment, we looked at two different sized seminal fluid stains, 1 μl and 20 μl that had been aging at room temperature in the dark for 16 days to 807 days (1 μl) or 791 days (20 μl). Table 3 shows that PRM2 expression is maintained in samples aged over 2 years. In fact, it appears as though the expression may even increase over time, perhaps due to the loss of an inhibitor.

Forensic samples are often exposed to a variety of environmental insults therefore, we stored parallel samples under ideal conditions (room temperature, dark) as well as in a sub-optimal condition (37˚C). By comparing gene amplification between the samples, we hoped to determine whether RNA profiling would be feasible for weathered forensic stains. In the first experiment, between 0.25 μl and 20 μl of blood was spotted onto cloth and dried at either room temperature in the dark, or in a 37˚C incubator. The samples were stored for 503 days prior to extraction with TRIzol®. cDNA was amplified using the HBB TaqMan® assay. Figure 4 shows that amplification of HBB is possible following storage at 37˚C for 503 days. In fact, there does not appear to be a considerable decrease in the degree of amplification after prolonged exposure to elevated temperatures.

Table 2. Time course of HBB expression in blood stains.

BLOOD	AGE (DAYS)	HBB Ct
1 μl	3	19.94
1 μl	23	19.99
1 μl	461	23.30
1 μl	496	23.52
1 μl	510	23.15
10 μl	3	17.68
10 μl	420	19.94
10 μl	490	20.27
10 μl	855	20.15
10 μl	864	20.51

Ct for the NTC was 35.15

Table 3. Time course of PRM2 expression in seminal fluid stains.

SEMEN	AGE (DAYS)	PRM2 Ct
1 μl	16	34.61
1 μl	461	30.35
1 μl	507	27.41
1 μl	699	30.95
1 μl	807	36.90
20 μl	16	27.92
20 μl	507	20.10
20 μl	607	22.19
20 μl	777	21.34
20 μl	791	22.94

Ct for NTC was not observed during the 50 cycles

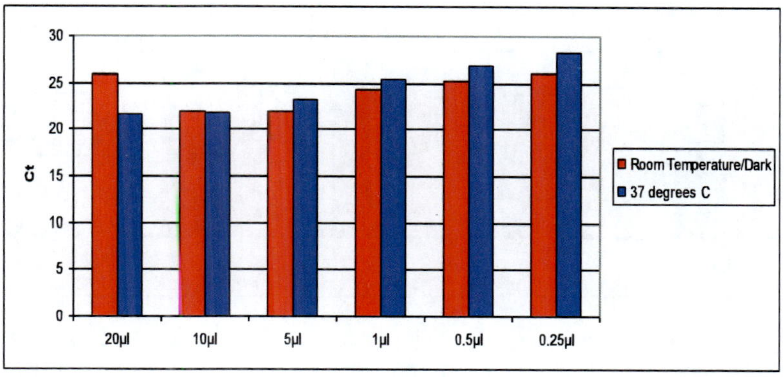

Figure 4. Effect of storage conditions on HBB amplification from blood stains. After room temperature or 37°C storage for 503 days, dried blood stains (0.25 – 20 µl) were extracted using the TRIzol® reagent. Six µl of RNA was reverse transcribed using the MessageSensor™ RT Kit. Two µl of the resulting cDNA was used with the HBB Assays-on-Demand™ Gene Expression Product. Real-time PCR reactions were performed on a Corbett Rotor-Gene 6000 and data analyzed using the Rotor-Gene Software.

One experiment we performed was to evaluate blood samples which had been stored under both conditions (room temp/dark, 37°C) for an extended period of time with a relatively new sample. One µl blood stains were stored for 544 days at room temperature (dark) or in a 37°C incubator and extracted with TRIzol®. For comparison purposes, a 1 µl blood stain was stored for 3 days at room temperature (dark) in order to see how much amplification of HBB decreases with prolonged storage at either of the two conditions. Regardless of the storage length or conditions, the resulting Cts were essentially the same with a difference of less than 1 Ct between all 3 samples (Figure 5). This result further demonstrates how stable HBB RNA is over time.

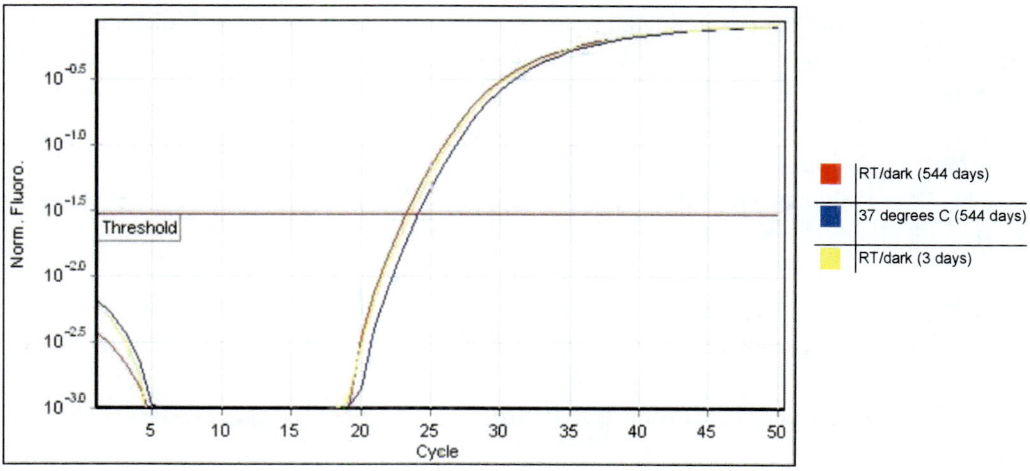

Figure 5. HBB expression in aged blood stains stored at various conditions. After room temperature or 37°C storage for 3 or 544 days, dried blood stains (1 µl) were extracted using the TRIzol® reagent. Six µl of RNA was reverse transcribed using the MessageSensor™ RT Kit. Two µl of the resulting cDNA was used with the HBB Assays-on-Demand™ Gene Expression Product. Real-time PCR reactions were performed on a Corbett Rotor-Gene 6000 and data analyzed using the Rotor-Gene Software.

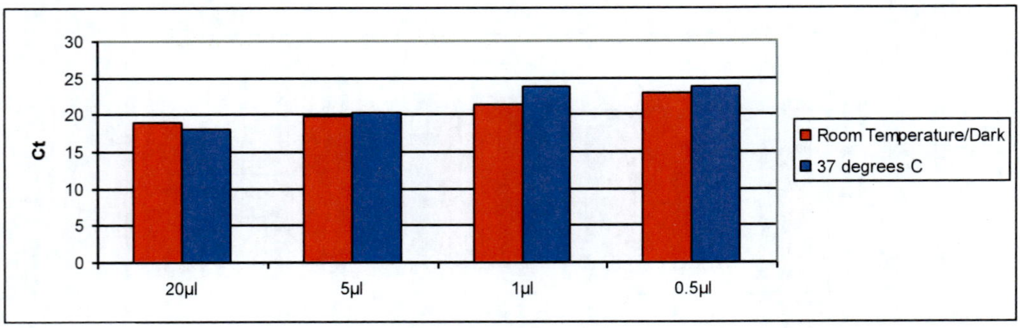

Figure 6. Effect of storage conditions on PRM2 amplification from seminal fluid stains. After room temperature or 37°C storage for 518 days, dried semen stains (0.5 - 20 µl) were extracted using the TRIzol® reagent. Six µl of RNA was reverse transcribed using the MessageSensor™ RT Kit. Two µl of the resulting cDNA were used with the PRM2 Assays-on-Demand™ Gene Expression Product. Real-time PCR reactions were performed on a Corbett Rotor-Gene 6000 and data analyzed using the Rotor-Gene Software.

In a similar experiment, between 0.5 µl and 20 µl of semen was spotted onto cloth and dried at either room temperature in the dark, or in a 37°C incubator. The samples were stored for 518 days prior to extraction with TRIzol®. cDNA was amplified using the PRM2 TaqMan® assay. Figure 6 shows that amplification of PRM2 is possible following storage at 37°C for 518 days. Similar to what was shown for blood and HBB expression, there does not appear to be a considerable decrease in the degree of amplification after prolonged exposure to elevated temperatures.

Finding genes specific for each stain type.

Survey of literature including PubMed, Gene and other databases have allowed for the selection of several genes that appear to be specific for a number of different tissues (Table 1). The following alphabetical list is a brief description of each gene target utilized in this chapter:

ACPP (acid phosphatase, prostate) - enzyme which catalyzes the conversion of orthophosphoric monoester to alcohol and orthophosphate; secreted by the epithelial cells of the prostrate gland

B2M (beta 2 microglobulin) - serum protein found in association with the major histocompatibility complex class I heavy chain on the surface of nearly all nucleated cells

CD3G (CD3G antigen, gamma polypeptide) - component of the T-cell receptor-CD3 complex

CRISP1 (cysteine-rich secretory protein 1) - expressed in the epididymis and plays a role at fertilization in sperm-egg fusion

Table 4. Sensitivities of TaqMan® blood assays.

BLOOD VOLUME	HBB	CD3G	SPTB
NTC	37.21	ND	ND
0.0001 µl	33.13	ND	ND
0.001 µl	29.92	ND	ND
0.01 µl	26.28	ND	ND
0.1 µl	24.31	ND	ND
1 µl	21.60	33.41	35.00
5 µl	19.08	29.63	34.35
10 µl	16.67	28.69	31.56
20 µl	15.84	28.62	31.24

ND = not detected

GAPDH (glyceraldehyde-3-phosphate dehydrogenase) - catalyzes an important energy-yielding step in carbohydrate metabolism, the reversible oxidative phosphorylation of glyceraldehyde-3-phosphate in the presence of inorganic phosphate and nicotinamide adenine dinucleotide

HBB (hemoglobin, beta) - iron containing oxygen-transport metalloprotein in red blood cells

HTN3 (histatin 3) - non-immunological, anti-microbial role in the oral cavity

MSMB (microseminoprotein, beta) - synthesized by the epithelial cells of the prostate gland and secreted into the seminal plasma

PRM2 (protamine 2) - major DNA-binding protein in the nucleus of sperm and packages DNA

PSA (kallikrein 3) - protease present in seminal plasma involved in liquefaction of seminal coagulum

SEMG1 (semenogelin I) - involved in formation of gel matrix that encases ejaculated spermatozoa

SPTB (spectrin, beta, erythrocytic) - major component of red blood cell membrane

STAT (statherin) - peptide that inhibits precipitation from calcium phosphate solutions, stabilizes saliva

TGM4 (transglutaminase 4) - catalyzes the cross linking of proteins and the conjugation of polyamines to specific proteins in the seminal tract

We tested the specificity, robustness and sensitivity of each assay by analysis of mRNA isolated from the body fluid of interest. Once the assay was been shown to be robust, we tested it on mRNA isolated from other fluids and tissues to demonstrate that the assay is specific. Additional fluids and tissues to those discussed previously included vaginal secretions and menstrual blood. For semen, we are interested in mRNAs specific for the sperm as well as for the prostatic components. In a number of cases, it is important to determine if semen is present even if the male is sterile or has had a vasectomy. In other cases, it is important to know if sperm are likely to be present. We obtained anonymous samples of a variety of tissues (kidney, colon, adipose, skin) from the Surgical Pathology Suite at Fletcher Allen Health Care. Lastly, control RNAs (liver, kidney, brain, heart, intestine) were purchased from Ambion to ensure that our assays are specific only for the intended tissue.

The one The sensitivity of the three candidate blood assays (HBB, CD3G, SPTB) was evaluated using a range of blood volumes spotted onto cotton cloth. Between 0.0001 µl and 20 µl of blood was allowed to air dry at room temperature for 18 hours prior to extraction using TRIzol®. cDNA was amplified using the three TaqMan® assays. The sensitivity of HBB was far superior to the other assays with amplification resulting from 0.0001 µl of blood, compared to a limit of 1 µl for the CD3G and SPTB assays (Table 4).

The sensitivity of the four candidate semen assays (SEMG1, PSA, MSMB, TGM4) and one sperm assay (PRM2) was evaluated using a range of semen volumes spotted onto cotton cloth. In one experiment (SEMG1, PSA, PRM2), between 0.01 µl and 20 µl of semen was allowed to air dry at room temperature for 8 days prior to extraction using TRIzol®. In a subsequent experiment (MSMB, TGM4), between 1 µl and 20 µl of semen was allowed to air dry at room temperature for 41 days prior to extraction using TRIzol®. cDNA was amplified using the five TaqMan® assays. The sensitivities of PSA and PRM2 were slightly higher than SEMG1 since both were able to amplify from 0.01 µl of semen (Table 5). However, since the Cts for PRM2 were significantly lower than those for PSA (and SEMG1) for most volumes, PRM2 appears to have a higher expression level, or is a more robust assay. Furthermore, the second experiment with MSMB and TGM4 suggests that those two candidates are expressed more than the other semen markers (SEMG1 and PSA) since the Cts are lower. But, since the two lowest volumes, namely 0.01 µl and 0.1 µl, were not analyzed with these two assays, the lower limits are unknown, but assumed to be lower than 1 µl based on the Cts.

The sensitivities for saliva candidate markers (STAT, HTN3) were evaluated using a range of saliva volumes spotted onto cotton cloth. Between 0.01 µl and 20 µl of saliva was allowed to air dry at room temperature for 18 hours prior to extraction using TRIzol®. cDNA was amplified using the two TaqMan® assays. As shown in Table 6, only the STAT assay was able to detect 1 µl of freshly dried saliva, whereas the HTN3 assay had a 5 µl limit of detection. The Cts were not as low as seen in the blood and semen/sperm assays, but were different than the absence of Cts observed for the negative controls.

Table 5. Sensitivities of TaqMan® semen and sperm assays.

SEMEN VOLUME	SEMG1	PSA	MSMB	TGM4	PRM2
NTC	ND	ND	ND	ND	ND
0.01 µl	ND	37.20	N/A	N/A	35.67
0.1 µl	33.79	34.15	N/A	N/A	32.26
1 µl	30.80	31.39	25.42	28.33	28.93
5 µl	31.43	32.48	28.90	30.69	27.57
10 µl	31.55	35.13	29.74	31.15	24.80
20 µl	31.86	34.02	28.35	30.79	27.24

ND = not detected

N/A = not determined

The specificity of the chosen genes for the tissue of interest was also examined. To test the TaqMan® sets (only the control B2M and tissue-specific sets should give amplification), mRNA was isolated from dried blood, semen, saliva, menstrual blood or vaginal secretions using TRIzol®. Human tissues (kidney, colon, adipose, skin) were extracted using the Absolutely RNA® Kit. Control human RNAs from Ambion (brain, heart, liver, kidney, intestine) were diluted to 100 ng/ul. The RNA samples were then reverse transcribed and PCR performed with each of the TaqMan® sets. The amplification results for the blood, semen, sperm, and saliva assays are depicted in Table 7.

For the three blood assays, amplification only occurred in the blood sample as expected. For the 6 semen assays there were mixed results. PSA, SEMG1, MSMB and TGM4 all appeared to be specific for semen when tested against blood, semen and saliva. However, SEMG1 cross-reacted with vaginal secretions, albeit at a much lower level (Ct of 29.40 for semen vs. Ct of 33.67 for vaginal secretions). Interestingly, MSMB amplified from a control intestine sample, but not with human colon tissues. ACPP and CRISP1, although ideal candidates based on literature searches, were not suitable assays for the specific detection of semen. The sperm marker PRM2 amplified from a seminal fluid sample, but nothing else that was tested. For saliva, amplification only occurred with the STAT assay as expected. The control B2M was detected in each of the fluids and tissues. Based on these results, several of these TaqMan® assays appear to be specific (HBB, SPTB, CD3G, PSA, STAT), however, screening against an expanded sample set would strengthen that claim. Alternatively, TGM4 and PRM2 have demonstrated their specificity for semen/sperm through the course of these studies.

Developing multiplex assays for the detection of stain types.

A major aim of stain identification using mRNA expression profiling is working towards multiplexing the real-time PCR assays once mRNAs are identified that clearly define specific types of stains. Since these assays are designed to function more qualitatively than quantitatively, a test of a single stain should typically give amplification with only one candidate. It is possible that a mixture could give several amplifications. However, it is of greater concern to know that a mixture does exist, than to know the exact amount of each fluid present.

Table 6. Sensitivities of TaqMan® saliva assays.

SALIVA VOLUME	STAT	HTN3
NTC	ND	ND
0.1 µl	ND	ND
1 µl	34.14	ND
5 µl	33.93	39.45
10 µl	37.80	37.93
20 µl	34.68	36.25

ND = not detected

Table 7. Specificity of blood, semen, sperm and saliva TaqMan® assays (grey = not tested).

	BLOOD			SEMEN						SPERM	SALIVA	CONTROL
	HBB	SPTB	CD3G	PSA	SEMG1	ACPP	CRISP1	MSMB	TGM4	PRM2	STAT	B2M
Blood	+	+	+	-	-	+	-	-	-	-	-	+
Semen	-	-	-	+	+	+	-	+	+	+	-	+
Saliva	-	-	-	-	-	-	-	-	-	-	+	+
Menstrual Blood					-	-	-	-	-	-		+
Vaginal Secretions					+	+	-	-	-	-		+
Kidney	-			-	-			-	-	-		+
Colon					-			-	-	-		+
Adipose					-			-	-	-		+
Skin					-			-	-	-		+
Ambion Brain					-			-	-	-		+
Ambion Heart					-			-	-	-		+
Ambion Liver					-			-	-	-		+
Ambion Kidney					-			-	-	-		+
Ambion Intestine					-			+	-	-		+

One methodology to achieve this goal is the Plexor® system from Promega. Depending on the dye-capability of the real-time instrument that is utilized, this system allows multiple mRNAs to be combined in one assay, thus reducing the amount of sample needed and time of analysis. For example, using a 4-dye channel instrument, eight mRNA targets could be analyzed; two per dye-channel, each of which has a distinctive melt curve. The Plexor® qRT-PCR system takes advantage of the specific interaction between two modified nucleotides to achieve quantitative PCR analysis. Promega's Plexor® design software allows the generation of primers which span an intron by designating a certain base to include in the primer design.

During the course of our work, we were contacted by Promega with an offer to collaborate on the development of stain identification assays. The combination of Promega's extensive knowledge of the Plexor® platform with our sample inventory and insight into the needs of the forensic laboratory, it was a strong partnership. The goal was to generate Plexor® Tissue Typing Systems; multiplexed qRT-PCR systems for determining the source and quantity of a variety of human stains and/or tissues. The systems would include Plexor® primers for the detection of tissue-specific mRNA transcripts associated with semen, sperm, blood, menstrual blood, saliva, etc. By limiting the initial system to two-color detection (i.e. FAM and HEX detection), it would be compatible with the majority of real-time thermal cyclers in forensic laboratories. We sought to explore the potential to include controls (e.g. a housekeeping gene) or multiple, tissue-specific targets using the ability of the Plexor® Data

Analysis Software to distinguish two different amplicons in the same dye channel, based upon thermal melt properties.

The Stain ID assay which we decided to first focus on was for detection of semen and saliva. Plexor® primers were designed to amplify the targets SEMG1 (semen), HTN3 (saliva) and GAPDH (housekeeper). Table 8 shows amplification of a variety of RNA samples using the triplex. The saliva primers (HTN3) show specific reactivity with saliva RNA (isolated in-house), and also with submaxillary gland RNA and salivary gland RNA that was purchased from a commercial vendor (BioChain). They show no cross-reactivity with other RNA samples tested, or with genomic DNA. The semen primers (SEMG1) show specific reactivity with semen RNA that was isolated in-house, and did not amplify any of the other RNA samples tested. However, they do show some minor reactivity with genomic DNA. This is likely due to some weak amplification across the intron in the genomic sequence. Therefore, new SEMG1 primers were designed that would hopefully address the genomic DNA amplification.

The modified SEMG1 primers were screened for the possibility of amplifying genomic DNA. The new primers did indeed eliminate cross-reactivity with genomic DNA, however the primers now yielded a product with RNA isolated from blood. The initial blood sample isolated showed no reactivity with the SEMG1 primers (old or new), but a commercial blood sample did show amplification with the SEMG1 primers. To confirm this, two additional RNA samples from blood were generated in-house and tested. Both of those showed amplification with the SEMG1 primers (Figure 7) with melt curves specific to the RNA product and not the product expected with genomic DNA. This indicates that the modified primers eliminated the amplification from genomic DNA, but not from RNA isolated from blood. Therefore, we assume that SEMG1 is expressed at low levels in blood. Although the article doesn't mention blood specifically, Lundwall et al. (2002) reported SEMG1 to be expressed in a number of non-genital tissues. As a result of these findings, we decided to replace SEMG1 with a different semen-specific target.

Protamine 2 was identified as a potential target in early experiments, so primers were designed and screened for PRM1 and PRM2 transcripts. The PRM1 primers were designed so that one primer sits on the junction of the two exons, however some amplification of genomic DNA was detected. Three different sets of PRM2 primers were designed. The first set was designed so that one primer sits on the junction of two exons, but some amplification of genomic DNA was detected, as well as amplification from other RNA transcripts. The second set was designed so that one primer is on either side of the intron. Some amplification of genomic DNA was detected; however, the DNA melt curve was distinct from the RNA melt (84°C for RNA vs. 88°C for DNA). The third set of PRM2 primers was also designed so that one primer is on either side of the intron. Some amplification of genomic DNA was detected, however, the DNA melt curve was distinct from the RNA melt curve. Unfortunately, minor amplification of skin and blood also occurred with these primers. The second set of PRM2 primers looked the most promising although in theory, PRM2 is a sperm-specific marker and therefore, may not be expressed in vaseoctomized males. We isolated RNA from a semen sample donated by a vasectomized male and not surprisingly, PRM2 was not detected (Figure 8). At this point, because the screening assay only includes one semen/sperm marker, it was decided that PRM2 would not be a viable candidate since the presence of semen would go undetected in samples from vasectomized males or donors with low-sperm counts.

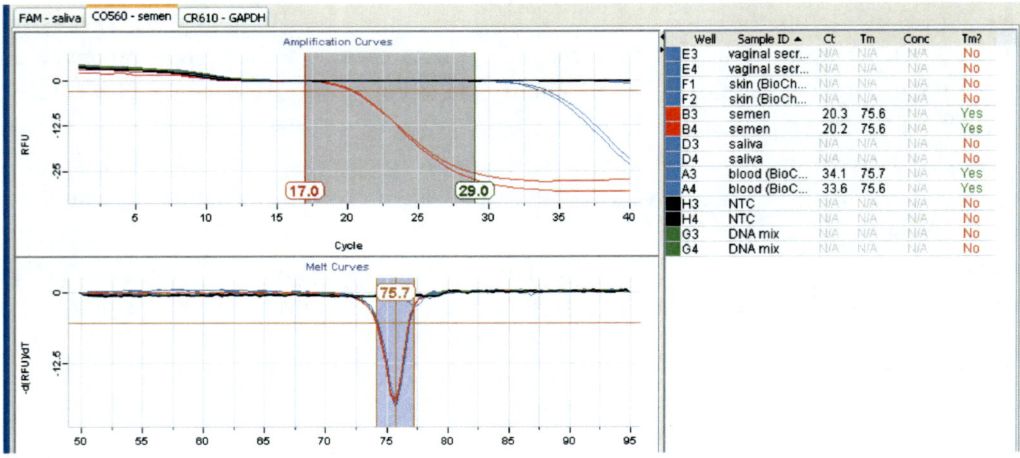

Figure 7. Cross-reactivity of SEMG1 Plexor® primers with blood. After room temperature storage for 18 hours, dried vaginal secretions, semen and saliva stains were extracted using phenol-chloroform or the Promega PureYield RNA midiprep kit (vaginal secretions). The RNA was quantitated and 100 ng of RNA was used in the Plexor® Stain ID assay. Control skin and blood RNA (BioChain) was also analyzed. Reactions were performed in duplicate on a RotorGene 6000 and data analyzed using the Plexor® analysis software.

Table 8. Detection of semen and saliva using a Plexor®-based Stain ID assay.

SAMPLE	HTN3 (FAM)	SEMG1 (HEX)	GAPDH (ROX)
NTC	ND	ND	ND
NTC	ND	ND	ND
DNA	ND	34.3	ND
DNA	ND	33.5	ND
vaginal secretion	ND	ND	22.1
vaginal secretion	ND	ND	22.3
vagina	ND	ND	21.4
vagina	ND	ND	21.1
submaxillary gland	**8.8**	ND	24.7
submaxillary gland	**8.2**	ND	24.8
salivary gland	**14.8**	ND	26.0
salivary gland	**14.8**	ND	25.8
saliva	**27.0**	ND	28.0
saliva	**26.9**	ND	27.2
skin	ND	ND	23.0
skin	ND	ND	23.0
semen	ND	**19.7**	25.1
semen	ND	**19.8**	25.5
blood	ND	ND	24.9
blood	ND	ND	25.5

ND = Not Detected

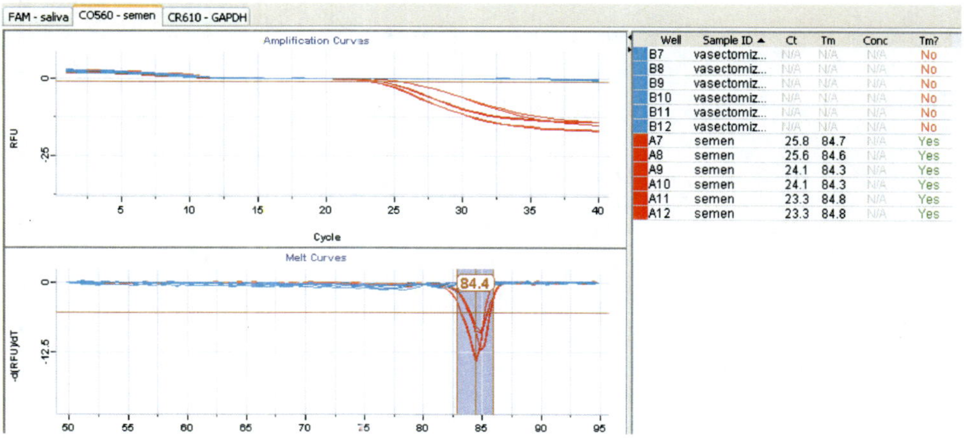

Figure 8. Absence of PRM2 expression in vasectomized males. After room temperature storage for 18 hours, dried semen stains from 6 individuals were extracted using phenol-chloroform. The RNA was quantitated and 100 ng of RNA was used in the Plexor® Stain ID assay. Reactions were performed in duplicate on a RotorGene 6000 and data analyzed using the Plexor® analysis software.

The next semen-specific candidate which was assessed was PSA. Initial studies with PSA primers showed non-specific amplification with all other RNAs that were tested, as well as genomic DNA and the no template control. Since all nonspecific amplification (including the no template control) was identical in the amplification and melt curves the cause was thought to be primer dimerization. However, additional experiments with variations in the PSA primer sequences yielded little improvement in the target specificity. Based on studies we had performed using the Gene Expression TaqMan® assays from Applied Biosystems, we switched our focus to MSMB and TGM4. Expression of MSMB occurred in both vasectomized and non-vasectomized semen samples, however, nonspecific amplification was observed with all the other RNAs as well as genomic DNA. Changing the fluorescent label to the opposite primer showed no improvement in target specificity. On the other hand, TGM4 was detected in both vasectomized and non-vasectomized semen samples without nonspecific amplification (Figure 9).

The preliminary experiments were not performed using sample types and sizes which are reflective of forensic-type samples. The RNA samples which were extracted in-house included vaginal secretions, whole blood, semen and saliva. In all cases, RNA was extracted from extremely large sample sizes (i.e. 2 mL of semen, 10 mL of saliva, 5 mL of whole blood) and the samples were all fresh. These were obviously not typical casework samples, but intended to isolate a large pool of RNA that experiments could be performed with. The semen, saliva and whole blood RNA samples were extracted with phenol-chloroform and the vaginal secretion was extracted using Promega's PureYield RNA midiprep kit. After extraction, the RNA was quantitated and 100 ng of RNA was used per reaction. Titration experiments were performed and indicated that semen had a detection threshold of approximately 10 pg of RNA whereas the threshold for saliva was approximately 1 ng of RNA. Although these studies gave an idea as to the relative sensitivities of the assay, it was important to assess how the assay would perform with typical forensic samples (i.e. various volumes of unknown RNA yields).

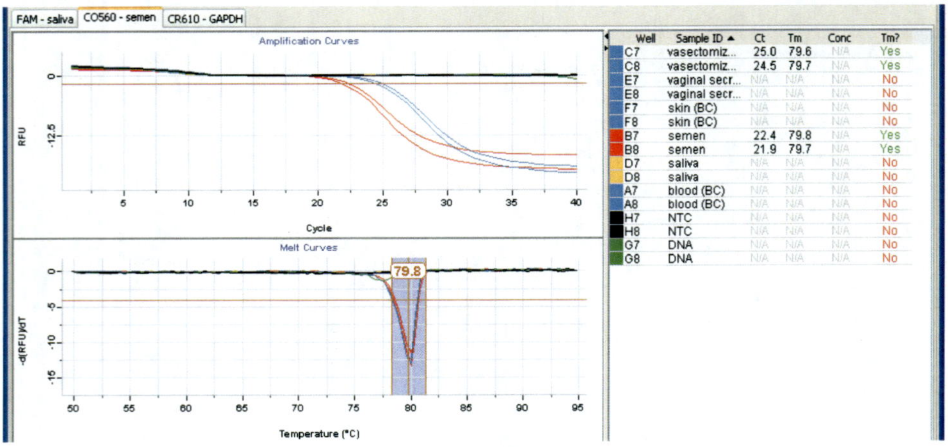

Figure 9. Specific expression of TGM4 in semen samples. After room temperature storage for 18 hours, dried vaginal secretions, semen and saliva stains were extracted using phenol-chloroform or the Promega PureYield RNA midiprep kit (vaginal secretions). The RNA was quantitated and 100ng of RNA was used in the Plexor® Stain ID assay. Control skin and blood RNA (BioChain) was also analyzed. Reactions were performed in duplicate on a RotorGene 6000 and data analyzed using the Plexor® analysis software.

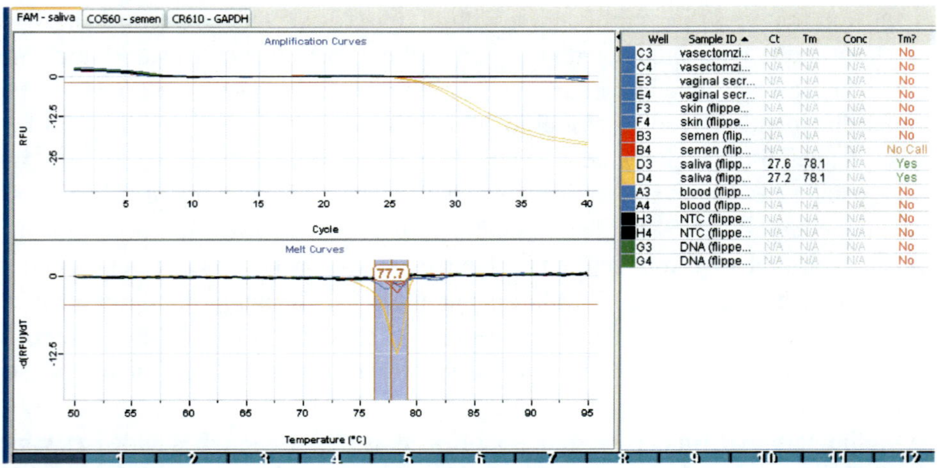

Figure 10. Modification of HTN3 primers increases specificity for saliva RNA. After room temperature storage for 18 hours, dried vaginal secretions, semen and saliva stains were extracted using phenol-chloroform or the Promega PureYield RNA midiprep kit (vaginal secretions). The RNA was quantitated and 100 ng of RNA was used in the Plexor® Stain ID assay. Control skin and blood RNA (BioChain) were also analyzed. Reactions were performed in duplicate on a RotorGene 6000 and data analyzed using the Plexor® analysis software.

A panel of samples was screened using the Stain ID assay which now consisted of HTN3, TGM4, and GAPDH. Samples included dried blood (20 µl aged 70 days), semen (1-20 µl aged 76 days), saliva (1-20 µl aged 4 days), menstrual blood (5x5 mm cutting aged 700 days) and vaginal secretions (1/2 swab aged 54 days) extracted with TRIzol®. Positive control samples of known RNA quantities were run along with the Plexor® Stain ID assay: semen (1 ng/reaction) and saliva (10 ng/reaction). The results are shown in Table 9. The HTN3 primers amplified from all four saliva samples ranging from 1-20 µl of dried saliva as well as the positive control sample. However, there was some nonspecific amplification in the vaginal secretions sample. The TGM4 primers only amplified extracts from the semen samples and positive control. Interestingly, GAPDH was not amplified in the four saliva samples even though HTN3 was clearly detected which suggests that the expression of the housekeeper in saliva is not as prevalent as the tissue-specific marker. Based on the HTN3 results, the primers were modified by moving the fluorescent label from one oligo to the other. The modified HTN3 primers no longer detect vaginal secretions (Figure 10).

A major selling point of the Stain ID assay was the development of a co-isolation method for RNA and DNA extraction. By utilizing one extraction step, a DNA sample would be ready and waiting for STR profiling if the RNA screening assay deemed it worthy of such analysis. In addition, obtaining RNA and DNA from a single stain would prevent the possibility of conclusions being drawn regarding the identity of one stain which may not hold true for a nearby stain. Therefore, a significant amount of time was spent optimizing a procedure which would co-extract the two nucleic acids so that they were of sufficient quality and quantity for downstream analyses.

An initial comparison of DNA yields using two different extraction techniques was made. One and ten μl of semen and saliva were extracted via the RNagents Total RNA Isolation System (Promega) or an organic method which used Tris-buffered phenol. Prior to extraction with these methods a pre-processing step consisting of incubation with 1.8 mg/mL proteinase K and 100 mM DTT at 56°C for 2 hours was performed. In parallel samples, this step was omitted and the samples went immediately into lysis with the guanidine denaturation solution. DNA yields were determined using the Plexor® HY quantitation assay. RNA yields were assessed using the saliva-semen screening assay. The Tris-buffered phenol extraction provided the best results since the DNA yield was better than for the RNagents extraction while the RNA yields were equivalent (Table 10). The DNA yields were significantly higher when the pre-processing step was included in the protocol, however the RNA was virtually destroyed and undetectable in the Stain ID assay (Table 11).

Table 9. Detection of semen and saliva with the Plexor® Stain ID assay.

SAMPLE	HTN3 (FAM)	TGM4 (HEX)	GAPDH (ROX)
NTC	ND	ND	ND
Blood (20ul)	ND	ND	30.1
Semen (20ul)	ND	**27.5**	34.9
Semen (10ul)	ND	**27.0**	35.9
Semen (5ul)	ND	**25.6**	33.5
Semen (1ul)	ND	**25.0**	35.3
Saliva (20ul)	**20.9**	ND	ND
Saliva (10ul)	**22.3**	ND	ND
Saliva (5ul)	**22.6**	ND	ND
Saliva (1ul)	**25.4**	ND	ND
Menstrual Blood	ND	ND	38.8
Vaginal Secretions	24.4	ND	24.4
Semen (+) Control	ND	**24.4**	36.5
Saliva (+) Control	**21.6**	ND	37.5

ND = Not Detected

Table 10. DNA yield comparison of Tris-buffered phenol and RNagents extractions.

	Pre-Processing Step		No Pre-Processing Step	
SAMPLE	Tris (ng)	RNagents (ng)	Tris (ng)	RNagents (ng)
10ul semen	2851.25	21.07	361.88	0.53
1ul semen	190.78	3.65	63.58	0.27
10ul saliva	166.45	0.56	89.38	0.94
1ul saliva	3.49	0.01	6.66	0.04

Table 11. RNA yield comparison of Tris-buffered phenol and RNagents extractions.

	Pre-Processing Step		No Pre-Processing Step	
SAMPLE	Tris (Ct)	RNagents (Ct)	Tris (Ct)	RNagents (Ct)
10ul semen	35.1	39.2	22.9	22.1
1ul semen	38.1	37.6	25.2	24.8
10ul saliva	ND	ND	29.6	30.3
1ul saliva	ND	ND	34.6	34.8

ND = not detected

Table 12. DNA yields and RNA expression of semen and saliva samples extracted with the Tris-buffered phenol procedure.

SAMPLE	DNA (ng/ul)	HTN3 RNA (Ct)	TGM4 RNA (Ct)	GAPDH RNA (Ct)
10ul semen	17.3	ND	21.9	23.7
1ul semen	2.0	ND	25.0	25.0
10ul saliva	13.8	29.0	ND	29.0
1ul saliva	1.3	32.9	ND	32.4

Experiments were undertaken to improve the sample lysis without compromising RNA integrity. Reducing agents were added to the guanidine-HCl denaturation solution which itself should protect the RNA. A protocol was optimized which yielded sufficient quality and quantity DNA and RNA for downstream analyses. The final optimized extraction method was used to extract 1 and 10 µl of semen and saliva. The resulting DNA was quantitated using the Plexor® HY assay with yields shown in Table 12. This extraction procedure yielded sufficient quantities of DNA to perform STR analysis. The RNA was used in the Stain ID assay to detect the presence of HTN3, TGM4 and GAPDH. The only amplification of HTN3 was in the saliva samples, whereas amplification of TGM4 only occurred with the semen samples. Alternatively, GAPDH was detected in both sample types. This experiment supports the use of Tris-buffered phenol extraction as a method to isolate sufficient quantities of quality nucleic acids for downstream analyses.

Promega generated a draft Technical Manual to accompany the Stain ID kit and subsequently distributed the prototype kits to several forensic laboratories for alpha-testing of the assay. As one of the testing sites, we extracted samples including blood, menstrual blood, saliva, semen, urine, vaginal secretions, buccal swabs on FTA®, kidney, adipose, colon and skin. In addition, mixtures of blood/saliva, blood/semen, and saliva/semen were tested. The samples ranged in age and size and were extracted using the Tris-buffered phenol protocol. In the multiplex Stain ID reaction we demonstrated that amplification of each fluid only occurred with the tissue-specific genes as expected; semen was amplified with TGM4 primers, whereas only saliva was amplified with the HTN3 primers (Table 13). Additionally, the artificial mixtures containing semen and saliva RNA extracts show that these primer sets are able to discriminate their target RNA in a heterogeneous sample. This experiment shows that the Stain ID assay can detect down to 1 μ of aged seminal fluid and saliva – the lowest volumes tested to date. Furthermore, there was no cross-reactivity with other commonly encountered bodily fluids such as blood, menstrual blood, vaginal secretions or urine. In order to compare the Stain ID assay with presumptive tests currently performed in forensic laboratories, we analyzed this set of samples using the acid phosphatase and amylase diffusion tests. As shown in Table 13, the acid phosphatase results mirrored the Stain ID results, however the amylase diffusion results were far inferior compared to the Stain ID assay. All of the samples containing saliva amplified with the HTN3 primers, but went undetected when subjected to the amylase diffusion test for the presence of saliva. These results suggest that the Stain ID assay is more sensitive and robust than at least one of the presumptive tests routinely used in the forensic laboratory.

In addition to the in-house mock forensic samples which were tested using the prototype kit from Promega, we also extracted and analyzed four swabs of unknown origin as a blind test of the assay's performance (Table 14). For swabs A-C, the results were as expected; swab A (10 μl semen) amplified from the TGM4 primers, swab B (10 μl saliva) amplified from the HTN3 primers, swab C (1 μl blood) did not amplify from either HTN3 or TGM4. Swab D was a vaginal swab to which 10 μl of semen was added. Each of these swabs was sent in duplicate and for swab D, one of the samples amplified from the TGM4 primers, but the other did not based on calls made by the software examining the melt curves. However, the raw data (Cts) indicates a positive reaction with the TGM4 primers for both swab D samples (Cts of ~25 for TGM4; equivalent to swab A). These results indicate the potential for false negatives if the melt curves are the only criteria used for making calls since samples containing a large amount of biological material (i.e. vaginal swabs) can lead to drift out of the melting temp range set based on the positive control samples. Therefore, both the raw data and calls made by the Stain ID software must be taken into account when analyzing data.

Table 13. Analysis of forensic samples using the Plexor® Stain ID assay.

Sample	Size	Age (months)	STAIN ID RESULTS HTN3 (FAM)	STAIN ID RESULTS TGM4 (HEX)	STAIN ID RESULTS GAPDH (ROX)	SEROLOGY RESULTS AP	SEROLOGY RESULTS Amylase
NTC	N/A	N/A	-	-	-	ND	ND
Positive Control	N/A	N/A	Yes	Yes	Yes	ND	ND
Blood	1ul	23	-	-	Yes	-	-
Blood	10ul	21	-	-	Yes	-	-
Blood/Saliva	5ul each	31	Yes	-	Yes	-	-
Blood/Semen	5ul each	28	-	Yes	Yes	Yes	-
Buccal (FTA)	5mm punch	27	-	-	Yes	-	-
Menstrual Blood	5x5mm cutting	27	-	-	Yes	-	-
Saliva	1ul	31	Yes	-	Yes	-	-
Saliva	10ul	29	Yes	-	Yes	-	-
Saliva/Semen	5ul each	31	Yes	Yes	Yes	Yes	-
Semen	1ul	7	-	Yes	Yes	Yes	-
Semen	10ul	7	-	Yes	Yes	Yes	-
Urine	1ul	31	-	-	No Call	-	-
Urine	10ul	31	-	-	Yes	-	-
Vaginal Secretions	1/4 swab	29	-	-	Yes	-	-
Kidney	10mg	17	-	-	Yes	ND	ND
Adipose	10mg	17	-	-	Yes	ND	ND
Colon	10mg	17	-	-	Yes	ND	ND
Skin	10mg	17	-	-	Yes	ND	ND
Neg Control	N/A	N/A	-	-	-	ND	ND

ND - not determined

Table 14. Blind analysis of mock forensic samples using the Plexor® Stain ID assay.

Sample	HTN3 (FAM)	TGM4 (HEX)	GAPDH (ROX)
NTC	-	-	-
Positive Control	Yes	Yes	Yes
Swab A	-	Yes	Yes
Swab B	Yes	-	Yes
Swab C	-	-	Yes
Swab D	-	-/Yes	-
Negative Control	-	-	-

(Key: Swab A – 10 µl semen, Swab B – 10 µl saliva, Swab C – 1 µl blood, Swab D – vaginal swab with 10 µl semen)

CONCLUSION

The classical biochemical analysis of evidence involved the determination of the nature of a biological stain. Through the use of simple enzymatic color reactions, crystal tests or microscopy, forensic examiners could probe a stain to determine if it contained blood, semen, saliva or some other tissue. The characterization of a stain was the basic question a forensic scientist was technically able to answer when presented with evidence from a homicide, sexual assault or other crimes where body fluid stains may be found. Through the years, researchers developed methods to permit some level of individualization of the evidence and knowledge of the material detected permitted the examiner to determine which typing tests were appropriate for the stain in question. For example, saliva stains contain a limited number of biological markers that could be typed, whereas semen allowed a better discrimination and blood even more so. Understanding the nature of the biological stain permitted the forensic scientist to select those appropriate stain specific typing methods. The identification of semen on clothing from a sexual assault victim or human blood on a suspect involved in a homicide, coupled with some limited typing information was seen in its day as useful information. Today DNA analysis can essentially pin point the donor of a stain and discussions within the forensic community debate the need to identify the tissue of origin of the stain when individual identity is assumed.

These discussions have merit since limiting the number of tests performed on an item would save time and resources especially given the backlogs existing in today's crime laboratories along with the need to meet judicial deadlines with limited resources. However, one could think of a number of crimes where stain identification may be crucial to a case. Finding a low level male DNA profile on a child's underwear from a father accused of sexual assault may not be truly meaningful given the sensitivity of DNA typing. In a court of law, a semen stain containing suspect DNA can have far more serious consequences than DNA deposited from shed epithelial cells due to incidental contact. In this case and similar cases, tissue identification would be important in the evaluation of the evidence. Prosecutors are seeking an answer to the CSI effect where juries expect forensic science to play a major role in solving cases. It is not unusual for a prosecutor to parade a number of experts in front of the jury to showplace forensics even if only limited or no useful forensic information was gleamed from the evidence. A forensic expert who offers court testimony about a DNA profile but says she didn't bother to identify a stain as seminal fluid in origin may not be appreciated by the members of a jury well versed in "CSI". The analysis of biological evidence without tissue identification is probably a procedural step backwards and should be viewed as a short term fix until technology catches up in the front-end analysis of crime scene evidence.

The work conducted in our laboratory and others concerning mRNA is an attempt to make improvements in tissue identification technology. Molecular biological testing has revolutionized the individualization of stains and research hopes to show these same improvements can be achieved with tissue identification. The first goal was to evaluate the stability of RNA in different types of stains. For over 5 years we have been collecting samples for the sole purpose of determining the stability of candidate mRNAs for stain identification. In order for a candidate to be implemented into a stain identification assay, it needs to be detectable for years following deposition of the sample. It was long believed that RNA was very unstable and difficult to work with, requiring dedicated equipment and laboratory space. However, due to the development of new techniques and the recent increase in knowledge concerning RNA, it has been shown to be relatively stable. Various groups have demonstrated the stability of RNA in forensic stains by traditional RT-PCR and more recently, RT reactions coupled with real-time PCR (Juusola and Ballantyne, 2003; Alvarez et al., 2004; Bauer et al., 2003; Nussbaumer et al., 2006). Furthermore, Setzer et al. (2008) recently addressed concerns on the stability, and hence recoverability, of RNA in forensic samples (blood, saliva, semen, and vaginal secretions) exposed to a range of environmental conditions from 1 to 547 days.

In support of these findings, we used amplification of HBB and PRM2 genes as an indicator of mRNA stability in blood and semen samples, respectively. Regardless of the sample size (i.e. down to 1 μl), we were able to detect these genes in samples aged well over 3 years. In fact, PRM2 was amplified from RNA extracted from the first sample collected for this project, a seminal fluid stain aged for 1666 days at room temperature in the dark. In fact, it appears as though PRM2 expression may even increase over time, perhaps due to the loss of an inhibitor.

Since forensic samples are often exposed to a variety of environmental insults, we stored parallel blood and semen samples under ideal conditions (room temperature, dark) as well as in a sub-optimal condition (37°C). By comparing gene amplification between the samples, we hoped to determine whether RNA profiling would be feasible for weathered forensic stains. There was no considerable decrease in the degree of amplification after prolonged exposure (>500 days) to elevated temperatures. These results further demonstrate how stable HBB and PRM2 RNA are over time. In conclusion, our results together with work published by other groups (Juusola and Ballantyne, 2003; Alvarez et al., 2004; Bauer et al., 2003; Nussbaumer et al., 2006; Setzer et al., 2008) prove that RNA can be recovered from forensically relevant biological stains in sufficient quantity and quality for mRNA analysis.

A major goal of ours was to identify 2-3 gene candidates which were specific for each tissue of interest. The gene candidates utilized through the course of this chapter were identified through surveys of the literature including PubMed, Gene and other databases. Initially we identified 2-3 genes that appeared to be specific for each tissue. These screening studies were performed using TaqMan® primer/probe sets from Applied Biosystems because they were pre-designed, inexpensive and already optimized for their intended target. For each target, the sensitivity and specificity for the body fluid of interest were assessed. Once the assay was shown to be robust, we tested it on mRNA isolated from other fluids and tissues to demonstrate that the assay was specific.

A sensitivity study of the three candidate blood assays (HBB, CD3G, SPTB) demonstrated how minute volumes of blood could be detected using mRNA profiling. Amplification with the HBB probe/primers occurred with as little as 0.0001 μl of blood. Although the seminal fluid genes were not as sensitive as the HBB assay, amplification using the PSA and PRM2 TaqMan® sets occurred with 0.01 μl of semen. The lower detection limit for TGM4 was not determined, but assumed to be lower than the 1 μl tested based on the Ct since it was far from nearing Cts typically observed in negative samples.

The sensitivities for saliva candidate markers (STAT, HTN3) were not as low as found for the blood and semen assays; 1 μl for STAT and 5 μl for HTN3. However, the result is significant because there are no Cts observed for the negative controls.

Our diverse sample bank was used to assess the specificity of the candidate tissue-specific genes. These samples included blood, semen, saliva, menstrual blood, vaginal secretions, kidney, colon, adipose, skin, and control RNAs (brain, heart, liver, kidney, intestine). The fluid results were fairly straight-forward. HBB, SPTB and CD3G (blood markers) only amplified in blood samples, whereas STAT (saliva marker) only amplified in saliva samples. There was some cross-reactivity with the semen candidates. Some minor amplification occurred with vaginal secretions (SEMG1 and ACPP) and blood (ACPP). But, for the most part, the assays only amplified from semen stains. Based on these studies PSA, TGM4 and PRM2 were the most specific.

A major aim of stain identification using mRNA expression profiling is working towards multiplexing the real-time PCR assays once mRNAs are identified that clearly define specific types of stains. Since these assays are designed to function more qualitatively than quantitatively, a test of a single stain should typically give amplification with only one candidate. It is possible that a mixture could give several amplifications. However, it is of greater concern to know that a mixture does exist, than to know the exact amount of each fluid present.

Work from Juusola and Ballantyne (2005) identified tissue-specific genes for the more common forensic casework samples: blood, saliva, semen and vaginal secretions. They developed a multiplex RT-PCR assay composed of eight genes (SPTB, prophobilinogen deaminase [PBGD], STAT, HTN3, PRM1, PRM2, human beta-defensin 1 [HBD1] and mucin 4 [MUC4]) which could be analyzed on a standard capillary electrophoresis platform. However, we sought to develop multiplex assays which could be performed more efficiently using real-time PCR; a technology currently used in most forensic laboratories.

A Plexor®-based multiplex assay arose out of collaboration with Promega to develop stain identification assays. The long-term goal of the collaboration is to generate a panel of Plexor® Tissue Typing Systems; multiplexed qRT-PCR systems for determining the source and quantity of a variety of human stains and/or tissues. The systems would include Plexor® primers for the detection of tissue-specific mRNA transcripts associated with semen, sperm, blood, menstrual blood, saliva, etc. By limiting the system to two-color detection (i.e. FAM and HEX detection), it is compatible with the majority of real-time thermal cyclers currently in place in forensic laboratories. We sought to include controls (e.g. a housekeeping gene) or multiple, tissue-specific targets using the ability of the Plexor® Data Analysis Software to distinguish two different amplicons in the same dye channel, based upon thermal melt properties.

The Stain ID assay which we have developed detects semen and saliva. Plexor® primers were designed to amplify the targets TGM4 (semen), HTN3 (saliva) and GAPDH (housekeeper). Through numerous optimization studies, we generated saliva primers that show specific reactivity with saliva RNA and no cross-reactivity with other RNA samples or with genomic DNA. The semen primers (TGM4) show specific reactivity with semen RNA and no cross-reactivity with other RNA samples or with genomic DNA. Although initial studies were performed using RNA extracted from large sample volumes, unlike what's encountered in routine forensic work, titration experiments were performed and indicated that semen had a detection threshold of approximately 10 pg RNA whereas the threshold for saliva was approximately 1 ng of RNA. Furthermore, we assessed how the Stain ID assay would perform with typical forensic samples (i.e. various volumes of unknown RNA yields). The only non-specific amplification that occurred was from the HTN3 primers and the vaginal secretions sample. Since this experiment was performed, the HTN3 primers have been modified and no longer detect vaginal secretions.

A major outcome of the Stain ID assay development was the generation of a co-isolation method for RNA and DNA extraction. A combination of the RNagents Total RNA Isolation System (Promega) with a Tris-buffered phenol protocol was optimized for simultaneous extraction of the nucleic acids. The final extraction method was shown to extract sufficient quantities of quality RNA and DNA from 1 and 10 µl of semen and saliva demonstrating its utility as a dual extraction technique. Although this method involves numerous hands-on steps, it is faster than the TRIzol® method, and produces significantly better DNA yields. To date, it's the best co-isolation method we've tested in terms of yields and amplifiability.

The research indicates that mRNA may have a significant impact in forensic analysis in the years to come. This change in the way we identify stain types may be ushered into forensics via mRNA analysis or perhaps other methods will become available and more appropriate for use. The National Institute of Justice (NIJ) has funded a number of projects that look at streamlining the screening of biological materials (http://www.dna.gov/audiences/researchers/r_funding). Our laboratory has been involved with a project to automate the microscopic search of sperm on slides from evidence retrieved from a sexual assault. An automated approach to scan and record the locations of sperm on a slide would reduce the time it takes an examiner to process a sexual assault case. Other NIJ projects include one from Towson University whose researchers will attempt to use DNA as the tissue determent for tissue identification with their project "Development of a DNA-Based Real-Time PCR Assay for Identification of Semen, Blood and Saliva". Additional NIJ grants are looking at proteomics as a means to provide a new way forward. The New York Medical Examiner's office has a project to evaluate proteomics as a method to determine body fluid cell type and researchers at the University of Denver have NIJ funding to identify proteins as markers for biological stains. A private company (MicroBioSystems, North Logan, UT) has performed research that has produced a hand held instrument with the qualities similar to that of the Star Trek "Tricorder". This device uses a variety of wavelengths of light to identify stains from crime scene materials and was developed under the NIJ funded project "Development of an Optical Handheld Biological Evidence Detection System".

The switch from a conventional to an mRNA or other non-traditional approach will not be an easy transition. Current testing methods are available for the majority of routinely encountered forensically important tissues. These tests are relatively fast and require limited resources and examiner training to perform. Each test does require a separate sampling consuming a portion of the specimen for each subsequent test. In addition, some of these tests have limited specificity and sensitivity. mRNA may be able to overcome these difficulties and with enough research one could imagine a streamlined extraction technology that would allow a simplified co-extraction of RNA and DNA. The extract could be used to identify a number of tissues in one test and then permit DNA analysis of that same extract, avoiding sampling question issues. More information could be derived from a single sample undergoing one test. This assay could be substituted for the numerous tests now required to identify semen, blood, saliva and others tissues with a single unified approach to the problem. Research and development of solutions for tissue identification is but one hurdle the forensic community must overcome to use mRNA in tissue typing. Training for non-DNA analysts currently performing only biological screening could be extensive. In the United States, those involved with DNA analysis must attain a certain level of training and experience as determined by the DNA Advisory Board which could be an obstacle for some laboratories. Identifying appropriate space to do the testing with concerns about contamination issues and resources to do the testing with high tech equipment would also need to be addressed. These are issues that the forensic community encountered when DNA typing was first introduced. The community was able to meet those challenges and in the years to come, as mRNA analysis matures, the community will once again need to face these concerns to see if the community will embrace a new technology to revolutionize how biological material is identified.

Acknowledgments

The authors would like to thank Doug Storts, Nadine Nassif and Len Goren at Promega Corporation for help in development of the Stain ID assay.

References

Akane, A; Shiono, H; Matsubara, K; Nakamura, H; Hasegawa, M; Kagawa M. Purification of forensic specimens for the polymerase chain reaction (PCR) analysis. *Journal of Forensic Science*, 1993, 38, 691-701.

Alvarez, M; Juusola, J; Ballantyne, J. An mRNA and DNA co-isolation method for forensic casework samples. *Analytical Biochemistry*, 2004, 335(2), 289-298.

Bauer, M. RNA in forensic science. *Forensic Science International: Genetics*, 2007 1, 69-74.

Bauer, M; Kraus, A; Patzelt, D. Detection of epithelial cells in dried blood stains by reverse transcriptase-polymerase chain reaction. *Journal of Forensic Science*, 1999, 44, 1232-1236.

Bauer, M; Patzelt, D. Evaluation of mRNA markers for the identification of menstrual blood. *Journal of Forensic Science*, 2002, 47, 1278-1282.

Bauer, M; Patzelt, D. A method for simultaneous RNA and DNA isolation from dried blood and semen stains. *Forensic Science International*, 2003, 136, 76-78.

Bauer, M; Patzelt, D. Identification of menstrual blood by real time RT-PCR: technical improvements and the practical value of negative test results. *Forensic Science International*, 2008, 174, 55-9.

Bauer, M; Polzin, S; Patzelt, D. Quantification of RNA degradation by semi-quantitative duplex and competitive RT-PCR: a possible indicator of the age of bloodstains? *Forensic Science International*, 2003, 138, 94-103.

Bourlet, T; Berthelot, P; Grattard, F; Genin, C; Lucht, F; Pozzetto, B. Detection of GB virus C/hepatitis G virus in semen and saliva of HIV type-1 infected men. *Clinical Microbiology and Infection*, 2002, 8, 352-357.

Cao, J; Cousins, R. Metallothionein mRNA in monocytes and peripheral blood on nuclear cells and in cells from dried blood spots increases after zinc supplementation of men. *Journal of Nutrition*, 2000, 130, 2180-2187.

Chomczynski, P. A reagent for the single-step simultaneous isolation of RNA, DNA, and proteins from cell and tissue samples. *Biotechniques*, 1993, 15(3), 532-537.

Daja, M; Aghmesheh, M; Ow, K; Rohde, P; Barrow, K; Russell, P. Beta-human chorionic gonadotropin in semen: a marker for early detection of prostate cancer? *Molecular Urology*, 2000, 4, 421-427.

Dulioust, E; Tachet, A; De Almeida, M; Finkielsztejn, L; Rivalland, S; Salmon, D; Sicard, D; Rouzioux, C; Jouannet, P. Detection of HIV-1 in seminal plasma and seminal cells of HIV-1 seropositive men. *American Journal of Reproductive Immunology*, 1998, 41, 27-40.

Forensic Science Research Training Center. Procedure for the detection of restriction length polymorphisms in Human DNA. Quantico VA: FBI Laboratory, 1989.

Fourney, R; DesRoches, A; Buckle, J. Recent Progress in Processing Biological Evidence and Forensic DNA Profiling. A Review: 2004 to 2007. 15th International Interpol Forensic Science Symposium. Lyon, France. October 23-26, 2007.

Goodwin, L; Karabinus, D; Pergolizzi, R; Benoff, S. L-type voltage-dependent calcium channel alpha-1C subunit mRNA is present in ejaculated human spermatozoa. *Molecular Human Reproduction*, 2000, 6, 127-136.

Jin, L; Vyse, A; Brown, D. The role of RT-PCR assay of oral fluid for diagnosis and surveillance of measles, mumps and rubella. *Bulletin of the World Health Organization*, 2002, 80, 76-77.

Juusola, J; Ballantyne, J. Messenger RNA profiling: a prototype method to supplant conventional methods for body fluid identification. *Forensic Science International*, 2003, 135, 85-96.

Juusola, J; Ballantyne, J. Multiplex mRNA profiling for the identification of body fluids. *Forensic Science International*, 2005, 152, 1-12.

Juusola, J; Ballantyne, J. mRNA profiling for body fluid identification by multiplex quantitative RT-PCR. *Journal of Forensic Science*, 2007, 52, 1252-1262.

Li, J; Wirtz, R; McCutchan, T. Analysis of malaria parasite RNA from decade-old Giemsa-stained blood smears and dried mosquitoes. *American Journal of Tropical Medicine and Hygiene*, 1997, 57, 727-731.

Lundwell, A; Bjartell, A; Olsson, A; Malm, J. Semenogelin I and II, the predominant human seminal plasma proteins, are also expressed in non-genital tissues. *Molecular Human Reproduction*, 2002, 8, 805-810.

Matsubara, Y; Ikeda, H; Endo, H; Narisawa, K. Dried blood spot on filter paper as a source of mRNA. *Nucleic Acids Research*, 1992, 20(8), 1998.

Miller, D. Analysis and significance of messenger RNA in human ejaculated spermatozoa. *Molecular Reproduction and Development*, 2000, 56, 259-264.

Nicklas, J; Buel, E. Simultaneous determination of total human and male DNA using a duplex real-time PCR assay. *Journal of Forensic Science*, 2006, 51(5), 1005-1015.

Noreault-Conti, T; Buel, E. The use of real-time PCR for forensic stain identification. *Profiles in DNA*, 2007, 10, 3-5.

Nussbaumer, C; Gharehbaghi-Schnell, E; Korschineck, I. Messenger RNA profiling: a novel method for body fluid identification by real-time PCR. *Forensic Science International*, 2006, 157, 181-186.

Richter, W; Dettmer, D; Glander, H. Detection of mRNA transcripts of cyclic nucleotide phosphodiesterase subtypes in ejaculated human spermatozoa. *Molecular Human Reproduction*, 1999, 5, 732-736.

Setzer, M; Juusola, J; Ballantyne, J. Recovery and stability of RNA in vaginal swabs and blood, semen, and saliva stains. *Journal of Forensic Science*, 2008, 53, 296-305.

Watanapokasin, Y; Winichagoon, P; Fuchareon, S; Wilairat, P. Relative quantitation of mRNA in beta-thalassemia/Hb E using real-time polymerase chain reaction. *Hemoglobin*, 2000, 24, 105-116.

Zhang, Y; McCabe, E. RNA analysis from newborn screening dried blood specimens. *Human Genetics*, 1992, 89, 311-314.

Zubakov, D; Hanekamp, E; Kokshoorn, M; van, I; Kayser, M. Stable RNA markers for identification of blood and saliva stains revealed from whole genome expression analysis of time-wise degraded samples. *International Journal of Legal Medicine*, 2008, 122, 135-42.

In: Horizons in DNA Research, Volume 1
Editor: Jason R. Chesterton, pp. 81-103

ISBN: 978-1-60876-968-1

Chapter 3

THE POTENTIAL ROLE OF ACTIVE METHYLATION CYCLE INTO LEUKEMIC CELL DIFFERENTIATION

Ioannis S. Vizirianakis*[*#], *Sotirios S. Tezias* and *Asterios S. Tsiftsoglou

Laboratory of Pharmacology, Department of Pharmaceutical Sciences, Aristotle University of Thessaloniki, GR-54124 Thessaloniki

ABSTRACT

Malignant cells exhibit metabolic alterations affecting growth, lack to respond to physiological stimuli and blockade of differentiation and apoptosis. Selective growth advantage on cells acquired by transformation influences biosynthetic enzymes and alters the internal pools of nucleotides. During the course of studying the induction of erythroid differentiation of murine erythroleukemia (MEL) cells, we observed that agents that inhibit the active methylation cycle, (N^6-methyladenosine, neplanocin A, 3-deazaneplanocin A and cycloleucine), blocked initiation of commitment, erased "memory" response to previous exposure to chemical inducers, altered the expression profile of specific genes and abrogated hemoglobin production in inducer-treated MEL cells presumably by interrupting a central process required for terminal erythroid maturation. In an effort to delineate the mechanism(s) by which such methylation inhibitors abrogate the developmental program of MEL cells, we have shown that exposure to these methylation inhibitors led to cytoplasmic accumulation of relatively small discrete RNA transcripts lacking poly(A) tail. These RNAs were hybridized with ^{32}P-labeled probes derived from 3'-end flanking sequences of the mouse β^{major} globin gene. Structural and functional analysis of this DNA region have revealed a B1 repeat element as well as consensus sequences recognized by transcription factors GATA-1, AP1/NF-E2 and erythroid Krüppel-like factor (EKLF). Successful cloning,

---

* Corresponding author: Laboratory of Pharmacology, Department of Pharmaceutical Sciences, Aristotle University of Thessaloniki, GR-54124 Thessaloniki, Greece; tel.: +30-2310-997658; Fax: +30-2310-997645; Email: <ivizir@pharm.auth.gr>.

The authors' experimental data presented in this chapter represent their work published earlier in different journals cited at the reference section.

characterization and identification of genomic DNA sequence encoding these RNA species would shed some light on how do the active methylation cycle inhibitors activate their biosynthesis and subsequently whether these RNA molecules contribute to the blockade of differentiation of MEL cells by hypomethylating agents.

Keywords: MEL cells, differentiation, hypomethylating agents, chemical inducers, globin genes, transcription, neplanocin A, 3-deazaneplanocin A, cycloleucine, silent DNA, DNA methylation, RNA methylation.

1. INTRODUCTION

Epigenetics is described as the heritable alterations in gene expression potential or functions that occur without any genetic change within the DNA sequence itself. Understanding the key epigenetic changes occurring *via* DNA methylation and histone post-translational modifications has clearly added new knowledge on vital processes in mammalian cells related to cell proliferation, differentiation and development (1-4). The machinery of active methylation cycle in mammals regulates specific methylation patterns in both DNA and RNA molecules that lead to differential gene expression profiles upon development and the production of specific proteins. Indeed, there is growing interest to reverse epigenetic abnormalities seen in pathophysiological disorders pharmacologically, especially in cases where the importance of keeping specific DNA methylation status has been emphasized and recognized. Unequivocally, aberrant DNA methylation has been correlated with many diseases including imprinting disorders and cancer (4-11). Specific methylation patterns are established and maintained through the function of DNA methyltransferases (DNMTs), enzymes that catalyze the transfer of a methyl group from S-adenosyl-methionine (SAM) to cytosine residues at position C5 in CpG dinucleotides organized in CpG-rich regions called "CpG islands". The "recognition and reading" of these methylation profiles has been succeeded by the methyl-CpG binding proteins (MBDs). It has been estimated that $5x10^7$ CpG islands exist in the human genome (1, 5-8, 12-15). In most cases, the CpG-islands are mapped within the promoter region spanning from 500bp to 5 kb. Evidence exists to support the notion that changes in the status of methylation of CpG islands at the promoter sequences can control gene expression. To this respect, the methylation of cytosine affects the interaction of promoters with transcription factors, chromatin remodeling enzymes or other DNA-binding proteins that ultimately leads to gene silencing. On the other hand, alterations in gene expression, cycle control, genome stability and genome reprogramming could be the result of modification of DNA methylation and chromatin remodeling patterns (16-20). Apart from CpG island methylation within the promoters, methylation also occurs in non-coding DNA regions including transposable elements, long interspersed elements (LINE) and DNA satellite sequences (16, 20-26). The potential role of methylation of these DNA regions it is not clear as yet.

The DNA sequences at the promoter regions of active genes under normal conditions in mammalian cells exhibit the open chromatin structure (eurochromatin), thus allowing histones acetyltransferases (HAT), transcription inducer co-activators (CA) and basic transcription factors to interact with their targeting sequences without permitting the interaction of DNMTs at the same time. The rest of the gene regions are accessible to

DNMTs for methylation of the appropriate CpG dinucleotides, to MBDs for binding to methylated CpG dinucleotides and finally to histone deacetylases (HDACs) for histone deacetylation. This cascade of events in specific DNA regions that contain sequences to control gene expression ultimately contributes in compact chromatin structure which in turn does not allow transcription factors to bind to their consensus DNA domains and activate gene expression (2, 5).

On the contrary, under the cancerous state, chromatin structure at the promoter region of genes exists in a more compact state (heterochromatin). To this end, the CpG islands in the promoter region of tumour suppressor genes, DNA repair genes, angiogenesis and metastasis inhibitory genes are heavily methylated rendering MBDs for binding on these sites. Furthermore, the HDACs make the chromatin structure more compact not allowing access and binding of transcription factors, thus leading to silencing of these genes (13, 27). At the same time, the remaining gene regions are totally unmethylated, a fact that is representative at least for the majority of cancer cells (2, 5). Overall, both global hypomethylation and regional hypermethylation confer a selective advantage to cancer cells by targeting different sets of genes with opposing roles in cellular transformation. As a consequence, global hypomethylation leads probably to the activation of genes which are required for different stages of the transformation process, in contrast to regional hypermethylation that seems to affect the silencing of genes which suppress tumorigenesis itself (10, 20). Although the changes of DNA methylation status in eukaryotic cells is heritable, they are also reversible, something that makes the modulation of DNA methylation profile a good therapeutic target. To this respect, many agents including methylation inhibitors, histone deacetylase inhibitors (HDACIs), antisense oligonucleotides and small interference RNAs (siRNAs) that target RNA transcripts encoding either specific DNMTs or MBDs are under investigation as potential anticancer drugs, a fact making the field of DNA methylation wider and more exciting (6, 17, 18, 24).

As far as RNA methylation is concerned, the methyl-groups are transferred from SAM into specific base residues on RNA molecules via the catalytic function of RNA methyltransferases. Mammalian cells contain several RNA species, e.g. snRNAs, tRNAs, rRNAs and mRNAs, most of which are methylated at specific nucleotides that, in some cases, are located at unique structures, like the 5′-cap in mRNA and the 5′-end of snRNAs (28-31). Unfortunately, little is known about the biological role of RNA methylation in cell proliferation, differentiation and development (32-34).

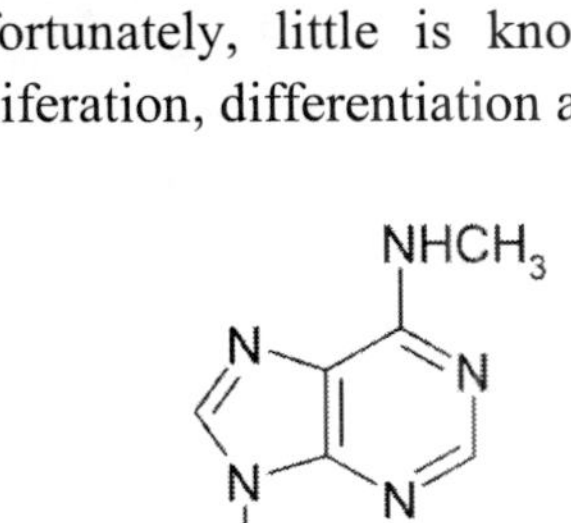

N^6-methyladenosine

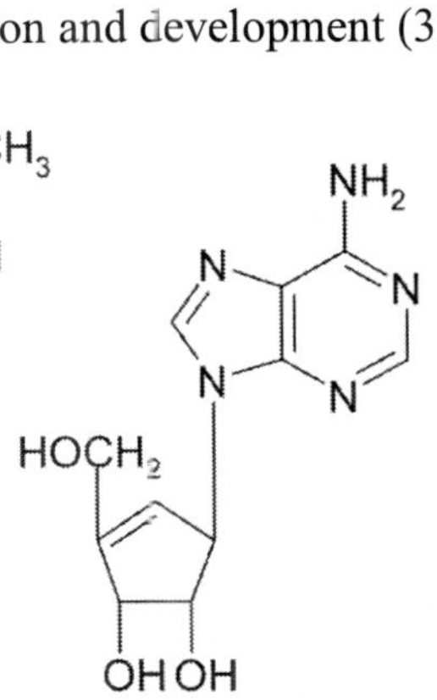

Neplanocin A

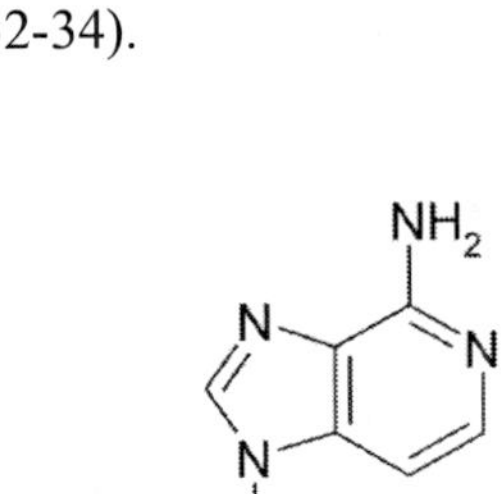

3-Deazaneplanocin A

Cycloleucine

Figure 1. Inhibitors of active methylation cycle that block the MEL cell differentiation program.

To this regard, we attempted to exploit the potential role of nucleic acid methylation and especially that of RNA in the initiation of differentiation of leukemia cells in culture by employing murine erythroleukemia (MEL or Friend) as a model system. Although the mechanisms governing MEL cell differentiation to terminal erythroid maturation are still not fully elucidated regardless the remarkable progress achieved in this field, earlier studies with metabolic inhibitors (e.g. cordycepin, cycloheximide) revealed that initiation of commitment to terminal maturation depends on the synthesis of new RNA transcripts and proteins (36-39). Biochemical studies with cordycepin (3'-deoxyadenosine), (an inhibitor of polyadenylation and methylation of RNA) have suggested that initiation of commitment of inducer-treated MEL cells to terminal erythroid maturation, may depend on both the synthesis and post-transcriptional modifications of RNA, such as methylation (36, 38). Furthermore, we assessed the effect of N^6-methyladenosine (N^6mAdo), [a methylated nucleoside found in RNAs (mRNA, rRNAs, snRNAs)], as well as that of neplanocin A, 3-deazaneplanocin A and cycloleucine on the *in vitro* differentiation program of MEL cells (Fig. 1). Neplanocin A, 3-deazaneplanocin A and N^6mAdo are inhibitors of SAH hydrolase (E.C. 3.3.1.1) (Ki values for the isolated SAH $2x10^{-9}$ M, $5x10^{-11}$ M and $2x10^{-4}$ M, respectively), whereas cycloleucine is an inhibitor of SAM synthetase (E.C. 2.5.1.6) (Ki value for the isolated SAM synthetase $4.8x10^{-5}$ M) (41-44). Interestingly enough, we have found that all these agents block the differentiation program of MEL cells induced by chemical inducers possibly *via* influencing the transmethylation of RNA in these cells. In particular, we assessed their effects: a) On both the synthesis and methylation of specific classes of RNA molecules; b) On the expression of "memory" to the previous exposure to inducers; c) On the accumulation of two active methylation cycle intermediates SAM and SAH; d) On the steady-state level of developmentally regulated genes; e) On the possibility of the transcriptional activation of silent non-coding DNA regions; and last, but not least, f) On the cytoplasmic accumulation of new poly(A)$^-$ RNA transcripts. The results obtained are presented and discussed extensively within this chapter.

2. The Developmental Program of MEL Cells

During normal haematopoiesis in bone marrow, stem cells are committed to early haematopoietic progenitors and differentiate into various mature blood cells along cell lineage restricted pathways (45, 46). The homeostasis in haematopoietic system is controlled by the interplay of growth factors acting on bone marrow cells to keep the balance steady, i.e. to control the number of differentiated progeny leaving the bone marrow compartment and entering the peripheral blood. Thus, haematopoietic stem cells (HSCs) replace the mature blood cells of limited life span with newly produced differentiated progeny via self-renewal and differentiation. The question whether this balancing act involves programmed cell death (apoptosis) of haematopoietic progenitors as part of the normal homeostatic process in addition to self-renewal and differentiation of HSCs remains under consideration (38, 39, 45-47). The concept however, that leukemias may arise primarily from individual stem or early haematopoietic progenitors that acquire immortality and fail to differentiate (48, 49) or undergo apoptosis (50), suggest that an impairment in apoptosis may be equally important element in leukemia development as the blockade for differentiation. A large number of studies carried out over the previous years have established that various cancer cells can be induced to differentiate both *in vitro* and *in vivo* by treatment with chemical, pharmacological or natural agents into cells which resemble their normal counterparts. This approach has led very early to the realization that the neoplastic phenotype, at least in certain tumors, can be reversed, a fact that was further strengthened by the observation that the differentiated neoplastic cells are not malignant any more and being unable to support tumor growth. The latter, led to the development of the so-called concept of "Differentiation Therapy of Cancer" with inducers of maturation as a new direction in anticancer therapeutics (39, 50-52). The use of permanent lines of leukemia cells as model systems over the years permitted an intensive investigation of the hematopoietic processes under the malignant state. Such approach gave valuable information on the basic cellular functions affected by these developmental state transitions that may be exploited therapeutically. Among the most widely studied systems for the erythroid lineage pathway is that of MEL cells (38).

MEL cells are virus-transformed erythroid precursor cells that although they retain the capacity of self-renewal they can also be differentiated in vitro by exposure to various chemical agents into cells resembling orthochromatophilic normoblasts. To this regard, MEL cells provide a suitable model system to uncover key-regulatory mechanisms responsible for specific gene expression during neoplastic cell differentiation as well as during normal haematopoiesis (38). The large number of chemical, pharmaceutical and natural agents found to induce commitment to terminal maturation of MEL cells includes purine and pyrimidine analogues, hybrid polar agents like bis-hydroxamic acid derivatives, bis-acetamides and substituted amides, cyclic ureas and thioureas, ureido-derivatives of pyridine (UDPs), benzodiazepines, sodium butyrate, phenyl-butyrate, HDAIs, interferons and many other agents (39, 53, 54). This highly diversified structure of the inducing agents suggests that there may be more than one mechanism to initiate commitment to terminal maturation of MEL cells. Differentiated cells are smaller in size and show a denser nucleus. In addition, under certain culture conditions, terminally differentiated MEL cells can extrude the nucleus yielding physically unstable reticulocyte-like cells. Commitment, as the central event in differentiation, leads to limitation of cellular proliferation to a few divisions and to expression of erythroid differentiation markers. Also, committed cells, in addition to producing highly detectable quantities of globin mRNA and haemoglobin, undergo a series of morphological and biochemical events that resemble those seen upon differentiation of normal early erythroid progenitors (38, 39, 53, 55). Upon this process, transcriptional activation of several genes and selective repression of many others has been observed. Indeed, over the past 35 years, the analysis of the commitment of MEL cells to terminal erythroid maturation has revealed that the differentiation program is associated with discrete highly-orchestrated patterns of gene expression. In particular, assessment of the steady-state level of several RNA transcripts by Northern blot hybridization, or RT-PCR analysis confirmed that some of them begin to decline prior to and after commitment, while in other cases the level is transiently increased and decline thereafter (38, 39). This kinetic analysis of gene expression in differentiating MEL cells indicated that initiation of commitment sets on discrete patterns of gene expression that contribute to the terminal differentiated phenotype (39). Alternatively, it is of great importance the fact that recently the detection, isolation and partial characterization of the primary structure of a ~38 kDa (UDP-p38) soluble post-mitochondrial protein(s) from

MEL cells with the ability to bind to the specific class of UDP inducers was reported by our laboratory (56). Evidence obtained from specific and competition binding studies has indicated that this binding protein(s) may also interact with other known inducers. Interestingly enough, these data shed light on how the execution of the developmental program of MEL cells may be triggered and regulated, since they impinge on the molecular mechanism(s) by which the UDP-p38 complexes contribute to the initiation of erythroid differentiation program.

3. Differentiation of MEL Cells in Vitro is Associated with Changes in SAH/SAM Ratio as well as in the Pattern of RNA Methylation

It has been shown that MEL cell differentiation is associated with hypomethylation of DNA (57), an event that may not be considered as a crucial initiative event, but still quite important for gene activation (4, 12). Also, earlier studies with cordycepin that synchronizes commitment of MEL cells (58) and blocks polyadenylation and methylation of RNA (36) indicated that induction of commitment depends on a methionine-sensitive event (40, 60). The latter, is further confirmed by the fact that MEL cell differentiation is blocked by agents that can inhibit methylation of RNA like cordycepin (58, 36), 3-deazaadenosine (61), S-5'-isobutylthioadenosine and 5'-methylthioadenosine (62), although the precise mechanism(s) of this inhibition is still not fully elucidated. Also, biochemical studies with cordycepin, and more recently with N^6mAdo, have suggested that initiation of commitment of inducer-treated MEL cells to terminal erythroid maturation may depends on both the synthesis and post-transcriptional modifications of RNA, such as methylation (36, 58-60, 63, 64).

In an effort to further explore the role of RNA methylation upon induction of MEL cell differentiation, we carried out experiments designed in a manner to particularly address the following questions: a) Do differentiating MEL cells undergo alterations in the synthesis and methylation of RNA? b) If yes, do these changes occur simultaneously or independently during growth and differentiation? c) Do polyA$^+$ and polyA$^-$ RNA transcripts undergo changes in methylation, to what extend and at which base residues? d) Do the changes in RNA methylation occur during differentiation as a result of alterations in the intracellular level of SAM and SAH? e) Are the changes in RNA methylation causally related to the induction of MEL cell differentiation and production of hemoglobin? f) Does inhibition of RNA methylation affect stability of polyA$^+$ RNA transcripts? h) Do specific methylation inhibitors block the MEL developmental program? g) Do methylation inhibitors converted into any intermediate products within the cells that affect the active methylation cycle? And finally, i) Do methylation inhibitors activate transcription of novel genes affecting the execution of MEL cell differentiation program? To approach these questions, various sets of experiments have been carried out. In particular, MEL cells were treated in culture with and without inducers and pulse-labeled with L-[methyl-^{3}H]-methionine. The ability of control and differentiating cells to methylate their total cytoplasmic RNA as well as polyA$^-$ and polyA$^+$ RNAs was assessed at different times with the use of a reversed-phase HPLC chromatographic method developed in our laboratory (60, 63). We analyzed [methyl-^{3}H]-

labeled polyA$^-$ and polyA$^+$ RNA fractions at the level of nucleosides by HPLC and assessed the intracellular levels of SAM and SAH, two active intermediates of methylation cycle. In a complementary series of experiments, MEL cells were exposed to both inducers of differentiation and inhibitors of RNA methylation (neplanocin A, 3-deazaneplanocin, cycloleucine) and examined whether such MEL-treated cells mature, produce globin mRNA and haemoglobin and accumulate other specific RNA transcripts, like those that play a critical role in growth and differentiation. In parallel, we investigated the effect of N^6mAdo on both synthesis and methylation of RNA, on the expression of "memory" and accumulation of SAM and SAH. Furthermore, the metabolic fate of methylation inhibitors in MEL cells was assessed by using radiolabeled compounds and HPLC analysis of cellular extracts.

The fact that RNA methylation occurs in MEL cells is not surprising since methylated bases have been detected in many RNA species (tRNAs, rRNAs, snRNAs, mRNAs) in eukaryotic cells (28-31). What is interesting in this case, however, is the fact that the rate of methylation of cytoplasmic RNA increased in differentiating MEL cells while the overall RNA synthesis markedly decreased. Kinetically, cells entering the plateau phase of growth after 60-72hr incubation exhibited a reduction in the rate of RNA methylation, while inducer-treated cells continued to methylate their RNA at higher rates. This finding suggests that hypermethylation of cytoplasmic RNA may be part of the differentiation process itself, rather than a cell-cycle related event. The increase in methylation of RNA was observed several hours after the reduction in RNA synthesis but not simultaneously. In addition, we observed that induction of differentiation by the inducer led to increased methylation of polyA$^+$ RNA after 36hr incubation, as compared to control untreated cells. During the same period, a transient increase in methylation of rRNA (28S, 18S and 5.8S) was also observed. The hypermethylation of polyA$^+$ and polyA$^-$ RNA was occurred at specific bases, as it has been seen by HPLC analysis (60, 63). In particular, in polyA$^-$ RNA the presence of 2'-O-methylcytidine, 2'-O-methyluridine, 5-methylcytidine (m^5C), 5-methyluridine (m^5U), N^6,N^6-dimethyladenosine ($m^6{}_2A$), 1-methylguanosine (m^1G), N^2-methylguanosine (m^2G) and 1-methylinosine (m^1I) in both control and inducer-treated cells was confirmed but in different quantities. The level of m^5C, m^1G and m^2G was higher in polyA$^-$ RNA fraction isolated from inducer-treated rather from untreated control cells. With respect to polyA$^+$ RNA, HPLC analysis revealed methylated constituents at 5′-cap structure [7-methylguanosine (m^7G), 2'-O-methylcytidine (C_m) and 2'-O-methyluridine (U_m)] and hardly detectable levels of N^6mAdo. The gradual accumulation of [methyl-^{3}H]-labeled 5'-cap structure constituents in inducer-treated cells indicates that mRNAs are relatively hypermethylated at their 5'-cap structure during differentiation, as well as at other base residues.

In order to delineate if the detected changes in RNA methylation in inducer-treated cells could be correlated with changes to intracellular levels of SAM and SAH, we assessed the accumulation of these substances in MEL cells. We observed that the level of SAM decreased in both control and differentiating MEL cells with different rates following an early increase. In contrast, the level of SAH remarkably increased in control cells, while dramatically decreased in differentiating cells from the first hours of incubation with the inducer. Overall, a lower ratio of SAH/SAM observed in differentiated cells (60, 63). These changes in the ratio SAH/SAM appear to affect the degree of RNA methylation, since RNA methyltransferases may no longer be inhibited by SAH (64, 65). In particular, control untreated cells growing from low to higher density towards the stationary (plateau) phase of growth exhibit a gradual increase in the ratio SAH/SAM. This finding suggests that may exist growth-dependent changes in SAH/SAM ratio. Comparing these changes with the rate of total RNA synthesis, we concluded that the value of SAH/SAM ratio reaches a relatively high value at the time when RNA synthesis declines in stationary phase cells. Therefore, an inverse relationship between the synthesis of RNA and the ratio SAH/SAM may exist. However, when cells were treated continuously with the inducer, the ratio SAH/SAM remained lower than that of control untreated cells, so did RNA synthesis. These data suggest that the changes in the level of both SAH/SAM ratio and RNA synthesis in inducer-treated cells occur as part of the differentiation process. In conclusion, we have observed two types of changes in the SAH/SAM ratio: growth-dependent as seen in control untreated cells where cells undergo changes from exponentially to stationary phase of growth and differentiation-dependent changes that appear to be affected by the inducer. Overall, these experimental data have suggested that MEL cell differentiation is associated with both qualitative and quantitative changes in RNA methylation at specific sites the precise role of these changes in maturation remains, however, elusive. Alternatively, it can be speculated that hypermethylation increases hydrophobicity of RNA and may thus facilitate mRNA transport from nucleus into cytoplasm as reported previously (66). On the other hand, the hypermethylation observed at 5′-cap structures of polyA$^+$ RNAs may affect the specific spatial configuration of 5′-end of mRNAs (67), thus protecting them against the action of 5′-nucleases (68, 69). Furthermore, hypermethylation may affect the tertiary structure and conformation of RNA in a way that renders it capable to interact with trans-acting proteins (70-73).

4. Inhibitors of Active Methylation Cycle Block "Memory" and Commitment of MEL Cells to Erythroid Maturation

In order to support the notion that there may be a causal relationship between methylation of RNA and specific events of MEL cell differentiation, we employed three known methylation inhibitors, (neplanocin A, 3-deazaneplanocin A and cycloleucine) (75-78), and asked whether selective inhibition of methylation of nuclear RNA and most importantly of cytoplasmic polyA$^+$ RNA leads to alterations in SAH/SAM ratio as well as changes in the steady-state accumulation and stability of RNA transcripts coded by various genes, thus preventing maturation. Although it can not be ruled out that these inhibitors could also affect methylation of other macromolecules in addition to RNA, it has been reasoned that their effect on DNA methylation must be neglegible, since this process is already suppressed in differentiating MEL cells (57). To this end, it has been shown that all these agents blocked the initiation of the erythroid maturation program of MEL cells promoted by various inducers like DMSO, HMBA, butyric acid, hypoxanthine and UDP-4 [2-(3-ethylureido)-6-methylpyridine] (63). In addition, as we expected, all three inhibitors increased the SAH/SAM ratio as compared to control and inducer-treated cells and inhibited methylation of polyA$^-$ and polyA$^+$ RNA fractions. At this point, we do not know whether treatment with inhibitors prevents the overall methylation of RNA or if these inhibitors act at specific sites of RNA and block particular methylated nucleosides accumulated in differentiating cells, as seen by HPLC analysis.

Exposure of MEL cells to an inducer for several hours and subsequently plating them in inducer-free medium leads to expression of "memory" response, a phenomenon describing the acquired ability of cells to respond and commit immediately upon re-challenge with the inducer. Furthermore, the fact that expression of "memory" response precedes initiation of commitment of MEL cells to maturation upon exposure to inducers, it was reasonable to ask if agents that block the differentiation program have also the capacity to modulate such early involved biochemical events during the execution of erythroid developmental program (39, 55, 58). This knowledge is considered very crucial in trying to understand molecular mechanisms governing the decisions of cells to grow and differentiate in culture. Thus, in an attempt to elucidate any possible correlation of RNA methylation with the "memory" ability of cells, the inverse approach was used. In particular, MEL cells were exposed to N^{6}mAdo, an agent that was seen to block both in vitro differentiation and RNA methylation of these cells, and asked if the expression of "memory" has also been affected. Indeed, it was observed that treatment of MEL cells with N^{6}mAdo erased the "memory" of cells to a previous exposure to inducer, an effect seen also to happen in the case of using cordycepin instead of N^{6}mAdo, another inhibitor of methylation and mRNA polyadenylation, as it has been already mentioned above (64, 79).

The experimental data presented thus far have indicated that N^{6}mAdo as well as neplanocin A, 3-deazaneplanocin and cycloleucine block a central process of MEL cell maturation. To further explore the mechanism(s) by which N^{6}mAdo inhibits initiation of commitment we investigated the possibility of its conversion into an active intermediate product within the cells. Indeed, the results obtained have confirmed that treatment of MEL cells with N^{6}mAdo resulted in the intracellular formation of S-(N^6-methyl)-adenosylhomocysteine (N^6-SAH) by the catalytic function of the enzyme SAH hydrolase (64). That treatment of MEL cells with N^{6}mAdo led to the intracellular accumulation of N^6-SAH is consistent with previous studies showing formation of N^6-SAH in mouse liver (80), lymphocytes (81) and after incubation of N^{6}mAdo with purified SAH hydrolase (82). Inhibition of SAH hydrolase by N^{6}mAdo may be responsible for accumulation of SAH in MEL cells and reversion in SAH/SAM ratio seen in inducer-treated cells. These data propose that N^{6}mAdo is converted into N^6-SAH in MEL cells, an intermediate that might be responsible for blocking methylation of RNAs involved in expression of "memory" and initiation of commitment. Alternatively, N^6-SAH can act as an inhibitor of RNA methyltransferases, enzymes involved in the methylation of RNA, as it has been reported earlier. The notion that SAH hydrolase was affected in MEL cells exposed to N^{6}mAdo in a way to influence the extent of RNA methylation during MEL cell differentiation is further supported by the fact that other known inhibitors of SAH hydrolase, like 3-deazaadenosine, S-5'-isobutylthioadenosine, 5'-methylthioadenosine, neplanocin A, and 3-deazaneplanocin A, or inhibitors of SAM synthetase, like cycloleucine, also inhibit MEL cell differentiation, as it has also been mentioned above (40, 61, 62, 83).

Moreover, by knowing that commitment of MEL cells depends on the synthesis of new RNA transcripts and proteins as well as on methylation and differential stability of polyA$^+$ RNAs, it is reasonable to assume that inhibition of RNA methylation may block initiation of differentiation by affecting stability of critical RNA transcripts. Alternatively, the increase in RNA methylation seen in differentiating MEL cells could be attributed to the observed alterations in the level of both SAH and SAM that result in a lower ratio SAH/SAM (63). Thus, the changes in SAH/SAM ratio would influence the extent of RNA methylation, since RNA methyltransferases will not be longer inhibited by SAH (83). Such changes in the extent of methylation of RNAs may affect the physicochemical behaviour of RNAs due to increased hydrophobicity and possible interactions with cytoplasmic proteins that influence RNA stability (40) as seen in other cases (70, 72, 84-88). If so, then we can assume that methylation of RNA may be a key-event in haematopoietic cell maturation, as we proposed elsewhere (40, 63).

5. Cytoplasmic Accumulation of Discrete Small Poly(A)$^-$ RNA Transcripts Induced by Treatment of MEL Cells with Methylation Inhibitors

As it has been already mentioned, the detailed mechanism by which RNA and DNA metabolic inhibitors block differentiation is still not fully delineated. It was very interesting, however, to observe in our studies for the first time that agents like N^{6}mAdo, neplanocin A, 3-deazaneplanocin A and cycloleucine besides the inhibition of RNA methylation led to cytoplasmic accumulation of discrete RNA transcripts of relatively small molecular weight detected by Northern blot hybridization analysis by using as probes mouse genomic DNA fragments derived from 3′-downstream sequences of the β^{major} globin gene locus (63, 89). Also, it has been shown that these discrete RNA transcripts lack poly(A) tail and accumulate in the cytoplasm of cells exposed to hypomethylating agents. The appearance of four to five different size RNAs (100 to 600 nt) lacking the poly(A) tail in MEL cells suggest that their accumulation is likely to result from unique, although unknown, effects of these agents. To this regard, preliminary structural analysis of the β^{major} globin gene 3′-end flanking DNA sequences revealed the presence within this region of one B1 repeat element as well as consensus sequences for the transcription factors GATA-1, AP1/NF-E2 and erythroid Krüppel-like factor (EKLF). However, it is not yet clear whether inducer/hypomethylating agent-treated cells become committed to terminal maturation without hemoglobin synthesis or both commitment to terminal maturation and hemoglobin production are blocked during this process. On the other hand, it is essential in trying to understand the function of these poly(A)$^-$ RNA transcripts to elucidate the fact if these RNAs are authentic products resulting from transcription of DNA sequences located within the β-globin gene family complex, or are abnormal products coming from mis-processed globin-like RNA precursors which are unable to be polyadenylated in the presence of hypomethylating agents. To this end, previous work has shown the presence of antisense RNA molecules of β^{major} globin gene in the cytoplasm of MEL cells as well as in erythroid spleen cells and reticulocytes from anemic mice (90, 91). These antisense globin RNA transcripts are detected in full size or as truncated molecules. They are complementary to the spliced β^{major} globin mRNA synthesized by the mouse erythroid cells, and they are implicated in a mechanism that gives rise to new sense-strand β^{major} globin mRNA by the involvement of an RNA-dependent RNA polymerase (91).

Unfortunately, no conventional transcriptional units have been detected in these 3'-flanking sequences of β^{major} globin gene. Therefore, it can not rule out the possibility that may be transcriptional units located somewhere else within the mouse genome that are activated by hypomethylating agents and produce such RNAs which share structural complementarities to the 3'-downstream sequences of β^{major} globin gene used as probes for detection. Recently, experiments done in our laboratory were designed in order firstly to investigate whether this DNA region exerts promoter-like activity by performing reporter gene assays and secondly to characterize the RNA transcripts through cloning and sequencing. Such data confirmed that specific DNA sequences derived from this region have indeed promoter-like activity, whereas the cloning of various poly(A)$^-$ RNA transcripts is in process to further characterize their structure and function (unpublished results). Thus, by moving forward, it would be very interesting to investigate whether these RNA species share any structural homology as well as to uncover their orientation (sense or antisense). The latter, could add new information by considering the fact that these RNA transcripts may be responsible for the blockade of MEL cell differentiation seen upon exposure of MEL cells with methylation inhibitors. The answers to these questions are still under experimental investigation in our laboratory, whereas additional work is needed to further characterize these RNAs and then investigate their role in differentiation. The fact, however, that the detected poly(A)$^-$ RNA species do not hybridize with DNA probes derived exclusively from the β^{major} globin coding sequences, it tends to suggest that they represent a different group of RNA molecules, although we do not know yet their orientation (sense or antisense). Also, the possibility that these short RNA transcripts may originate from other regions of the mouse genome that contain sequences structurally homologous to those being present within the 3´-end flanking sequences of β^{major} globin gene is under thorough examination.

In order to gain more insight, electrophoretic mobility shift-band assays (EMSAs) with DNA fragments derived from 3'-downstream sequences of β^{major} globin gene were carried out and the data obtained thus far confirmed the binding of several transcription factors within this region (unpublished results) This finding tends to suggest that the identified sequences and the B1 element share somehow a functional role, although unknown, during the transcription process. Alternatively, the presence of the consensus binding sequences for GATA-1 and AP1/NF-E2 transcription factors near the B1 repeat element may suggest some specific role in the expression pattern of β^{major} gene during development (93, 94). The B1 repeat elements are the mouse equivalent of the Alu element seen in human DNA (92, 93). This is an attractive direction, since it is well known that these two transcription factors have crucial roles in the regulation of globin gene expression through binding in consensus DNA sequences located within the locus control region (LCR) and the promoter of these genes (95-97). This hypothesis is further strengthened by recent findings that implicate the function of B1 repeat family in complex regulatory elements involved in posttranscriptional control of gene expression in the mouse genome (98). In particular, the insertion of a B1 repeat element in the 3'-untranslated region of the rabbit β-globin gene conferred growth- and transformation-dependent transcriptional regulation of this gene after its transfection into the cells. This effect was attributed to the function of B1 repeat element that cooperated with other regulatory sequences in a way that somehow modulated the RNA polymerase II transcription machinery (98).

6. Nucleic Acid Methylation, Chromatin Remodeling and Erythroid Differentiation

Nowadays, epigenetic regulation of gene expression via chromatin structural modification, including DNA methylation and histone acetylation status, has still considered as a key developmental process in higher mammals. The β-globin locus represents a well-studied system for the examination of interplay between chromatin structure and transcriptional regulation. Specific structural elements and modifications like locus control region (LCR), DNase I hypersensitivity sites, ATP-dependent chromatin remodeling, localized H3 hyperacetylation and intergenic transcription are considered very crucial by linking chromatin and β-globin locus regulation (99). In the context of such studies, the MEL model system has been extensively used over the previous years. Nuclear condensation and irreversible cessation of DNA replication are among the most prominent events occurring in terminal maturation of MEL cells (100). Previous studies examining the structure of chromatin by digestion with nucleases have shown that induction of differentiation of MEL cells with chemical inducers correlates with an increase of DNAse I hypersensitivity in DNA sequences located near α- and β-globin genes (101, 102). These changes in chromatin structure, however, have not been correlated with changes in the pattern of DNA methylation of these genes. Although the alterations in chromatin structure of globin genes are still elusive, it is reasonable to postulate that such changes could reveal cis-acting elements in DNA that affect the function of trans-acting factors related to globin gene transcription. To this respect, recent evidence has shown that the progressive loss of specific cyclin-dependent kinase (cdk) activities associated with MEL cell differentiation may lead to the accumulation of dephosphorylated linker histones and restricted chromatin flexibility (103). Furthermore, the presence of phosphorylated H1, -(H1 histones are reversibly phosphorylated primarily at cdk consensus sites)-, has a dominant effect on the binding capacity of other linker histones, a fact that may be crucial in chromatin folding and modeling process (103). To this end, recent data have also proposed that the methyl-CpG binding protein (MeCP2) interacts with the histone deacetylase-involved complex, thus providing evidence on the connection between DNA methylation, chromatin structure and gene transcription silencing (104, 105). Further experimental support to this notion has been recently provided in studies where a role for DNA methylation and MBD2 in the silencing of the human fetal γ-globin gene has been postulated (106). In particular, the use of 5-azacytidine indicated that DNA methylation acts primarily through MBD2 to maintain γ-globin suppression in adult erythroid cells. Furthermore, in studies using transgenic mice it has been demonstrated that DNA methylation plays a local role in preventing the activation of γ-globin gene, and this effect is mediated by histone deacetylation and the modulation of binding of specific transcription factors (107).

Over the past several years, extensive experimental analysis has established the notion that agents which hypomethylate DNA in CpG islands alter the superfine structure of chromatin. Moreover, chromatin structure has been shown to be affected by histone acetylation that acts in synergy with inhibition of topoisomerases to regulate gene transcription. As a matter of fact, specific classes of agents like hypomethylating agents, HDACIs and inhibitors of topoisomerases have been developed as modulators of cell growth and differentiation. But how DNA methylation and chromatin structure can cooperate in the execution of gene expression patterns seen during erythropoiesis? The observations that discrete patterns of gene expression operate in differentiating MEL cells suggest that the steady-state cytoplasmic accumulation of RNA transcripts result either from transcriptional activation, repression, or posttranscriptional processes (38, 39). In fact, previous studies have considered DNA hypomethylation in differentiating MEL cells responsible for activation of transcription of globin genes (57, 108). It is also known that sodium butyrate, an inhibitor of HDACs, can induce or inhibit differentiation of MEL cells in vitro depending on the concentration used in the culture (109). Another interesting point that emerged recently is the fact that c-myc has been involved in chromatin remodeling machinery like that of histone acetyltransferase (HATs) and ATP-dependent chromatin remodeling complexes (71, 105). On the other hand, c-myc plays an important role in MEL cell differentiation and regulates the activation of the expression of histone H1 gene very early upon induction of differentiation of MEL cells with chemical inducers (110-112). C-myc also is considered to be a crucial factor for haemopoietic homeostasis in decision making switches of progenitor cells towards proliferation, apoptosis and/or differentiation (113). Since histones are considered crucial components of the DNA condensation process, it is reasonable to correlate these changes with chromatin structure alterations that could result in gene expression modulation. Furthermore, these data have made more attractive the hypothesis of how finally c-myc regulates gene transcription and exerts its multiple biological functions on enforcement of cell growth, blockade of differentiation and/or sensitization of apoptosis.

Treatment of MEL cells with the DNA methylation inhibitor 5-azacytidine resulted in a small but substantial accumulation of differentiated cells, suggesting also the involvement of DNA methylation in gene activation or silencing processes (105). On the other hand, it is also known that DNA hypomethylating agents like 5-azacytidine, 5-aza-2'-deoxycytidine (decitabine) and 5-fluoro-2'-deoxycytidine can modulate differentiation signals in other leukemia cell models. In particular, 5-azacytidine triggers differentiation of HL-60 (114), K-562 (115) in addition to MEL cells, but with different potencies (116). Furthermore, the combination of 5-azacytidine with either retinoic acid (117) or vitamin D_3 analogues (118) caused extensive cell growth inhibition and induction of differentiation of HL-60 and NB-4 cells. In addition, specific inhibitors of SAH hydrolase like 3-deazaadenosine, aristeromycin, 3-deaza-(±)-aristeromycin, neplanocin A and MDL28842 have been reported to induce differentiation of various cell lines (119), instead to inhibit differentiation as seen in our studies with MEL cells.

The precise mechanisms involved in the interplay between c-myc, HDCAs, HATs and DNMTs to modulate chromatin structure, DNA methylation and gene expression in MEL cell differentiation, need further investigation for uncovering key components upon the switch of normal eryhtropoiesis to erythroleukemia. Undoubtedly, however, these observations provide new mechanistic insights on how DNA methylation and chromatin structure could work for affecting gene transcription during differentiation of MEL cells on the light of recent evidence indicating a broader role of chromatin structure and DNA methylation in physiological and pathological processes during development of higher organisms (2, 4, 120).

7. Methylation and HDACs Inhibitors Developed as Epigenetic Drugs for Heamatological Disorders

The silencing of gene expression through DNA methylation could be achieved by various ways. For example, except of DNMTs that function as the main enzymes contributing to DNA methylation, there is another category of proteins that of MBDs which recognize methylated DNA, thus preventing transcription factors to reach their DNA binding sites. This is undoubtedly a fact that result to gene expression silencing (4). To this end, a number of data accumulated over the previous years have suggested that DNMTs possess gene-silencing activities independent of their catalytic DNA methylation activities. In particular, DNMT1 interacts with HDACs as well as histone methyltransferases, a result that may inhibit gene expression by affecting protein-protein interactions rather than inhibiting DNA methylation (10). Alternatively, MBD proteins recognize methylated spots on the CpG islands, bind on them and through this way can inhibit the binding capacity of transcription factors rendering them capable to interact with specific consensus DNA sequences. The latter, can finally lead to specific gene expression silencing (4, 17, 20, 21, 24). Another possibility could be gene silencing to occur via the non-catalytic action of DNMTs, which inhibits the transcription process. To this regard, it has been shown that these proteins have the ability to reconstitute complexes of histone deacetylases and other inhibitory proteins on the DNA (24). On the other hand, the gene silencing procedure is a very complex phenomenon, in which, except the involvement of DNA methylation, other processes participate, like the deacetylation of H3 and H4 histones and the methylation of H3 histone at lysine 9, just to mention a few. In any case, however, all these processes result in the transformation of euchromatin to heterochromatin, a fact that leads to the inhibition of the binding of transcription factors onto their DNA targeting sequences and ultimately to the gene silencing (3, 6, 12, 13).

Structural analogs of cytosine that act as non-competitive antagonists and inhibit DNMT1 include agents like 5-azacytidine, 5-aza-2'-deoxy-cytidine (DAC, decitabine), arabinosyl-5-azacytosine (fazarabine) and dihydro-5-azacytidine (DHAC). In particular, these molecules function as pro-drugs being converted into their tri-nucleotide active derivatives and act as inhibitors of DNMT1 through their incorporation into DNA and/or RNA. The existence of a nitrogen atom at C5 instead of a carbon into their structure results to the creation of a homopolar binding with DNMTs, a result that prevents their regeneration and finally contributes to less methylation reactions (2, 4, 9, 10, 17, 19). In addition, these agents can exert their action by leading to a rapid and selective degradation of DNMT1 by the proteasomal protein-degradation pathway, as it has been recently proposed (6, 9). On the other hand, it has also been observed that such molecules seem to promote the acetylation process and the methylation of histone H3 at Lys 4 causing the alteration of chromatin structure to its more compact form (3, 15). Moreover, it has been shown that these demethylating agents can restore the sensitivity to a range of chemotherapeutic agents including cisplatin, epirubicin and temozolomide through regulation of MDR1 (multi-drug resistance) gene expression (19).

Nowadays, decitabine is used for the treatment of chronic myelogenous leukemia and myelodysplastic syndrome (6, 8, 19). Although this substance was the first agent that was shown to induce HbF production in patients with β-thalassemia and sickle cell disease, concerns about its potential carcinogenic effects halted its further development as a therapeutic agent in patients with hemoglobin disorders (9). Moreover, methylation inhibitors are used for experiments that try to correlate differentiation, apoptosis and cell development with DNA methylation. Such experiments have shown that these compounds act synergistically with HDACIs (eg. trichostatin) and induce the expression of silenced tumor suppressor genes due to modulation of the DNA methylation machinery (21, 24).

Agents that inhibit HDACs can alter the transcription of genes through a genome-wide effect on histone acetylation state, but since some HDACs associate with the transcriptional machinery their inhibition can also lead to more generalized gene-specific effects. The latter, can be attributed to the fact that DNA methylation represses gene expression in part due to modulation of histone deacetylation. Over the past years, HDACIs have been extensively used to activate the expression of methylated genes in different experimental model systems. Although HDACIs when used alone fail to activate the expression of genes with a dense methylated pattern, they can do it synergistically with demethylating agents (6, 19). Such inhibitors can reactivate several genes involved in apoptotic signaling pathways and the control of cell cycle, thereby inhibiting proliferation and displaying anti-invasive and anti-angiogenic potential. Furthermore, treatment with these inhibitors can sensitize cancer cells against other chemotherapeutic compounds, irradiation or even during gene therapy. Indeed, HDACIs like phenylbutyrate, sodium valproate, hydralazine, procainamide, retinoic acid, trichostatin, depsipeptide and NVP-LAQ824, a hydroxamic acid derivative have been used in experimental models for the treatment of haematological malignancies as well as in vivo (6, 8, 19). Recently, the use of antisense oligonucleotides, siRNAs or even ribozymes targeting DNMTs has been proposed as an alternative way to be applied in order to exploit therapeutically the modulation of DNA methylation and histone posttranslational status in cancer cells (6, 8, 10, 24).

8. Conclusion and Perspectives

Overall, the data presented strongly support our hypothesis that MEL cell differentiation is associated with both qualitative and quantitative changes in the pattern of both polyA$^-$ and polyA$^+$ RNA methylation seen at specific sites, as well as with lower SAH/SAM ratio. N^6mAdo, neplanocin A, 3-deazaneplanocin, cycloleucine and related agents may block initiation of commitment in MEL cells via inhibition of RNA and DNA methylation by affecting SAH/SAM ratio, processes that may be responsible for the induction of differentiation, the specific pattern of gene expression and the differential RNA stability observed in differentiating MEL cells. Experiments in our laboratory are now underway to clone and characterize RNA transcripts that accumulate in the cytoplasm of cells treated with methylation inhibitors and demonstrate if the identified consensus sequences for the GATA-1, AP1/NF-E2 and EKLF found at the 3′-end flanking sequences of β^{major}-globin gene are indeed functional. In particular, we have cloned DNA fragments derived from this region into pGL3 vector bearing the luciferase gene as a reporter gene, in order to uncover any potential promoter-like function activated by hypomethylating agents (unpublished results). Moreover, it would be of great interest to delineate how DNA regions which are being silent in untreated MEL cells are somehow activated upon treatment with methylation inhibitors, and how this process leads to the blockade of erythroid differentiation program. Since the appearance of discrete poly(A)⁻ RNA transcripts is seen exclusively in cells exposed to hypomethylating agents, but not in control cells, we tend to conclude that their production may be due to the direct action of hypomethylating agents on chromatin transcriptional state, or alternatively these agents may exert their action indirectly on chromatin by modulating active methylation cycle. Although at present we have no clue what do these RNA transcripts do, we may ask whether these RNAs act at transcriptional or even posttranscriptional level to abrogate erythroid differentiation. Such a direction may lead to unraveling novel mechanisms that regulate gene expression posttranscriptionally. Alternatively, the detected poly(A)⁻ RNAs may be encoded by genes that belong to the family of non-coding RNAs (ncRNAs) that it is well known they have important roles in processes like transcriptional regulation, RNA maturation, stability and translation as well as protein degradation and translocation. Such a direction is expected to enrich our knowledge on how the methylation inhibitors exert their mode of action in heamatological disorders and in a way that could be further exploited pharmacologically toward the development of improved epigenetic drugs.

9. References

Jones, PA; Takai, D. The role of DNA methylation in mammalian epigenetics. *Science*, 2001, 293, 1068-1070.

Jones, PA; Baylin, SB. The fundamental role of epigenetic events in cancer. *Nature Rev. Genet.*, 2002, 3, 415-428.

Laird, PW. Cancer epigenetics, *Hum Mol Genet*, 2005, 14 Spec No 1, R65-76.

Robertson, KD. DNA methylation and human disease, *Nat Rev Genet*, 2005, 6, 597-610.

Baylin, SB; Herman, JG. DNA hypermethylation in tumorigenesis: epigenetics joins genetics, *Trends Genet*, 2000, 16, 168-174.

Das, PM; Singal, R. DNA methylation and cancer, *J Clin Oncol*, 2004, 22, 4632-4642.

Ehrlich, M. DNA methylation in cancer: too much, but also too little, *Oncogene*, 2002, 21, 5400-5413.

Esteller, M. Relevance of DNA methylation in the management of cancer, *Lancet Oncol*, 2003, 4, 351-358.

Fathallah, H; Atweh, GF. DNA hypomethylation therapy for hemoglobin disorders: Molecular mechanisms and clinical applications, *Blood Reviews*, 2006, 20, 227-234.

Szyf, M; Pakneshan, P; Rabbani, SA. DNA methylation and cancer: therapeutic implications, *Cancer Letters*, 2004, 211, 133-143.

Doerfler, W. DNA methylation: site-specific methylations cause gene inactivation, *Angew. Chem.*, 1984, 23, 12, 919-931.

Robertson, KD; Wolffe, AP. DNA methylation in health and disease, *Nat. Rev., Genet.*, 2000, 1, 11-19.

Garinis, GA; Patrinos, GP; Spanakis, NE; Menounos, PG. DNA hypermethylation: when tumour suppressor genes go silent, *Hum Genet*, 2002, 111, 115-127.

Yaneva, J; Zacharieva, S. DNA methylation and cancer: an overview, *Exp. Pathol. Parasitol.*, 2002, 45-54.

Laird, PW. The Power and the Promise of DNA Methylation Markers, *Nat. Rev. Cancer*, 2003, 3, 253-266.

Eng, C; Herman, JG; Baylin, SB. A bird's eye view of global methylation, *Nat. Genet.* 2000, 24, 101-102.

Momparler, RL; Bovenzi, V. DNA methylation and cancer, *J. Cell Physiol.*, 2000, 183, 145-154.

Preston, RJ. Epigenetic processes and cancer risk assessment, *Mutat. Res.*, 2007, 616, 7-10.

Theodoridis, JM; Strathdee, G; Brown, R. Epigenetic silencing mediated by CpG island methylation: potential as a therapeutic target and as a biomarker, *Drug Resistance Updates*, 2004, 7, 267-278.

Lorincz, MC; Dickerson, DR; Schmitt, M; Groudine, M. Intragenic DNA methylation alters chromatin structure and elongation efficiency in mammalian cells, *Nat. Struct. Mol. Biol.*, 2004, 11, 1068-1075.

Curradi, M; Izzo, A; Badaracco, G, Landsberger, N. Molecular mechanisms of gene silencing mediated by DNA methylation, *Mol Cell Biol*, 2002, 22, 3157-3173.

Chan, SWL; Zhang, X; Bernatavichute, YV; Jacobsen, SE. Two-Step Rcruitment of RNA-Directed DNA Methylation to Tandem Repeats, *PlosBiology*, 2006, 4, 11, 1923-1933.

El-Osta, A. DNMT cooperativity-the developing links between methylation, chromatin structure and cancer, *BioEssays*, 2003, 25, 1071-1084.

Karpf, AR; Jones, DA. Reactivating the expression of methylation silenced genes in human cancer, *Oncogene*, 2002, 21, 5496-5503.

Labuda, D; Sinnett, D; Richer, C; Deragon, JM; Striker, G. Evolution of mouse B1 repeats: 7SL RNA folding pattern conserved, *J Mol Evol*, 1991, 32, 405-414.

Lorincz, MC; Groudine, M. C(m)C(a/t)GG methylation: a new epigenetic mark in mammalian DNA?, *Proc Natl Acad Sci U S A*, 2001, 98, 10034-10036.

Sontag, LB; Lorincz, MC; Luebeck, EG. Dynamics, stability and inheritance of somatic DNA methylation imprints, *J. Theor. Biol.*, 2006, 242, 890-899.

Banerjee, AK. 5'-Terminal cap structure in eukaryotic messenger ribonucleic acids. *Microbiol Reviews*, 1980, 44, 175-205.

Choi, YC; Reddy, R; Busch, H. Methylation of snRNAs and 18S rRNA of Novikoff hepatoma. In: Usdin E, Borchardt RT and Creveling CR (eds), Biochemistry of S-adenosylmethionine and related compounds, 1982, pp. 313-320, Macmillan Press, London.

Kersten, H. On the biological significance of modified nucleosides in tRNA. *Prog Nucleic Acid Res Mol Biol*, 1984, 31, 59-114.

Maden, BEH. The numerous modified nucleotides in eukaryotic ribosomal RNA. *Prog Nucleic Acid Res Mol Biol,* 1990, 39, 241-303.

Alix, JH. Relationship between methylation and maturation of ribosomal RNA in prokaryotic and eukaryotic cells. In: Borchardt RT, Creveling CR and Ueland PM (eds), Biological methylation and drug design, 1986, pp. 175-187, Humana Press, Clifton, New Jersey.

Rottman, FM; Narayan, P; Goodwin, R; Camper, S; Yao, Y; Horowitz, S; Albers, R; Ayers, D; Maroney, P; Nilsen, T. Distribution of m^6A in mRNA and its biological role. In: Borchardt RT, Creveling CR and Ueland PM (eds), Biological methylation and drug design, 1986, pp. 189-200, Humana Press, Clifton, New Jersey.

Munholland, JM; Nazar, RN. Methylation of ribosomal RNA as a possible factor in cell differentiation. *Cancer Res,* 1987, 47, 169-172.

Scarpa, S; Renzo, LD; Strom, R. In: Clawson GA, Willis DB, Weissbach A, Jones PA, (eds), Nucleic acid methylation, 1990, pp 103-112, Alan R. Liss, New York.

Kredich, NM. Inhibition of nucleic acid methylation by cordycepin. In vivo synthesis of S-deoxyadenosylmethionine by WI-L2 human lymphoblasts. *J Biol Chem*, 1980, 255, 7380-7385.

Levenson, R; J. Kernen, D; Housman, D. Synchronization of MEL cell commitment with cordycepin. *Cell*, 1979, 18, 1073-1078.

Tsiftsoglou AS, Pappas, IS; Vizirianakis, IS. The developmental program of murine erythroleukemia cells, *Oncol Res*, 2003, 13, 339-346.

Tsiftsoglou, AS; Pappas, IS; Vizirianakis, IS. Mechanisms involved in the induced differentiation of leukemia cells, *Pharmacol Ther*, 2003, 100, 257-290.

Vizirianakis, IS; Wong, W; Tsiftsoglou, AS. Analysis of inhibition of commitment of murine erythroleukemia (MEL) cells to terminal maturation by N^6-methyladenosine. *Biochem Pharmacol,* 1992, 44, 927-936.

Glazer, RI; Hartman, KD; Knode, MC; Richard, MM; Chiang, PK; Tseng, CKH; Marquez, VE. 3-Deazaneplanocin: a new and potent inhibitor of S-adenosylhomocysteine hydrolase and its effects on human promyelocytic leukemic cell line HL-60. *Biochem Biophys Res Comm*, 1986, 135, 688-694.

Tseng, CKH; Marquez, VE; Fuller, RW; Goldstein, BM; Haines, DR; McPherson, H; Parsons, JL; Shannon, WM; Arnett, G; Hollingshead, M; Driscoll, JS. Synthesis of 3-deazaneplanocin A, a powerful inhibitor of S-adenosylhomo-cysteine hydrolase with potent and selective in vitro and in vivo antiviral activities. *J Med Chem*, 1989, 32, 1442-1446.

Chiang, PK; Miura, GA. S-Adenosylhomocysteine hydrolase, In: Borchardt RT, Creveling CR, Ueland PM, (eds), Biological methylation and drug design, 1986, pp. 239-251, Humana Press, Clifton, New Jersey.

Sufrin, JR; Lombardini, JB; Kramer, DL; Alks, V; Bernacki, RJ; Porter, CW. Methionine analog inhibitors of S-adenosylmethionine biosynthesis as potential antitumor agents. in: R.T. Borchardt, C.R. Creveling, P.M. Ueland, (Eds.), Biological Methylation and Drug Design, Humana Press, Clifton, New Jersey, (1986) 373-384.

Orkin, SH. Diversification of hematopoietic stem cells to specific lineages, *Nat. Rev Genet*, 2000, 1, 57-64.

Domen, J. The role of apoptosis in regulating hemopoietic stem cell numbers, *Apoptosis* 2001, 6, 239-252.

Pierce, GB. Neoplasms, differentiation and mutations, *Am J Pathol*, 1974, 77, 103-118.

Sachs, L. Constitutive uncoupling of pahtways of gene expression that control growth and differentiation in myeloid leukemia: a model for the origin and progression of malignancy, *Proc Natl Acad Sci USA*, 1980, 77, 6152-6156.

McKenna, SL; Cotter, TG. Functional Aspects of apoptosis in hematopoiesis and consequences of failure, *Adv Cancer Res*, 1997, 71, 121-164.

Reis, M; Gamba-Vitalo, C; Sartorelli, AC. Induction of tumor cell differentiation as a therapeutic approach: Preclinical models for hematopoietic and solid neoplasms, *Cancer Treat Rep,* 1986, 70, 201-218.

Lotan, R; Francis, GE; Freeman, CS; Waxman, S. Differentiation Therapy, *Cancer Res*, 1990, 50, 3453-3464.

Leszczyniecka, M; Roberts, T; Dent, P; Grant, S; Fischer, PB. Differentiation therapy of human cancer: Basic science and clinical applications, *Pharmacol Ther*, 2001, 90, 105-156.

Tsiftsoglou, AS; Robinson, SH. Differentiation of leukemic cell lines: A review focusing on murine erythroleukemia and human HL-60 cells, *Int J Cell Cloning* 1985, 3, 349-366.

Marks, PA; Richon, VM; Breslow, R; Rifkind, RA. Histone deacetylase inhibitors as new cancer drugs, *Curr Opin Oncol*, 2001, 13, 477-483.

Tsiftsoglou, AS; Wong, W. Molecular and cellular mechanisms of leukemic hematopoietic cell differentiaion: An analysis of the Friend system. *Anticancer Res*, 1985, 5, 81-99.

Pappas, IS; Lambris, J; Vizirianakis, IS; Winter, MS; Tsiftsoglou, AS. Mechanisms of action of differentiation inducers: Detection of an inducer-binding protein in murine erythroleukemia cells, *Oncol Res*, 2005, 15, 21-37.

Christman, JK; Weich, N; Schoenbrun, B; Schneiderman, N; Acs, G. Hypomethylation of DNA during differentiation of Friend erythroleukemia cells. *J Cell Biol*, 1980, 86, 366-370.

Levenson, R; Housman, D. Memory of MEL cells to a previous exposure to inducer, *Cell*, 1979, 17, 485-490.

Hatse, S; De Clercq, E; Balzarini, J. Role of antimetabolites of purine and pyrimidine nucleotide metabolism in tumor cell differentiation, *Biochem Pharmacol*, 1999, 58, 539-555.

Vizirianakis, IS; Tsiftsoglou, AS. Hypermethylation of RNA as part of the differentiation program of murine erythroleukemia (MEL) cells. In: S. Waxman (Edr.), Challenges of Modern Medicine: Differentiation Therapy, 1995, vol. 10, pp. 263-268, Ares-Serono Symposia Publications, Rome.

Sherman, ML; Shafman, TD; Spriggs, DR; Kufe, DW. Inhibition of murine erythroleukemia cell differentiation by 3-deazaadenosine, *Cancer Res*, 1985, 45, 5830-5834.

Di Fiore, PP; Grieco, M; Pinto, A; Attadia, V; Porcelli, M; Cacciapuoti, G; Carteni-Farina, M. Inhibition of dimethyl sulfoxide-induced differentiation of Friend erythroleukemic cells by 5'-methylthioadenosine. Cancer Res. 1984, 44, 4096-4103.

Vizirianakis, IS; Tsiftsoglou, AS. Induction of murine erythroleukemia cell differentiation is associated with methylation and differential stability of poly(A)+ RNA transcripts, *Biochim Biophys Acta*, 1996, 1312, 8-20.

Vizirianakis, IS; Tsiftsoglou, AS. N^6-methyladenosine inhibits murine erythroleukemia cell maturation by blocking methylation of RNA and memory via conversion to S-(N^6-methyl)-adenosylhomocysteine, *Biochem Pharmacol*, 1995, 50, 1807-1814.

Taylor, SM; Constantinides, PA; Jones, PA. 5-Azacytidine, DNA methylation, and differentiation, Cur Topics Microbiol Immunol, 1984, 108, 115-127.

Camper, SA; Albers, RJ; Coward, JK; Rottman, FM. Effect of undermethylation on mRNA cytoplasmic appearance and half-life. *Mol Cell Biol*, 1984, 4, 538-543.

Kim, CH; Sarma, RH. Spatial configuration of the bizarre 5′ terminus of mammalian mRNA, *J Am Chem Soc,* 1978, 100, 1571-1590.

Coutts, M; Brawerman, G. A 5′ exoribonuclease from cytoplasmic extracts of mouse sarcoma 180 ascites cells *Biochem Biophys Acta*, 1993, 1173, 57-62.

Coutts, M; Krowczynska, A; Brawerman, G. Protection of mRNA against nucleases in cytoplasmic extracts of mouse sarcoma ascites cells, *Biochem Biophys Acta,* 1993, 1173, 49-56.

Belasco, J; Brawerman, G; (eds) Control of messenger RNA stability, 1993, Academic Press Inc. San Diego.

Amati, B; Frank, SR; Donjerkovic, D; Taubert, S. Function of the c-Myc oncoprotein in chromatin remodeling and transcription, *Biochim Biophys Acta*, 2001, 1471, M135-M145.

Peltz, SW; Brewer, G; Bernstein, P; Hart, A; Ross, J. Regulation of mRNA turnover in eukaryotic cells, *Critical Rev. Eukaryotic Gene Expression* 1991, 1, 99-126.

Klausner, RD; Rouault, TA; Harford, JB. Regulating the fate of mRNA: the control of cellular iron metabolism *Cell* 1993, 72, 19-28.

Volloch, V; Schweitzer, B; Rits, S. Messenger RNA changes during differentiation of murine erythroleukemia cells, *Exp Cell Res,* 1987, 173, 38-48.

Tsiftsoglou, AS; Wong, W; Volloch, V; Gusella, J; Housman, D. Commitment of murine erythroleukemia (MEL) cells to terminal differentiation is associated with coordinated expression of globin and ribosomal genes. *Prog Clin Biol Res* 1982, 102A: 69-79.

Glazer, RI; Knode, MC; Neplanocin, A. A cyclopentenyl analog of adenosine with specificity for inhibiting RNA methylation, *J Biol Chem*, 1984, 259, 12964-12969.

Pugh, CSG; Borchardt, RT; Stone, HO. Inhibition of Newcastle disease virion messenger RNA (guanine-7-)-methyltransferase by analogues of S-adenosyl-homocysteine, *Biochemistry,* 1977, 16, 3928-3932.

Glazer, RI; Knode, MC; Tseng, CK; Haines, DR; Marquez, VE. 3-Deazaneplanocin A: a new inhibitor of S-adenosylhomocysteine synthesis and its effects in human colon carcinoma cells, *Biochem Pharmacol*, 1986, 35, 4523-4527.

Levenson, R; Kernen, J; Housman, D. Synchronization of MEL cell commitment with cordycepin, 1979, *Cell*, 18, 1073-1078.

Hoffman, JL. Biosynthesis of S-N^6-methyladenosylhomosysteine an inhibitor of RNA methyltransferases *J Biol Chem.*, 1978, 253, 2905-2907.

Zimmerman, TP; Wolberg, G; Duncan, GS; Elion, GB. Adenosine analogues as substrates and inhibitors of S-adenosylhomocysteine hydrolase in intact lymphocytes, *Biochemistry*, 1980, 19, 2252-2259.

Guranowski, A; Montgomery, JA; Cantoni, GL; Chiang, PK. Adenosine analogues as substrates and inhibitors of S-adenosylhomocysteine hydrolase, *Biochemistry*, 1981, 20, 110-115.

Cantoni, GL. The centrality of S-adenosylhomocysteinase in the regulation of the biological utilization of S-adenosylmethionine, In: Borchardt RT, Creveling CR, Ueland PM, (eds), Biological methylation and drug design, 1986, pp. 227-238, Humana Press, Clifton, New Jersey.

Costello, JF; Vertino, PM. Methylation matters: a new spin o maspin, *Nat Genet*, 2002, 31, 123-124.

Müllner, EW; Neupert, B; Kühn, LCA. A specific mRNA binding factor regulates the iron-dependent stability of cytoplasmic transeffin receptor mRNA, *Cell*, 1989, 58, 373-382.

Hentze, MW; Caughman, SW; Rouault, TA; Barriocanal, JG; Dancis, A; Harford, JB; Klausner, RD. Identification of the iron-responsive element for the translational regulation of human ferritin mRNA, *Science*, 1987, 238, 1570-1573.

Müllner, EW; Kühn, LCA. A stem-loop in the 3′ untranslated region mediates iron-dependent regulation of transferring receptor mRNA stability in the cytoplasm, *Cell*, 1988, 815-825.

Burd, CG; Dreyfuss, G. Conserved structures and diversity of functions of RNA-binding proteins, *Science*, 1994, 265, 615-621.

Vizirianakis, IS; Tsiftsoglou, AS. Blockade of murine erythroleukemia cell differentiation by hypomethylating agents causes accumulation of discrete small poly(A)- RNAs hybridized to 3'-end flanking sequences of beta(major) globin gene, *Biochim Biophys Acta*, 2005, 1743, 101-114.

Volloch, V. Cytoplasmic synthesis of globin RNA in differentiated murine erythroleukemia cells: Possible involvement of RNA-dependent RNA polymerase, *Proc Natl Acad Sci USA,* 1986, 83, 1208-1212.

Volloch, V; Schweitzer, B; Rits, S. Antisense globin RNA in mouse erythroid tissues, Structure, origin, and possible function, *Proc Natl Acad Sci USA*, 1996 93, 2476-2481.

Coggins, LW; Vass, JK; Stinson, MA; Lanyon, WG; Paul, J. A B1 repetitive sequences near the mouse beta-major globin gene, *Gene*, 1982, 17, 113-116.

Labuda, D; Sinnett, D; Richer, C; Deragon, JM; Striker, G. Evolution of mouse B1 repeats: 7SL RNA folding pattern conserved, *J Mol Evol*, 1991, 32, 405-414.

Shehee, WR; Loeb, DD; Adey, NB; Burton, FH; Casavant, NC; Cole, P; Davies, CJ; McGraw, RA; Schichman, SA; Severynse, DM; Voliva, CF; Weyter, FW; Wisely, GB; Edgell, MH;. Hutchison III, CA. Nucleotide sequence of the BALB/c mouse beta-globin complex, *J Mol Biol*, 1989, 205, 41-62.

Cantor, AB; Orkin, SH. Transcriptional regulation of erythropoiesis: an affair involving multiple partners, *Oncogene*, 2002, 21, 3368-3376.

Choe, KS; Radparvar, F; Matushansky, I; Rekhtman, N; Han, X; Skoultchi, AI. Reversal of tumorigenicity and the block to differentiation in erythroleukemia cells by GATA-1, *Cancer Res,* 2003, 63, 6363-6369.

Sawado, T; Igarashi, K; Groudine, M. Activation of beta-major globin gene transcription is associated with recruitment of NF-E2 to the beta-globin LCR and gene promoter *Proc Natl Acad Sci USA*, 2001, 98, 10226-10231.

Vidal, F; Mougneau, E; Glaichenhaus, N; Vaigot, P; Darmon, M; Cuzin, F. Coordinated posttranscriptional control of gene expression by modular elements including Alu-like repetitive sequences. *Proc Natl Acad Sci* USA, 1993, 90, 208-212.

Fu X-H, Liu D-P, Liang C-C, Chromatin structure and transcriptional regulation of the β-globin locus, *Exp Cell Res*, 2002, 278, 1-11.

Scher, W; Scher, BM; Waxman, S. Nuclear events during differentiation of erythroleukemia cells. In: Dunn CDR, (ed), Current Concepts of Erythropoiesis, 1983, pp. 301-388, John Wiley & Sons Ltd., New York.

Sheffery, M; Rifkind, RA; Marks, PA. Murine erythroleukemia cell differentiation: DNAse I hypersensitivity and DNA methylation near the globin genes *Proc Natl Acad Sci USA*, 1982, 79, 1180-1184.

Yu, J; Smith, RD. Sequential alterations in globin gene chromatin structure during erythroleukemia cell differentiation. *Proc Natl Acad Sci USA*, 1985, 260, 3035-3040.

Rupon JW, Wang SZ, Gaensler K, Lloyd J, Ginder GD, Methyl binding domain protein 2 mediates γ-globin gene silencing in adult human βYAC transgenic mice, *Proc Natl Acad Sci USA*, 2006, 103, 6617-6622.

Ng, HH; Bird, A. DNA methylation and chromatin modification. *Curr Opin Genet Dev*, 1999, 9, 158-163.

Robertson, KD. DNA methylation and chromatin-unraveling the tangled web, *Oncogene*, 2002, 21, 5361-5379.

Yellajoshyula D, Brown DT, Global modulation of chromatin dynamics mediated by dephosphorylation of linker histone H1 is necessary for erythroid differentiation, *Proc Natl Acad Sci USA*, 2006, 103, 18568-18573.

Goren A, Simchen G, Fibach E, Szabo PE, Tanimoto K, Chakalova L, Pfeifer GP, Fraser PJ, Engel JD, Cedar H, Fine tuning of globin gene expression by DNA methylation, *PLoS ONE*, 2006, 1, e46. doi:10.1371/journal.pone.0000046.

Christman, JK; Price, P; Pedrinan, L; Acs, G. Correlation between hypomethylation of DNA and expression of globin genes in Friend erythroleukemia cells. *Eur J Biochem*, 1977, 81, 53-61.

Corin, RE; Haspel, HC; Peretz, AM; Sonenberg, M; Rifkind, RA. Antagonistic effect of butyrate on hexamethylene bisacetamide induced differentiation of murine eryhtroleukemia cells. *Cancer Res*, 1986, 46, 1136-1141.

Lachman, HM; Skoultchi, AI. Expression of c-myc changes during differentiation of mouse erythroleukaemia cells, *Nature*, 1984, 310, 592-594.

Prochownik, EV; Kukowska, J. Deregulated expression of c-myc by murine erythroleukaemia cells prevents differentiation, Nature 1986, 322, 848-850.

Cheng, GH; Skoultchi, AI. Rapid induction of polyadenylated H1 histone mRNAs in mouse erythroleukemia cells is regulated by c-myc, Mol Cell Biol, 1989, 9, 2332-2340.

Muschel, RJ; McKenna, WG. Alterations in cell cycle control during tumor progression: effects on apoptosis and the response to therapeutic agents, *Curr Top Microbiol Immunol*, 1996, 213, 197-213.

Christman, JK; Mendelsohn, N; Herzog, D; Schneiderman, N. Effect of 5-azacytidine on differentiation and DNA methylation in human promyelocytic leukemia cells (HL-60), *Cancer Res*, 1983, 43, 763-769.

Gambari, R; del Senno, L; Barbieri, R; Viola, L; Tripodi, M; Raschella, G; Fantoni, A. Human leukemia K-562 cells: induction of erythroid differentiation by 5-azacytidine, Cell Differ, 1984, 14, 87-97.

Creusot F, Acs G, Christman JK, Inhibition of DNA methyltransferase and induction of Friend erythroleukemia cell differentiation by 5-azacytidine and 5-aza-2'-deoxycytidine, J Biol Chem, 1982, 257, 2041-2048.

Momparler RL, Dore BT, Momparler LF, Effect of 5-aza-2'-deoxycytidine and retinoic acid on differentiation and c-myc expression in HL-60 myeloid leukemic cells, Cancer Lett, 1990, 54, 21-28.

Dore BT, Chomienne C, Momparler RL, Effect of 5-aza-2'-deoxycytidine and vitamin D3 analogs on growth and differentiation of human myeloid leukemic cells, Cancer Chemother Pharmacol, 1998, 41, 275-280.

Niitsu N, Honma Y, Adenosine analogs as possible differentiation-inducing agents against acute myeloid leukemia. Leuk Lymphoma, 1999, 34, 261-271.

Brown, R; Strathdee, G. Epigenomics and epigenetic therapy of cancer, *Trends Mol Med*, 2002, 8, S43-S48.

In: Horizons in DNA Research, Volume 1
Editor: Jason R. Chesterton, pp. 105-127

ISBN: 978-1-60876-968-1

Chapter 4

MITOCHONDRIAL DNA VARIATION, MITOCHONDRIAL HAPLOGROUPS AND NEURODEGENERATION

M. Mancuso, C. Carlesi, D. Orsucci and G. Siciliano
Department of Neuroscience, Neurological Clinic, University of Pisa, Italy.

ABSTRACT

Mitochondria, the powerhouse of the cell, play a critical role in multiple metabolic processes as well as in the apoptotic pathways. Many lines of evidence suggest that mitochondria and its DNA (mtDNA) have a central role in ageing-related neurodegenerative diseases. Moreover, there is a long history of investigations on mitochondria aimed at identifying genetic markers that would relate to aging or neurodegenerative diseases. In this chapter, we highlight some of the major neurodegenerative disorders and discuss the role of the mtDNA variations and haplogroups in the pathogenetic cascade leading to these diseases.

Keywords: Mitochondria, mtDNA, mtDNA haplogroups, neurodegenerative disorders.

INTRODUCTION

Mitochondria, which are cytoplasmic organelles, are the result of a symbiotic association between a glycolytic protoeukaryctic cell and oxidative bacterium about 1.5 billion years ago (Reddy and Beal 2005). Mitochondrial DNA (mtDNA) has lost much of its independence and keeps functioning under the overarching control of the nuclear genome (nDNA). Mitochondria are present virtually in all eukaryotic cells, including neurons, and are essential for survival. Their primary function is to support aerobic respiration and to provide energy and heat. Mitochondria also play other important roles, including cell signalling for apoptotic cell death.

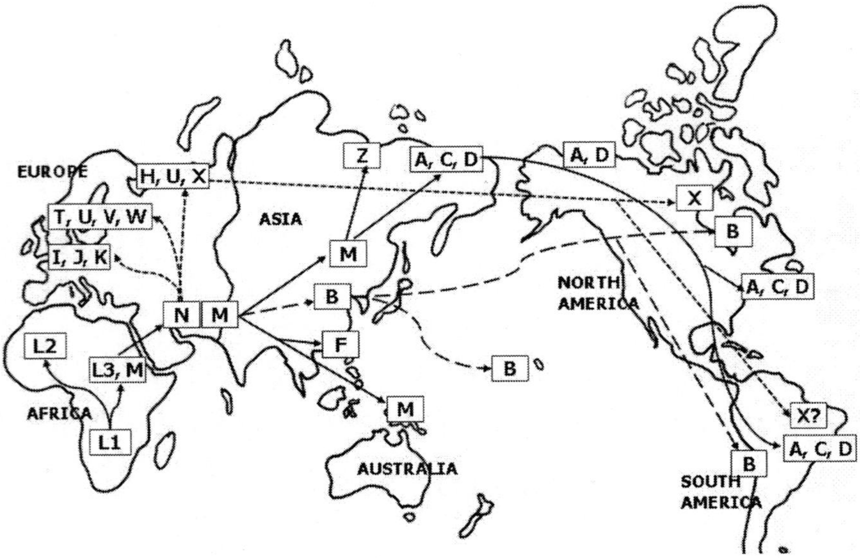

Legend to figure: mtDNA haplogroup migration patterns (reproduced with permission from Mancuso, M; Filosto, M; Orsucci, D; Siciliano, G. Mitochondrial DNA sequence variation and neurodegeneration. *Hum Genomics*, 2008 Sep;3,71-78 (c) Henry Stewart Publications). The map shows the migration patterns of the main mtDNA haplogroups. The first common ancestor is postulated to have lived in Africa about 200,000 years ago. From Africa, the arrows show the mtDNA migration route.

By convention, the term "mitochondrial diseases" refers to disorders of the mitochondrial respiratory chain, thus excluding dysfunction in other metabolic pathways located in the mitochondria as pyruvate metabolism, Krebs cycle and fatty acid oxidation. The respiratory chain is the only metabolic pathway in the cell that is under the dual control of the mtDNA and the nDNA.

In this chapter, after a brief reminding of the mitochondrial biology and genetics, we will focus on the role of the mtDNA haplogroups in selected neurodegenerative disorders.

Mitochondrial Compartment

Mitochondria are highly dynamic and pleomorphic organelles. Mitochondria are compartmentalized with two lipid membranes: the inner mitochondrial membrane and the outer mitochondrial membrane (Reddy, 2007). The inner mitochondrial membrane houses the mitochondrial respiratory chain and provides a highly efficient barrier to the flow of ions. The inner mitochondrial membrane covers the mitochondrial matrix, which contains tricarboxylic acid and beta-oxidation. The outer mitochondrial membrane is basically porous and allows the passage of low molecular weight substances between the cytosol and the intermembrane space (Reddy, 2008).

Although mitochondria contain their own genome and protein synthesizing machinery, the majority of mitochondrial polypeptides are encoded in the nuclear genome, synthesized in the cytosol and imported into the mitochondria post-transcriptionally.

Mitochondria are the power houses of cells, providing energy for several cellular functions, including ATP production, intracellular calcium regulation, the release of proteins that activate the caspase family of proteases, the alteration of the reduction-oxidation potential of cells and free radical scavenging, but they are also involved in other metabolic processes including the biosynthesis of amino acids, vitamin cofactors, fatty acids, iron-sulphur clusters, cell signalling and programmed cell death (Reddy, 2007).

ATP molecules are generated via glycolysis or by oxidation of glucose to ethanol or lactic acid. In aerobic conditions, the complete oxidation of carbohydrates, proteins, and fats to CO2 and H2O, via the efficient mitochondrial respiratory chain and oxidative phosphorylation system, generates an ATP yield 15 times higher than glycolysis (Fontanesi et al, 2006). Electrons from oxidative substrates are transferred to oxygen, via a series of redox reactions, to generate water (Elston et al, 1998). In the process, protons are pumped from the matrix across the mitochondrial inner membrane through the electron transport chain (ETC), which consists of four multimeric complexes -I to IV- plus two small electron carriers, coenzyme Q -or ubiquinone- and cytochrome *c*. When protons return to the mitochondrial matrix down their electrochemical gradient, ATP is synthesized via complex V (ATP synthase) (Noji and Yoshida, 2001).

MITOCHONDRIAL GENOME AND COMMON MTDNA HAPLOGROUPS

Mitochondria are controlled by both nuclear and mitochondrial genomes. MtDNA consists of a 16.5 kb circular DNA molecule that is maternally inherited (Anderson et al. 1981). MtDNA has 2 strands: a guanine-rich heavy strand and cytosine-rich light strand (DiMauro and Schon, 2003; Filosto and Mancuso, 2007). MtDNA carries 37 genes encoding 22 transfer RNAs (tRNAs), two ribosomal RNAs (12S and 16S) and 13 polypeptides (mRNAs). The 13 mtDNA-encoded polypeptides are part of the respiratory chain system and are assembled together with nuclear-encoded subunits; seven of them belong to complex I or NADH dehydrogenase, NADH: ubiquinone oxidoreductase (ND1, ND2, ND3, ND4, ND4L, ND5, ND6), one to complex III or ubiquinol: ferricytochrome c oxidoreductase, three to complex IV or cytochrome c oxidase (COX I, COX II and COX III) and two to complex V or ATP synthase (ATPase6 and ATPase8). Nuclear genes encode the remaining mitochondrial proteins, including all the complex II subunits, metabolic enzymes, DNA and RNA polymerases, ribosomal proteins, and mtDNA regulatory factors, such as the mitochondrial transcription factor A. Nuclear mitochondrial proteins are synthesized in the cytoplasm and are subsequently transported into mitochondria. The import of the majority of mitochondrial proteins is accomplished by membranespanning, multi-subunit translocators of the outer and inner mitochondrial membranes (Reddy and Beal, 2005).

MtDNA gives a view of diversity present in the human gene pool, first because mutations accumulate in this DNA several times faster than in nucleus, second because mtDNA is inherited maternally and does not recombine so it is a tool for relating individuals to one another and third because there are about 10 molecules of mtDNA within a typical human and they are usually identical to one another. MtDNA is used to complement information provided by Y chromosome studies on the evolution of modern humans. We can theoretically follow the transmission of mtDNA from the original "ancestor mother" by identifying common polymorphisms that have accumulated with time. Specific and unique sets of those common polymorphisms define groups of mtDNA, named haplogroups, that have evolved from the same ancestor. Phylogeographic studies identified the first common ancestor named "mitochondrial Eve", (Cann et al, 1987) postulated to have lived in Africa about 200,000 years ago.

The earliest mitochondrial DNA research began by digesting the entire molecule with just a few restriction enzymes. As early as 1980, Brown's pioneering study of 21 humans from diverse ethnic and geographic backgrounds indicated that mtDNA restriction-enzyme fragment length polymorphism (RFLP) patterns could be used to trace human genetic history (Brown, 1980). On the basis of the observed diversity in a worldwide sample, he obtained a surprisingly recent coalescence age estimate of 180 000 years for global mtDNA variation.

Direct sequencing and RFLP analysis allowed to identify the mtDNA haplogroups tree and the mtDNA migration route (Figure).

The basal branching structure of mtDNA variation in most parts of the world is now well understood (Torroni et al, 2006). African haplogroups fall into seven major families (L0, L1, L2, L3, L4, L5, L6). About 80000-90000 years ago, probably in the Horn of Africa, the root of haplogroup L3 gave rise to many (successful) descendant haplogroups perhaps reflecting some colonization event or local population growth which might have been triggered by the glacial interstadial phase 21 and, possibly, followed by cultural adaptation in response to the changing climate (Torroni et al, 2006). Two splinters of this basal L3 variation, haplogroups M and N, cover the mtDNA pool of all non-Africans (excluding the descendants of migrations from Africa within the past few thousand years, such as those engendered by various slave trades). Haplogroup N almost immediately diverged further to give rise to haplogroup R. There is now growing evidence that the richest basal variation in the three founder haplogroups M, N and R is found along the southern stretch of Eurasia, particularly in the Indian subcontinent (Palanichamy et al, 2004; Sun et al, 2006). Although Southeast Asia has been less comprehensively analysed to date, recent first results indicate a similarly high basal diversification in this region (Macaulay et al, 2005). This suggests a rapid colonization along the southern coast of Asia, about 60000 years ago. The expansions northwords occurred later, about 45000 years ago, when technology and climatic conditions enabled the exploration of the interior of Eurasia. Over 30 subclades of the haplogroup M are present in Asia.

Haplogroups A, B, C, D and X have been found in the Americas, coming mainly from Asia. In Europeans and Near Easterners (who share a rather recent common ancestor) nine different mitochondrial haplogroups have been identified (H, I, J, K, T, U, V, W, X). The variation in the basal European mtDNA haplogroups dates to about 45000 years ago. Complete mtDNA sequencing and the increasing number of samples analyzed allow subdividing haplogroups in smaller groups identifying younger branches on the mtDNA evolution tree. Therefore, sub-haplogroups classification is continuously evolving (Torroni et al, 2006).

MtDNA Haplogroups and Neurodegeneration

It has been speculated that polymorphisms in mtDNA may cause subtle differences in the encoded proteins and, thus, minimum changes in oxidative phosphorylation activity and to increased amount of reactive oxygen species (ROS), both resulting in cell damage. In fact this could predispose an individual, or a population sharing the same mtDNA genotype, to an earlier onset of apoptotic processes, such as accumulation of somatic mtDNA mutations and oxidative phosphorylation impairment. The opposite could be true for different polymorphism(s), which could be beneficial increasing oxidative phosphorylation activity and/or reducing ROS production (Mancuso et al, 2006).A series of experimental evidence have been published suggesting that specific mtDNA haplogroups are associated with longevity (Tanaka et al. 1998; De Benedictis et al. 1999; Niemi et al. 2003). Thus, if they can be associated with longevity, the same or other haplogroups could be involved in the other side of the spectrum of the life's circle, the neurodegeneration and, thus, death. Because of the sensitivity of mtDNA as a marker for human migration patterns, all studies of mtDNA haplogroup association with disease must pay rigorous attention to the ethnic matching of cases to controls.

Alzheimer's Disease

Alzheimer's disease (AD) is a late-onset, progressive, age-dependent neurodegenerative disorder which results in irreversible loss of neurons, particularly in the cortex and hippocampus. It is clinically characterized by progressive memory impairment, disordered cognitive function, altered behaviour including paranoia, delusions and loss of social appropriateness, and a progressive decline in language function. Its pathological hallmarks are the presence of intracellular neurofibrillary tangles, composed of a hyperphosphorylated form of the microtubular protein tau (Selkoe, 2001), extracellular beta amyloid (Aβ) plaques, a loss of neuronal subpopulations, synaptophysin immunoreactivity of presynaptic terminals, cholinergic fibers, and the proliferation of reactive astrocytes and microglia (Tanzi and Bertram, 2001).

The Aβ cascade hypothesis remains the main pathogenetic model of familiar AD with mutation in amyloid precursor protein and presenilin genes, (Hardy and Selkoe, 2002) but its role in the majority of sporadic AD cases without mutations in these genes (accounting for 90-95% of AD cases) is still unclear.

Several studies on the morphological and morphometric estimation of mitochondria in AD, by electron microscopy, showed substantial morphological and morphometric changes in the neurons of the hippocampus, the acoustic cortex, the frontal cortex, the cerebellar cortex, the climbing fibers, the thalamus, the globus pallidus, the red nucleus and the locus coeruleus. The morphological alterations consisted of considerable changes of the mitochondrial cristae, accumulation of osmiophilic material, and decrease of their size, in comparison with the normal controls. The majority of the mitochondria, small, round, or elongated, presented disruption of the cristae or osmiophilic inclusions (Baloyannis et al, 2004; Baloyannis, 2006).

Stewart and collegues (1992) in a morphometric studies of the mitochondria in AD revealed a significant reduction in mitochondria density in endothelial cells as well as in fibroblasts obtained from patients with AD. Furthermore, apparently normal dendrites from the frontal cortex of seven patients with AD showed mitochondria with increased-density matrices and paracrystalline inclusions in the intercristal space (Lovell et al, 1999). Several studies indicates that abnormalities of cerebral metabolism are common in neurodegenerative disease, including AD (Gibson et al, 2000). Decreased glucose metabolism in AD precedes clinical diagnosis and correlates closely with the clinical state (Minoshima et al 1997). The decline in the Mini-Mental State Examination scores in AD correlated highly to reductions in glucose metabolism as measured by positron emission tomography in the temporoparietal, frontal and occipital cortices (Mielke et al, 1994). This suggests that the clinical deterioration and metabolic impairment in AD are related closely.

Since mitochondria are the powerhouse of all cells damage to mitochondria will inevitably impair energy metabolism. Measurement of mitochondrial enzyme activities indicate inherent damage to mitochondria in AD brain (Sorbi et al, 1983; Mastrogiacomo and Kish, 1994; Simonian and Hyman, 1994). Deficiency of two cycle tricarboxylic acid key enzyme has been documented in AD, suggesting defects in glucose metabolism in the AD brains (Elson et al, 2006; Sorbi et al, 1983).

Abnormal mitochondrial function in AD was first shown in 1992, when Kish and colleagues demonstrated a marked reduction (-50%) of cytochrome oxidase activity (COX; complex IV) in platelets of AD patients. Several authors also evidenced a decrease of COX activity in different brain regions (Maurer et al, 2000; Mutsya et al, 1994; Simonian and Hyman et al 1994) as well as in platelets and fibroblasts (Curti et al, 1997) of AD patients. Cardoso and collaborators (1997) found that COX activity is reduced in platelets of both early and early-onset sporadic AD subjects and that this defect is associated with increased ROS generation and ATP depletion. Our group therefore evidenced a significantly increased blood resting levels of lactate in AD and a decreased of COX but not of F1F0-ATPase activity in hippocampus and platelets of sporadic AD cases (Mancuso et al 2003), with dysfunction of energy metabolism (Bosetti et al, 2002), particularly at the synapse level, where high metabolic activity is present (Cassarino and Bennett, 1999). Further, cybrid models, obtained transfecting mtDNA from platelets of AD patients into neuronal cultured cells deprived of their own mtDNA, revealed decreased COX activity, increased ROS generation, increased Aβ production and morphological abnormalities (Khan et al, 2000; Trimmer et al, 2004). This suggests a mitochondrial genomic contribution to mitochondria dysfunction in AD (Onyango et al, 2006).

Studies attempting to identify mtDNA mutations in brains of AD patients had limited success. In 2006 Elson and colleagues sequenced the complete coding regions of 145 autoptic AD brain samples and 128 normal controls; they observed that for both synonymous and non-silent changes the overall numbers of nucleotide substitutions were the same for the AD and control sequences (Elson et al, 2006).

For this reason, several studies analyzed the frequencies of polymorphisms and/or mutations in mtDNA in AD patients, with somehow conflicting results.

The study conducted by Chagnon and coworkers showed that haplogroup T is under-represented in AD patients, while haplogroup J seems to be over-represented (Chagnon et al, 1999). Carrieri and colleague (2001), by studying an Italian sample of subjects, found that haplogroups K and U were present at a lower frequency in *Apolipoprotein* (*Apo*) *E4* carriers than in non-carriers AD patients, while, in controls, there were independence between *E4* allele and mtDNA haplogroups. This finding suggests that some haplogroups are able to modulate susceptibility to the disease. Indeed, some haplogroups seem to neutralize the harmful effect of the *E4* allele by lowering of the *E4* allele odds ratio from statistically significant to non-significant values in patients with haplogroups K and U. The statistical interaction observed between APOE alleles and mtDNA haplogroups in AD could be related to the antioxidant role exerted by APOE. It has been shown that mtDNA haplogroups are significantly associated with oxidative phosphorylation performance (Ruiz-Pesini et al. 2000), and that oxidative phosphorylation performance is correlated with reactive ROS production (Esposito et al. 1999). This suggests that the low antioxidant effect of the *E4* isoform with respect to E2 and E3 is compensated by a moderate ROS production by mtDNA types included in haplogroups K and U.

Another report showed, in contrast with Carrieri results, that (independently from *ApoE* genotype) significant association of haplogroup U was only revealed when data were stratified by sex: males classified as haplogroup U had a significant increase in risk of AD, while females demonstrated a significant decrease in risk with the same U haplogroup. Thus, the inheritance of haplogroup U may have negative effect on aging in Caucasian males (Van der Walt et al, 2004).

Two other studies (including only neuropathologically proven cases of AD of European descent) showed that mtDNA haplogroups were not associated with AD, either individually or by grouping together closely related haplogroups (Elson et al, 2006; Chinnery et al, 2000).

In our laboratory we did not find any evidence for an etiological role of haplogroup-associated polymorphisms. We have genotyped the frequency of the European mtDNA haplogroups in 209 patients with AD and 191 matched controls, both with clear Tuscan origin in order to minimize the risk of false associations between gene markers and disease (with at least three generations of Tuscanborn relatives). The frequency of haplogroups H, I, J, K, T, U, V, W, and X did not differ between the two groups, and no correlation with gender, age of onset or disease status was observed. *ApoE4* allele was confirmed as a risk factor for AD, as it was found at a significant higher frequency in patients than in controls (23,3% versus 8,11%), but no association between *ApoE* alleles and mtDNA haplogroups was observed (Mancuso et al, 2007).

In conclusion, it has been suggested that inherited haplogroups K and U may influence AD risk in Caucasians, but this is still an unresolved question. To date, mtDNA haplogroups do not seem to play a major role in AD.

Parkinson Disease

Parkinson's disease (PD) is the second most common neurodegenerative movement disorder, afflicting approximately 6 million people worldwide (Dauer and Przedborski, 2003). The cardinal features of PD include resting tremor, rigidity, bradykinesia and postural instability. These symptoms are a result of the degeneration of the dopaminergic nigrostriatal pathway originating in the substantia nigra pars compacta (SNpc). While SNpc neurons are preferentially lost in PD, they are by no means the only neuronal population affected. Degeneration in other areas, locus ceruleus, where cell loss may exceed that seen in SNpc (Zarow et al, 2003), dorsal raphe nuclei, nucleus basalis of Meynert, postganglionic sympathetic neurons, likely contributes to many of the non-motor symptoms of PD, such as depression, dementia, and autonomic dysfunction (Braak et al, 2003). Pathologically, in addition to neuronal loss, surviving neurons contain Lewy bodies, which are cytoplasmic, eosinophilic inclusions (Dauer and Przedborski, 2003). PD exists as both familial and idiopathic forms. The aetiology of sporadic PD remains largely unclear, but accumulating evidence suggesting that mitochondrial dysfunction occurs in brain and peripheral tissues of PD patients (Lean and Beal, 2006; Fukae et al, 2007). Since discovery of 1-methyl-4-phenyl-1,2,3,6-tetrahydropyridine (MPTP) as environmental factor, a number of studies have revealed that mitochondrial dysfunctions play important roles in the selective dopaminagic neuronal loss (Langston et al., 1983; Heikkila et al., 1984). In selected cases, mitochondrial genetic abnormalities can directly cause PD (Lean et al, 2006; Fukae et al, 2007). Cells fused with mtDNA from PD patients showed 25% decreased complex I activity (Gu et al, 1998).

Some reports indicate the occurrence of primary pathogenic mtDNA mutations in PD, both in the form of large scale mtDNA rearrangements (Chalmers et al, 1996), and as point mutations in complex I (Kosel et al, 1998) and tRNA genes (Grasbon-Frodl et al, 1999). Further, co-existence of parkinsonism and mutations in the mitochondrial polymerase gamma nDNA gene (*POLG1*) was recently described in several families, suggesting that when defective, this gene can be responsible for some of the mendelian transmission of parkinsonism (Davidzon et al, 2006; Luoma et al, 2004). Clustering of rare variants of the *POLG1* CAG-repeat, encoding a polyglutamine tract, have been found in Finnish patients with idiopathic PD as compared to controls (Luoma et al, 2007).

On the other hand, primary mtDNA point mutations causing parkinsonism have been reported only rarely, typically as one feature of a larger syndrome, with sensorineural deafness, levodopa-responsive parkinsonism and neuropathy. This heteroplasmic, maternally inherited 12SrRNA point mutation (T1095C) caused a significant reduction in cytochrome c oxidase activity, disrupted a highly conserved loop in the small subunit ribosomal RNA, which is important in the initiation of mitochondrial protein synthesis (Thyagarajan et al, 2000).

Recently, Horvath (2007) observed the A8344G myoclonus epilepsy, ataxia, and myopathy with ragged red fibers (MERRF) mutation in the tRNALys gene of the mtDNA in a 66-year-old man with dopa-responsive parkinsonism and reduced muscle strength. No other clinical signs typical of MERRF were noted. The A8344G mutation was present in a virtually homoplasmic state in the patient's muscle, and 80% mutant was detected in blood DNA (Horvath et al, 2007).

In order to see if the A8344G can be considered a common cause of hydiopathic sporadic PD, our group analysed 159 Italian patients with PD; none of the patients carried the A8344 mutation in DNA extracts from peripheral blood lymphocytes (Mancuso et al, 2008a). This indicates that the screening for this mutation should be restricted to those patients with parkinsonisms and other canonical features of mitochondrial encephalomyopathies, in whom screening for POLG1 gene mutations has been previously conducted with negative results (Mancuso et al, 2008a).

The role of the mtDNA variants in PD has been extensively studied. Van der Walt and coworkers (2003) genotyped 10 single-nucleotide polymorphisms (SNPs) that define the European mtDNA haplogroups in 609 white PD patient and 340 white controls. These authors observed that Haplotype J and K reduced the incidence of PD by 50%. A further analysis revealed that the SNP at 9055A of *ATP6* (haplogroup K) reduced the risk in women, and the SNP at 13708A of *ND5* gene (haplogroup J) was protective in individuals older than

70-yrs (probably either increasing the performance of complex I within the brain and other tissues in individuals derived from J and K haplogroup lineages, or reducing ROS generation) (Van der Walt et al, 2003). Consistently with the described study, it has been reported that haplotype cluster UKJT was associated with a 22% reduction in population-attributable risk for PD (Pyle et al, 2005).

Autere and colleagues (2004) hypothesize that the risk of PD is conveyed by the total number of nonsynonymous substitutions in the mtDNA-encoded complex I genes. To test this possibility, they determined the number of nonsynonymous substitutions of the seven mtDNA-encoded complex I genes from 183 Finns. They observed that the supercluster JTIWX increased the risk of both PD and PD with dementia; this cluster was associated with a twofold increase in nonsynonymous substitutions in the mtDNA genes encoding complex I subunits (Autere et al, 2004).

Ghezzi and colleagues (2005) evaluated the distribution of the different mtDNA haplogroups in a large cohort of 620 Italian patients with adult-onset idiopathic PD *vs* two groups of ethnic-matched controls. They found that haplogroup K was associated with a lower risk and reported also that the 10398G polymorphism resulted protective against PD (Ghezzi et al, 2005). This last finding, but not the association between haplogroup K and lower risk, was confirmed in 271 Spanish PD patients *vs* 230 healthy controls by Huerta and coworkers (2005).

Latsoudis and collegues determined the distributions of ten mtSNPs that define the nine major European haplogroups among 224 PD patients and 383 controls from Crete, an island of 0.6 million inhabitants who share a similar genetic background and a common environment. The recruitment of patients and controls was restricted to individuals of Cretan origin for at least three generations from both parental sides in order to avoid population admixture and subsequent genetic heterogeneity. They found no mtSNP or mtDNA haplogroup that predisposes to PD (Latsoudis et al, 2008)

Further Takasaki examined the relations between three classes of people (96 Japanese centenarians, 96 Japanese AD patients and 96 Japanese PD patients) and their mtSNP frequencies at individual mtDNA positions of the entire mt-genome. These mtSNPs showed distinct differences for three classes of people. Individual classes of people were characterized by unique mtSNPs: Japanese centenarians were closely associated with haplogroup D4, Japanese AD patients with haplogroup G2a, and Japanese PD patients with haplogroup M7a. These characteristics of mtSNPs were different from those of previously reported works. As the amino acid replacement mtSNPs were at four mtDNA positions, it is indicated that mtSNPs of synonymous nucleotide substitutions as well as those of nonsynonymous nucleotide substitutions may play important roles in mitochondrial functions (Takasaki, 2008).

Gaweda-Walerych and colleagues evaluated the impact of the mitochondrial haplogroups, subhaplogroups, coding and non-coding single-nucleotide polymorphisms on PD risk in 241 Polish PD patients and 277 control subjects. After stratification by gender, they found that haplogroup J was associated with a lower PD risk in males. Unexpectedly, subhaplogroup analysis based on the control region polymorphisms demonstrated that subcluster K1a was more prevalent in healthy controls, while K1c was more frequent in PD patients. Additionally, they confirmed the hypothesis that sublineages (U4 + U5a1 + K+J1c + J2), previously proposed to partially uncouple oxidative phosphorylation, decrease PD risk (Gaweda-Walerych et al, 2008).

Khusnutdinova and colleagues (2008) in their study analysed single-nucleotide polymorphisms on PD Russian Tatar population. They found that polymorphisms associated with haplogroup H mtDNAs increased PD risk, whereas those associated with haplogroup UK cluster mtDNAs were protective. Moreover, mtDNA sequencing revealed that PD patients with either haplogroup H or UK cluster mtDNAs can harbour additional recent variants that might further modulate PD risk. Therefore, the complexity of PD genetics may reflect the complex mitochondrial genetics (Khusnutdinova et al, 2008).

Amyotrophic Lateral Sclerosis

Amyotrophic lateral sclerosis (ALS) is a neurological disorder characterized by a progressive degeneration and death of upper and lower motor neurons. The mean survival time is not longer than 3 years from the diagnosis (Yoshida et al, 1986). About 90% of ALS patients are sporadic (sALS), whereas 10% are familial (fALS). In this group, a mutation in the gene encoding the antioxidant enzyme Cu/Zn superoxide dismutase (SOD1), on chromosome 21 has been found in a subset of 20% of patients. Since the first SOD1 missense mutations were reported in 1993 (Rosen et al, 1993), the number of known mutations has increased to more than 100. Furthermore, in the past few years additional genes responsible for familial forms of ALS have been isolated, such as *TDP-43* and *FUS* (Lagier-Tourenne C and Cleveland DW 2009).

The aetiology and pathogenesis of the sporadic form of the disease are poorly understood; mitochondrial dysfunction and oxidative stress are likely involved (Mancuso et al, 2006).

Swerdlow et al. (1998) created cybrids by fusing platelets from ALS patients, as mitochondrial donor, with neuroblastoma cells that have been depleted of mitochondrial DNA, as mitochondrial acceptor. They have observed a reduction of complex I activity in ALS cybrids and a trends toward low complex III and IV activities were also observed. This result is associated with an impairment of respiratory chain function, an increase of free radical production, and an altered calcium homeostasis, suggesting the presence of mtDNA defects.

However, this findings were not replicated by Gajewsky et al. (2003). They generated hybrid cells by using a human osteosarcoma cell line and patients' haematic cells. They also stressed the cybrids with low doses of KCN that should have elicited latent respiratory defects. Any statistically significant decrease in respiratory function in ALS cybrids compared to controls was observed.

Further, increased levels of mtDNA single "common deletion" and multiple deletions have been detected in brain (Dhaliwal and Grewal, 2000) and skeletal muscle (Ro et al, 2003) of patients with sporadic ALS.

Three patients with ALS and primary pathogenic mtDNA mutations have been reported. Comi et al (1998) described a first case of a man with an initially spastic paraparesis, early evolved in a spastic tetraparesis. He also presented mild dysphagia and dysarthria and he died at 34 years for bulbar impairment. Muscle biopsy analysis evidenced a marked COX deficiency. MtDNA analysis evidenced a 5-bp microdeletion in the 5' region of the subunit I of COX gene. This mutation was heteroplasmic out-of-frame mutation (leading to premature termination of the translation) and was absent in muscle mtDNA from control subjects (Comi et al, 1998).

The second one was an homoplasmic mutation in the mitochondrial tRNACys, in a 60-year-old Caucasian male suffering from asymmetrical pure lower motor neuron variant of ALS and temporal lobe epilepsy (Kirches et al, 1999).

Further, Borthwick et al (2006) described a man of 73 years with a slowly deterioration of motor function and presence of upper and lower limb weakness. In few time, he also developed bulbar muscle involvement with dysphonia and dysarthria. Neurological examination showed generalized muscle atrophy, reduced deep tendon reflexes, absence of clinically evident fasciculations. A diagnosis of progressive muscular atrophy was made and post-mortem examination of brain and spinal cord was performed. In this patient motor neurons in the anterior horn of lumbar spinal cord were COX deficient and molecular genetic analysis showed high levels of the pathogenic T4274C mutation, previously observed in a patient with chronic progressive external ophtalmoplegia (Chinnery et al, 1997).

To investigate if specific genetic polymorphisms within the mtDNA could act as susceptibility factors and contribute to the clinical expression of sporadic ALS, our group genotyped predefined European mtDNA haplogroups in 222 patients of clear Italian origin with sporadic ALS and 151 matched controls (Mancuso et al, 2004). All patients did not present an inherited form (either recessive or dominant) of ALS, and mutations on the entire *SOD1* gene were excluded by SSCP analysis and/or by direct sequencing. Individuals classified as haplogroup I demonstrated a significant decrease in risk of ALS versus individuals carrying the most common haplogroup, H.

In a comparison developed to test what makes this haplogroup I different from the other haplogroups tested, two nucleotides at positions 16391 and 10034 were identified; we found that only the difference in 16391A and 10034C alleles frequency between ALS patients and control subjects was highly significant (Mancuso et al, 2004). This finding, however, was not confirmed by a subsequent study (Chinnery et al., 2007) which studied an UK cohort of 504 ALS patients and 493 controls.

Multiple Sclerosis

Multiple sclerosis (MS) is a chronic and progressive disease characterised by central nervous system inflammation and axonal damage from the demyelination process, affecting young people, mostly females. This disease affects a significant part of life and is a significant health problem that causes limitations in individuals' physical, emotional, social and cognitive functioning (Bethoux et al. 2001, Bakshi 2003).

Ebers and coworkers (2004) demonstrated a maternal parental-of-origin effect in MS. The involvement of the mtDNA in determining susceptibility to MS has been hypothesized on detection of more frequency Leber's hereditary optic neuropathy (LHON) mutations in MS patients than expected by chance (Flanigan and Johns, 1993; Harding et al, 1992; Olsen et al, 1995; Hanfeld et al, 1994; Kalman et al, 1995).

In a recent study, Mihailova and coworkers (2007) analysed 58 unrelated Bulgarian patients with relapsing remitting form of MS and 104 randomly selected healthy individuals for the presence of 14 mtDNA polymorphisms determining major European haplogroups as well as three (4216, 14 798, 13 708) secondary LHON mutations. They found that 21 of the 58 patients (36.2%) resulted positive for T4216C mutation, while only 11.3% of the controls carried this secondary LHON mutation ($P < 0.01$). This finding indicated that 4216C base substitution could be considered as a predisposing marker for MS and supported the hypothesis that particular mtDNA variants could contribute to genetic susceptibility of MS (Mihailova et al, 2007) .

Kalman and colleagues reported a large scale screening of mtDNA in 77 Caucasian patients with relapsing-remitting or secondary progressive form of MS and in 84 Caucasian controls: they found that haplogroups K and J showed association with MS at a *P*-value of 0.001 (Kalman et al, 1999).

Two recent works noted a SNP (RS660339) in the *UCP2* nuclear gene as a MS risk factor; *UCP2* is a member of the mitochondrial proton transport family, which uncouples proton entry in the mitochondrial matrix from ATP synthesis. RS660339 SNP is correlated with lower UCP2 expression levels *in vitro* and *in vivo*, which is an MS risk factor (Vogler et al, 2005). In the second study, authors also found that none of the mitochondrial haplogroups were related with the disease; however there seemed to be a protective trend in the "JT + the protective allele of *UCP2*" combination (Otaegui et al, 2007). These genetic variables are independently hereditary, but there may be a synergetic effect between them.

Recently, in order to investigate the role of mtDNA variations in MS, Yu and co-workers investigated six European MS case-control cohorts comprising 5000 individuals. They found that the SNP nt13708 G/A was significantly associated with MS susceptibility (Yu et al, 2008). The frequency of haplogroup J, a haplogroup constructed with nt10398 and nt13708 SNPs, was higher in patients than in controls. Subsequent sequencing of the mtDNA of 50 individuals revealed that the nt13708A variant itself, rather than SNPs linked to it, was responsible for the association. Thus, this mtDNA variant might represent a susceptibility allele to MS.

In the same year Ban and colleagues re-sequenced the mtDNA from 159 patients with multiple sclerosis and completed a haplogroup analysis including a further 835 patients and 1506 controls. A trend towards overrepresentation of super-haplogroup U was the only evidence for association with mtDNA that they identified in these samples (Ban et al, 2008).

In parallel their analysis of nuclear encoded variants showed a trend suggesting the possible involvement of the NDUFS2 gene from Complex I. This complex is the main mediator of the production of mitochondrial super oxide (Pitkanen and Robinson, 1996) and is involved in ATP synthesis through the generation of a proton gradient. It is the first and largest subunit involved in the electron transport chain (Ugalde et al, 2004). Mutations in the NDUFS2 gene have been shown to result in a decreased activity of Complex I due to defects in the assembly and stability of this complex, and these also affect the stability of Complex III but without reducing its activity (Ugalde et al, 2004). Aside from altered energy production, a disruption in Complex I may also affect the elimination of autoreactive T cells. Mitochondrial encoded proteins are essential for activation induced T cell death. Specifically Complex I is involved in the generation of ROS and thus the induction of activation induced T cell death, an important regulator of autoimmunity, via increased expression of CD95L (Kaminski et al, 2007). Given the role of CD95L in the development of T cells and induction of activation induced T cell death, mitochondrial damage could result in impaired thymocyte development and the inability to eliminate auto-reactive T cells.

Very recently, Ghabaee and coworkers assessed possible involvement of mtDNA point mutation in MS patients. They genotyped fifty-two Iranian MS patients. They found that haplotype A was more prevalent in patients with younger age of onset and high proportion of haplogroup H was associated with optic nerve involvement. No motor symptoms were seen in haplogroup H patients (Ghabaee et al, 2009).

Huntington Disease

Huntington's disease (HD) is a genetic disease characterized by psychiatric disturbances, progressive cognitive impairment, choreiform movements, and death 15 to 20 years after the onset of symptoms (Walker, 2007). Psychiatric and behavioural symptoms do not show stepwise progression with disease severity. In most patients the onset of the disease occurs in the fourth to fifth decade of life. The pathological hallmark is the selective degeneration of medium spiny GABAergic neurons in the striatum, which results in the progressive atrophy of the caudate nucleus, putamen and globus pallidus (Walker, 2007). HD is inherited as an autosomal dominant disorder, and the mutation is highly penetrant. It is one of the diseases that involve trinucleotide expansion. In particular, HD is caused by an expansion of a CAG trinucleotide in the *IT15* gene on chromosome 4. This gene encodes a large cytoplasmatic protein called huntingtin, the function of which is as yet unknown (The Huntington's Disease Collaborative Research Group, 1993). The CAG expansion results in abnormally long stretches of polyglutamine (poly-Q) in the protein.

Various lines of evidence demonstrate the involvement of mitochondrial dysfunction in the pathogenesis of HD. However, the precise role of mitochondria in the neurodegenerative cascade leading to HD is still unclear. The length of CAG repeat seems to be related to the clinical features of HD, such as age of onset and progression of motor impairment. The basis for the impaired cognitive functions and for the mood changes is less clear. In order to determine whether mtDNA polymorphism(s) could play the role of "modifier gene(s)" in HD, in our laboratory we genotyped predefined European mtDNA haplogroups in 51 patients with HD and 181 matched controls. The frequency of the haplogroups and haplogroup clusters did not differ between the two groups, and no correlation with gender, age of onset and disease status was observed. No significant difference was observed between different haplogroups and haplogroup clusters in the cognitive or motor progression of the disease. Our study did not support any association between mtDNA haplogroups and HD (Mancuso et al, 2008b).

Friedreich's Ataxia

Friedreich's ataxia (FA), an autosomal recessive neurodegenerative disorder, is the most common hereditary ataxia among white people (Harding, 1981).1 The disease is characterised by gait and limb ataxia, dysarthria, absent tendon reflexes, Babinski sign, impairment of position and vibratory senses, and pes cavus (Filla et al, 2000). Cardiac manifestations are prominent in some cases (Isnard et al, 1997). Diabetes mellitus or carbohydrate intolerance is frequently described (Harding, 1981; De Michele et al, 1994). The molecular defect that occurs in FA is the trinucleotide GAA expansion in the first intron of the X25 gene. This gene encodes for a 210 aa mitochondrial protein called frataxin (Campuzano et al, 1996). Levels of frataxin mRNA and protein are severely reduced as a result of this expansion (Campuzano et al, 1996). Although the exact physiological function of frataxin is not known, its involvement in iron–sulphur (Fe–S) cluster biogenesis has been suggested (Yoon and Cowan, 2003).

The FA phenotype shows a variable expression with respect to the age of onset and the presence of some signs or symptoms. Some FA phenotype features are strongly correlated with GAA expansion size. An inverse relationship has been demonstrated between GAA repeat size and the age of onset (Durr et al, 1996; Filla et al, 1996).

Modifier genes are defined on the basis of their ability to modulate the clinical phenotype of individuals with monogenic and multigenic diseases. MtDNA could be considered a candidate modifier factor for FA, since mitochondrial oxidative stress is thought to be involved in the pathogenesis of this diseases (Orth and Schapira, 2001).

For this reason Giachetti and coworkers to analysed the role of mtDNA variations in FA, genotyped mtDNA haplogroups in 99 patients with FA and 48 control individuals, all from southern Italy. They found haplogroup U seems to have a protective effect on the development of the cardiomyopathy in Friedreich's ataxia. Patients belonging to this mtDNA class seem to have a delay in the disease onset and a lower rate of cardiomyopathy (Giachetti et al, 2004). This study has not been replicated so far.

Conclusions

In the past 15 years, research has been directed at clarifying the involvement of mitochondria and defects in mitochondrial oxidative phosphorylation in late-onset neurodegenerative disorders. A critical role for mitochondrial dysfunction and oxidative damage in neurodegenerative diseases has been greatly strengthened by recent findings. However, despite the evidence of morphological and biochemical abnormalities of the mitochondria in various tissues of patients with neurodegenerative disorders, the role of the mitochondrial genome and of its haplogroups as a risk factor is still controversial. MtDNA haplogroups have been associated with a number of different neurodegenerative diseases, but to date, the only disease consistently associated with a different mtDNA haplogroup frequency is PD. Because very large cohorts are required to reliably detect an association with complex human diseases (Samuels et al, 2006), further analysis with larger samples will be required to more definitively reveal the contribution of mtDNA mutations or haplogroups, if any, to the pathogenesis of neurodegenerative diseases.

References

Anderson, S; Bankier, AT; Barrell, BG; de Bruijn, MH; Coulson, AR; Drouin, J; Eperon, IC; Nierlich, DP; Roe, BA; Sanger, F; Schreier, PH; Smith, AJ; Staden, R; Young, IG. Sequence and organization of the human mitochondrial genome. *Nature,* 1981 Apr 9, 290, 457-465.

Autere, J; Moilanen, JS; Finnilä, S; Soininen, H; Mannermaa, A; Hartikainen, P; Hallikainen, M; Majamaa, K. Mitochondrial DNA polymorphisms as risk factors for Parkinson's disease and Parkinson's disease dementia. *Hum Genet*, 2004 Jun; 115, 29-35.

Bakshi, R. Fatigue associated with multiple sclerosis: diagnosis, impact and management. *Mult Scler*, 2003 Jun; 9, 219-227.

Baloyannis, SJ. Mitochondrial alterations in Alzheimer's disease. *J Alzheimers Dis*, 2006 Jul, 9, 119-126.

Baloyannis, SJ; Costa, V; Michmizos, D. Mitochondrial alterations in Alzheimer's disease. *Am J Alzheimers Dis Other Demen*, 2004 Mar-Apr, 19, 89-93.

Ban, M; Elson, J; Walton, A; Turnbull; D; Compston, A; Chinnery, P; Sawcer, S. Investigation of the role of mitochondrial DNA in multiple sclerosis susceptibility. *PLoS ONE*, 2008 Aug 6, 3, e2891.

Bethoux, F; Miller, DM; Kinkel, RP. Recovery following acute exacerbations of multiple sclerosis: from impairment to quality of life. *Mult Scler*, 2001 Apr, 7, 137-142.

Borthwick, GM; Taylor, RW; Walls, TJ; Tonska, K; Taylor, GA; Shaw, PJ; Ince, PG; Turnbull, DM. Motor neuron disease in a patient with a mitochondrial tRNAIle mutation. *Ann Neurol*, 2006 Mar, 59, 570-574.

Bosetti, F; Brizzi, F; Barogi, S; Mancuso, M; Siciliano, G; Tendi, EA; Murri, L; Rapoport, SI; Solaini, G. Cytochrome c oxidase and mitochondrial F1F0–ATPase (ATP synthase) activities in platelets and brain from patients with Alzheimer's disease. *Neurobiol Aging*, 2002 May-Jun, 23, 371-376.

Braak, H; Del Tredici, K; Rub, U; de Vos, RA; Jansen Steur, EN; Braak, E. Staging of brain pathology related to sporadic Parkinson's disease. *Neurobiol Aging*, 2003 Mar-Apr, 24, 197-211.

Brown, WM. Polymorphism in mitochondrial DNA of humans as revealed by restriction endonuclease analysis. *Proc Natl Acad Sci USA*, 1980 Jun, 77, 3605–3609.

Campuzano, V; Montermini, L; Moltò, MD; Pianese, L; Cossée, M; Cavalcanti, F; Monros, E; Rodius, F; Duclos, F; Monticelli, A; Zara, F; Cañizares, J; Koutnikova, H; Bidichandani, SI; Gellera, C; Brice, A; Trouillas, P; De Michele, G; Filla, A, De Frutos, R; Palau, F; Patel, PI; Di Donato, S; Mandel, JL; Cocozza, S; Koenig, M; Pandolfo, M. Friedreich's ataxia: autosomal recessive disease caused by an intronic GAA triplet repeat expansion. *Science*, 1996 Mar 8, 271, 1423-1437.

Cann, RL; Stoneking, M; Wilson, AC. 'Mitochondrial DNA and human evolution', *Nature*, 1987 Jan 1-7, 325, 31-36.

Cardoso, SM; Santana, I; Swerdlow, RH; Oliveira, CR. Mitochondria dysfunction of Alzheimer's disease cybrids enhances Abeta toxicity. *J Neurochem*, 2004 Jun, 89, 1417-1426.

Carrieri, G; Bonafè, M; De Luca, M; Rose, G; Varcasia, O; Bruni, A; Maletta, R; Nacmias, B; Sorbi, S; Corsonello, F; Feraco, E; Andreev, KF; Yashin, AI; Franceschi, C; De Benedictis, G. Mitochondrial DNA haplogroups and ApoE4 allele are non-independent variables in sporadic Alzheimer's disease. *Hum Genet*, 2001 Mar, 108, 194-198.

Cassarino, D.S. Bennett, J.P. Jr. (1999). An evaluation of the role of mitochondria in neurodegenerative diseases: mitochondrial mutations and oxidative pathology, protective nuclear responses, and cell death in neurodegeneration. *Brain Res Rev,* 29, 1-25.

Chagnon, P; Gee, M; Filion, M; Robitaille, Y; Belouchi, M; Gauvreau, D. Phylogenetic analysis of the mitochondrial genome indicates significant differences between patients with Alzheimer disease and controls in a French-Canadian founder population. *Am J Med Genet*, 1999 Jul 2, 85, 20-30.

Chalmers, RM; Brockington, M; Howard, RS; Lecky, BR; Morgan-Hughes, JA; Harding, AE. Mitochondrial encephalopathy with multiple mitochondrial DNA deletions: a report of two families and two sporadic cases with unusual clinical and neuropathological features. *J Neurol Sci*, 1996 Nov, 143, 41-45.

Chinnery, PF; Johnson, MA; Taylor, RW; Durward, WF; Turnbull, DM. A novel mitochondrial tRNA isoleucine gene mutation causing chronic progressive external ophthalmoplegia. *Neurology*, 1997 Oct, 49, 1166-1168.

Chinnery, PF; Mowbray, C; Elliot, H; Elson, JL; Nixon, H; Hartley, J; Shaw, PJ. Mitochondrial DNA haplogroups and amyotrophic lateral sclerosis. *Neurogenetics*, 2007 Jan, 8, 65-67.

Chinnery, PF; Taylor, GA; Howell, N; Andrews, RM; Morris, CM; Taylor, RW; McKeith, IG; Perry, RH; Edwardson, JA; Turnbull, DM. Mitochondrial DNA haplogroups and susceptibility to AD and dementia with Lewy bodies. *Neurology*, 2000 Jul 25, 55, 302-304.

Comi, GP; Bordoni, A; Salani, S; Franceschina, L; Sciacco, M; Prelle, A; Fortunato, F; Zeviani, M; Napoli, L; Bresolin, N; Moggio, M; Ausenda, CD; Taanman, JW; Scarlato, G. Cytochrome c oxidase subunit I microdeletion in a patient with motor neuron disease. *Ann Neurol*, 1998 Jan, 43, 110-116.

Curti, D; Rognoni, F; Gasparini, L; Cattaneo, A; Paolillo, M; Racchi, M; Zania, L; Bianchetti, A; Trabucchi, M; Bergamaschi, S; Covoni, S. Oxidative metabolism in cultured fibroblasts derived from sporadic Alzheimer's disease (AD) patients. *Neurosci Letters*, 1997 Oct 24, 236, 13-16.

Dauer, W. & Przedborski, S. (2003). Parkinson's disease: mechanisms and models. *Neuron*, 39, 889-909.

Davidzon, G; Greene, P; Mancuso, M; Klos, KJ; Ahlskog, JE; Hirano, M; DiMauro, S. Early-onset familial parkinsonism due to POLG mutations. *Ann Neurol*, 2006 May, 59, 859-862.

De Benedictis, G; Rose, G; Carrieri, G; De Luca, M; Falcone, E; Passarino, G; Bonafe,` M; Monti, D; Baggio, G; Bertolini, S; Mari, D; Mattace, R; Franceschi, C. Mitochondrial DNA inherited variants are associated with successful aging and longevity in humans. *FASEB J*, 1999 Sep, 13, 1532-1536.

De Michele, G; Filla, A; Cavalcanti, F; Di Maio, L; Pianese, L; Castaldo, I; Calabrese, O; Ponticelli, A; Marrone, S; Campanella, G; et al. Late onset Friedreich's disease: clinical features and mapping of mutation to the FRDA locus. *J Neurol Neurosurg Psychiatry*, 1994 Aug, 57, 977-979.

Dhaliwal, G.K. & Grewal, R.P. (2000). Mitochondrial DNA deletion mutation levels are elevated in ALS brains. *Neuroreport*, 11, 2507-2509.

DiMauro, S. & Schon, E.A. (2003). Mitochondrial respiratory-chain diseases. *N Engl J Med*, 348, 2656-2668.

Durr, A; Cossee, M; Agid, Y; Campuzano, V; Mignard, C; Penet, C; Mandel, JL; Brice, A; Koenig, M. Clinical and genetic abnormalities in patients with Friedreich's ataxia. *N Engl J Med*, 1996 Oct 17, 335, 1169-1175.

Ebers, GC; Sadovnick, AD; Dyment, DA; Yee, IM; Willer, CJ; Risch, N. Parent-of-origin effect in multiple sclerosis: observations in half-siblings. *Lancet*, 2004 May 29, 363, 1773-1774.

Elson, JL; Herrnstadt, C; Preston, G; Thal, L; Morris, CM; Edwardson, JA; Beal, MF; Turnbull, DM; Howell, N. Does the mitochondrial genome play a role in the etiology of Alzheimer's disease? *Hum Genet*, 2006 Apr, 12, 1-14.

Elson, JL; Herrnstadt, C; Preston, G; Thal, L; Morris, CM; Edwardson, JA; Beal, MF; Turnbull, DM; Howell, N. Does the mitochondrial genome play a role in the etiology of Alzheimer's disease? *Hum Genet*, 2006 Apr, 119, 241-254.

Elston, T; Wang, H; Oster, G. Energy transduction in ATP synthase. *Nature*, 1998 Jan 29, 391, 510-513.

Esposito, LA; Melov, S; Panov, A; Cottrell, BA; Wallace, DC. Mitochondrial disease in mouse results in increased oxidative stress. *Proc Natl Acad Sci USA*, 1999 Apr 27, 96, 4820-4825.

Filla, A; De Michele, G; Cavalcanti, F; Pianese, L; Monticelli, A; Campanella, G; Cocozza, S. The relationship between trinucleotide (GAA) repeat length and clinical features in Friedreich ataxia. *Am J Hum Genet*, 1996 Sep, 59, 554-560.

Filla, A; De Michele, G; Coppola, G; Federico, A; Vita, G; Toscano, A; Uncini, A; Pisanelli, P; Barone, P; Scarano, V; Perretti, A; Santoro, L; Ponticelli, A; Cavalcanti, F; Caruso, G; Cocuzza, S. Accuracy of clinical diagnostic criteria for Friedreich's ataxia. *Mov Disord*, 2000 Nov, 15, 1255-1258.

Filosto, M. & Mancuso, M. (2007). Mitochondrial diseases: a nosological update. *Acta Neurol Scand*, 115, 211-221.

Flanigan, K.M. & Johns, D.R. (1993). Association of the 11,778 mitochondrial DNA mutation and demyelinating disease. *Neurology*, 43, 2720-2722.

Fontanesi, F; Soto, IC; Horn, D; Barrientos, A. Assembly of mitochondrial cytochrome c-oxidase, a complicated and highly regulated cellular process. *Am J Physiol Cell Physiol*, 2006 Jun 7, 291, C1129-C1147.

Fukae, J; Mizuno, Y; Hattori, N. Mitochondrial dysfunction in Parkinson's disease. *Mitochondrion*, 2007 Feb-Apr, 7, 58-62.

Gajewski, CD; Lin, MT; Cudkowicz, ME; Beal, MF; Manfredi, G. Mitochondrial DNA from platelets of sporadic ALS patients restores normal respiratory functions in rho(0) cells. *Exp Neurol*, 2003 Feb, 179, 229-235.

Gaweda-Walerych, K; Maruszak, A; Safranow, K; Bialecka, M; Klodowska-Duda, G; Czyzewski, K; Slawek, J; Rudzinska, M; Styczynska, M; Opala, G; Drozdzik, M; Canter, JA; Barcikowska, M; Zekanowski, C. Mitochondrial DNA haplogroups and subhaplogroups are associated with Parkinson's disease risk in a Polish PD cohort. *J Neural Transm*, 2008 Nov; 115, 1521-1526.

Ghabaee, M; Omranisikaroudi, M; Amrisaroukolaei, S; Meysamie, A; Sahraian, MA; Bayati, A; Sanati, MH; Houshman, M; Sadeghian, H; Vajihazaman, K. Mitochondrial mutation in Iranian patients with multiple sclerosis, correlation between haplogroups H, A and clinical manifestations. *Cell Mol Neurobiol*, 2009 May, 29, 341-346.

Giacchetti, M; Monticelli, A; De Biase, I; Pianese, L; Turano, M; Filla, A; De Michele, G; Cocozza, S. Mitochondrial DNA haplogroups influence the Friedreich's ataxia phenotype. *J Med Genet*, 2004 Apr, 41, 293-295.

Gibson, GE; Haroutunian, V; Zhang, H; Park, LC; Shi, Q; Lesser, M; Mohs, RC; Sheu, RK; Blass, JP. Mitochondrial damage in Alzheimer's disease varies with apolipoprotein E genotype. *Ann Neurol*, 2000 Sep; 48, 297-303.

Grasbon-Frodl, EM; Kösel, S; Sprinzl, M; von Eitzen, U; Mehraein, P; Graeber, MB. Two novel point mutations of mitochondrial tRNA genes in histologically confirmed Parkinson disease. *Neurogenetics*, 1999 Apr, 2, 121-127.

Grezzi, D; Marelli, C; Achilli, A; Goldwurm, S; Pezzoli, G; Barone, P; Pellecchia, MT; Stanzione, P; Brusa, L; Bentivoglio, AR; Bonuccelli, U; Petrozzi, L; Abruzzese, G; Marchese, R; Cortelli, P; Grimaldi, D; Martinelli, P; Ferrarese, C; Garavaglia, B; Sangiorgi, S; Carelli, V; Torroni, A; Albanese, A; Zeviani, M. Mitochondrial DNA haplogroup K is associated with a lower risk of Parkinson's disease in Italians. *Eur J Hum Genet*, 2005 Jun, 13, 748-752.

Gu, M; Cooper, JM; Taanman, JW; Schapira, AH. Mitochondrial DNA transmission of the mitochondrial defect in Parkinson's disease. *Ann Neurol*, 1998 Aug, 44, 177-186.

Hanefeld, FA; Ernst, BP; Wilichowski, E; Christen, HJ. Leber's hereditary optic neuropathy mitochondrial DNA mutations in childhood multiple sclerosis. *Neuropediatrics*, 1994 Dec, 25, 331.

Harding, AE. Friedreich's ataxia: a clinical and genetic study of 90 families with an analysis of early diagnostic criteria and intrafamilial clustering of clinical features. *Brain*, 1981 Sep, 104, 589-620.

Harding, AE; Sweeney, MG; Miller, DH; Mumford, CJ; Kellar-Wood, H; Menard, D; McDonald, WI; Compston, DA. Occurrence of multiple sclerosis-like illness in women who have a Leber's hereditary optic neuropathy mitochondrial DNA mutation. *Brain*, 1992 Aug, 115, 979-989.

Hardy, J. & Selkoe, D.J. (2002). The amyloid hypothesis of Alzheimer's disease: progress and problems on the road to therapeutics. *Science*, 297, 353-356.

Heikkila, RE; Hess, A; Duvoisin, RC. Dopaminergic neurotoxicity of 1-methyl-4-phenyl-1,2,5,6-tetrahydropyridine in mice. *Science*, 1984 Jun 29, 224, 1451-1453.

Horvath, R; Kley, RA; Lochmüller, H; Vorgerd, M. Parkinson syndrome, neuropathy, and myopathy caused by the mutation A8344G (MERRF) in tRNALys. *Neurology*, 2007 Jan 2, 68, 56-58.

Huerta, C; Castro, MG; Coto, E; Blázquez, M; Ribacoba, R; Guisasola, LM; Salvador, C; Martínez, C; Lahoz, CH; Alvarez, V. Mitochondrial DNA polymorphisms and risk of Parkinson's disease in Spanish population. *J Neurol Sci*, 2005 Sep 15, 236, 49-54.

Isnard, R; Kalotka, H; Durr, A; Cossee, M; Schmitt, M; Pousset, F; Thomas, D; Brice, A; Koenig, M; Komajda, M. Correlation between left ventricular hypertrophy and GAA trinucleotide repeat length in Friedreich's ataxia. *Circulation*, 1997 May 6, 95, 2247-2249.

Kalman, B; Li, S; Chatterjee, D; O'Connor, J; Voehl, MR; Brown, MD; Alder, H. Large scale screening of the mitochondrial DNA reveals no pathogenic mutations but a haplotype associated with multiple sclerosis in Caucasians. *Acta Neurol Scand*, 1999 Jan, 99, 16-25.

Kalman, B; Lublin, FD; Alder, H. Mitochondrial DNA mutations in multiple sclerosis. *Mult Scler*, 1995 Apr, 1, 32-36.

Kaminski, M; Kiessling, M; Suss, D; Krammer, PH; Gulow, K. Novel role for mitochondria: protein kinase Ctheta-dependent oxidative signaling organelles in activation-induced T-cell death. *Mol Cell Biol*, 2007 May, 27, 3625-3639.

Khan, SM; Cassarino, DS; Abramova, NN; Keeney, PM; Borland, MK; Trimmer, PA; Krebs, CT; Bennett, JC; Parks, JK; Swerdlow, RH; Parker, WDJr; Bennett, JPJr. Alzheimer's disease cybrids replicate beta-amyloid abnormalities through cell death pathways *Ann. Neurol*, 2000 Aug, 48, 148-155.

Khusnutdinova, E; Gilyazova, I; Ruiz-Pesini, E; Derbeneva, O; Khusainova, R; Khidiyatova, I; Magzhanov, R; Fallace, DC. A mitochondrial etiology of neurodegenerative diseases: evidence from Parkinson's disease. *Ann N Y Acad Sci*, 2008 Dec, 1147, 1-20.

Kirches, E; Winkler, K; Vielhaber, S; Michael, M; Warich-Kirches, M; von Bossanyi, P; Plate, I; Kunz, WS; Szibor, R; Feistner, H; Dietzmann, K. Mitochondrial tRNA(Cys) mutation A5823G in a patient with motor neuron disease and temporal lobe epilepsy. *Pathobiology*, 1999 Jul-Aug, 67, 214-218.

Kish, SJ; Bergeron, C; Rajput, A, Dozic, S; Mastrogiacomo, F; Chang, LJ; Wilson, JM; DiStefano, LM; Nobrega, JN. Brain cytochrome oxidase in Alzheimer's disease. *J Neurochem*, 1992 Aug, 59, 776-779.

Kösel, S; Grasbon-Frodl, EM; Mautsch, U; Egensperger, R; von Eitzen, U; Frishman, D; Hofmann, S; Gerbitz, KD; Mehraein, P; Graeber, MB. Novel mutations of mitochondrial complex I in pathologically proven Parkinson disease. *Neurogenetics*, 1998 Mar, 1, 197-204.

Lagier-Tourenne, C; Cleveland, DW. Rethinking ALS: the FUS about TDP-43. *Cell*. 2009 Mar 20, 136, 1001-1004.

Langston, JW; Ballard, P; Tetrud, JW; Irwin, I. Chronic Parkinsonism in humans due to a product of meperidine-analog synthesis. *Science*, 1983 Feb 25, 219, 979-980.

Latsoudis, H; Spanaki, C; Chlouverakis, G; Plaitakis, A. Mitochondrial DNA polymorphisms and haplogroups in Parkinson's disease and control individuals with a similar genetic background. *J Hum Genet*, 2008 Feb 20, 53, 349-356.

Lean, M.T. & Beal, M.F. (2005). Mitochondrial dysfunction and oxidative stress in neurodegenerative diseases. *Nature*, 443, 787-795.

Lovell, MA; Gabbita, SP; Markesbery, WR. Increased DNA oxidation and decreased levels of repair products in Alzheimer's disease ventricular CSF. *J Neurochem*, 1999 Feb, 72, 771-776.

Luoma, P; Melberg, A; Rinne, JO; Kaukonen, JA; Nupponen, NN; Chalmers, RM; Oldfors, A; Rautakorpi, I; Peltonen, L; Majamaa, K; Somer, H; Suomalainen, A. Parkinsonism, premature menopause, and mitochondrial DNA polymerase mutations: clinical and molecular genetic study. *Lancet*, 2004 Sep 4-10, 364, 875-882.

Luoma, PT; Eerola, J; Ahola, S; Hakonen, AH; Hellström, O; Kivistö, KT; Tienari, PJ; Suomalainen, A. Mitochondrial DNA polymerase gamma variants in idiopathic sporadic Parkinson disease. *Neurology*, 2007 Sep 11, 69, 1152-1159.

Macaulay, V; Hill, C; Achilli, A; Rengo, C; Clarke, D; Meehan, W; Blackburn, J; Semino, O; Scozzari, R; Cruciani, F; Taha, A; Shaari, NK; Raja, JM; Ismail, P; Zainuddin, Z; Goodwin, W; Bulbeck, D; Bandelt, HJ; Oppenheimer, S; Torroni, A; Richards, M. Single, rapid coastal settlement of Asia revealed by analysis of complete mitochondrial genomes. *Science*, 2005 May 13, 308, 1034-1036.

Mancuso, M; Cardini, M; Micheli, D; Rocchi, A; Nesti, C; Figlioli, NJ; Petrozzi, L; Rossi, C; Ceravolo, R; Bacci, A; Choub, A; Ricci, G; Rognoni, G; Manca, ML; Siciliano, G; Murri, L. Lack of association between mtDNA haplogroups and Alzheimer's disease in Tuscany. *Neurol Sci*, 2007 Jun, 28, 142-147.

Mancuso, M; Conforti, FL; Rocchi, A; Tessitore, A; Muglia, M; Tedeschi, G; Panza, D; Monsurrò, M; Sola, P; Mandrioli, J; Choub, A; DelCorona, A; Manca, ML; Mazzei, R; Sprovieri, T; Filosto, M; Salviati, A; Valentino, P; Bono, F; Caracciolo, M; Simone, IL; La Bella, V; Majorana, G; Siciliano, G; Murri, L; Quattrone, A. Could mitochondrial haplogroups play a role in sporadic amyotrophic lateral sclerosis? *Neurosci Lett*, 2004 Nov 23, 371, 158-162.

Mancuso, M; Coppede, F; Migliore, L; Siciliano, G; Murri, L. Mitochondrial dysfunction, oxidative stress and neurodegeneration. *J Alzheimers Dis*, 2006 Sep, 10, 59-73.

Mancuso, M; Filosto, M; Bosetti, F; Ceravolo, R; Rocchi, A; Tognoni, G; Manca, ML; Solaini, G; Siciliano, G; Murri, L. Decreased platelet cytochrome c oxidase activity is accompanied by increased blood lactate concentration during exercise in patients with Alzheimer disease. *Exp Neurol*, 2003 Aug, 182, 421-426.

Mancuso, M; Kiferle, L; Petrozzi, L; Nesti, C; Rocchi, A; Ceravolo, R; Orsucci, D; Maluccio, MR; Bonuccelli, U; Filosto, M; Siciliano, G; Murri, L. Mitochondrial DNA haplogroups do not influence the Huntington's disease phenotype. *Neurosci Lett*, 2008b Oct 17, 444, 83-86.

Mancuso, M; Nesti, C; Petrozzi, L; Orsucci, D; Frosini, D; Kiferle, L; Bonuccelli, U; Ceravolo, R; Murri, L; Siciliano, G. The mtDNA A8344G "MERRF" mutation is not a common cause of sporadic Parkinson disease in Italian population. *Parkinsonism Relat Disord*, 2008a, 14, 381-382.

Mastrogiacomo, F. & Kish, S.J. (1994). Cerebellar alpha-ketoglutarate dehydrogenase activity is reduced in spinocerebellar ataxia type 1. *Ann Neurol*, 35, 624-626.

Maurer, I; Zierz, S; Moller, HJ. A selective defect of cytochrome c oxidase is present in brain of Alzheimer disease patients. *Neurobiol Aging*, 2000 May-Jun, 21, 455-462.

Mielke, R; Herholz, K; Grond, M; Kessler, J; Heiss, WD. Clinical deterioration in probable Alzheimer's disease correlates with progressive metabolic impairment of association areas. *Dementia*, 1994 Jan-Feb, 5, 36-41.

Mihailova, SM; Ivanova, MI; Quin, LM; Naumova, EJ. Mitochondrial DNA variants in Bulgarian patients affected by multiple sclerosis. *Eur J Neurol*, 2007 Jan, 14, 44-47.

Minoshima, S; Giordani, B; Berent, S; Frey, KA; Foster, NL; Kuhl, DE. Metabolic reduction in the posterior cingulate cortex in very early Alzheimer's disease. *Ann Neurol*, 1997 Jul, 42, 85-94.

Mutisya, EM; Bowling AC; Beal, MF. Cortical cytochrome oxidase activity is reduced in Alzheimer's disease. *J Neurochem*, 1994 Dec, 63, 2179-2184.

Niemi, AK; Hervonen, A; Hurme, M; Karhunen, PJ; Jylha, M; Majamaa, K. Mitochondrial DNA polymorphisms associated with longevity in a Finnish population. *Hum Genet*, 2003 Jan, 112, 29-33.

Noji, H. & Yoshida, M. (2001). The rotary machine of the cell, ATP synthase. *J Biol Chem*, 276, 1665-1668.

Olsen, NK; Hansen, AW; Nørby, S; Edal, AL; Jørgensen, JR; Rosenberg, T. Leber's hereditary optic neuropathy asseociated with a disorder indistinguishable from multiple sclerosis in a male harboring the mitochondrial DNA 11.778 mutation. *Acta Neurol Scand*, 1995 May, 91, 326-329.

Onyango, I; Khan, S; Miller, B; Swerdlow, R; Trimmer, P; Bennett, PJr. Mitochondrial genomic contribution to mitochondrial dysfunction in Alzheimer's disease. *J Alzheimers Dis*, 2006 Jul, 9, 183-193.

Orth, M. & Schapira, A.H. (2001). Mitochondria and degenerative disorders. *Am J Med Genet*, 106, 27-36.

Otaegui, D; Saenz, A; Ruiz-Martinez, J; Olaskoaga, J; López de Munain, A. UCP2 and mitochondrial haplogroups as a multiple sclerosis risk factor. *Mult Scler*, 2007 May, 13, 454-458.

Palanichamy, MG; Sun, C; Agrawal, S; Bandelt, HJ; Kong, QP; Khan, F; Wang, CY; Chaudhuri, TK; Palla, V; Zhang, YP. Phylogeny of mitochondrial DNA macrohaplogroup N in India, based on complete sequencing: implications for the peopling of South Asia. *Am J Hum Genet*, 2004 Dec, 75, 966-978.

Pitkanen, S. & Robinson, B.H. (1996). Mitochondrial complex I deficiency leads to increased production of superoxide radicals and induction of superoxide dismutase. *J Clin Invest*, 98, 345-351.

Pyle, A; Foltynie, T; Tiangyou, W; Lambert, C; Keers, SM; Allcock, LM; Davison, J; Lewis, SJ; Perry, RH; Barker, R; Burn, DJ; Chinnery, PF. Mitochondrial DNA haplogroup cluster UKJT reduces the risk of PD. *Ann Neurol*, 2005 Apr, 57, 564-567.

Reddy, P.H. & Beal, M.F. (2005). Are mitochondria critical in the pathogenesis of Alzheimer's disease? *Brain Res Brain Res Rev*, 49, 618–632.

Reddy, PH. Mitochondrial dysfunction in aging and Alzheimer's disease: Strategies to protect neurons. *Antioxid Redox Signal*, 2007 Oct, 9, 1647–1658.

Reddy, PH. Mitochondrial medicine for aging and neurodegenerative diseases. *Neuromolecular Med*, 2008 Jun 20, 10, 291-315.

Ro, LS; Lai, SL; Chen, CM; Chen, ST. Deleted 4977-bp mitochondrial DNA mutation is associated with sporadic amyotrophic lateral sclerosis: a hospital-based case-control study. *Muscle Nerve*, 2003 Dec, 28, 737-743.

Rosen, DR; Siddique, T; Patterson, D; Figlewicz, DA; Sapp, P; Hentati, A; Donaldson, D; Goto, J; O'Regan, JP; Deng, HX; Rahmani, Z; Krizus, A; McKennaYasek, D; Cayabyab, A; Gaston, SM; Berger, R; Tanzi, RE; Halperin, JJ; Herzfeldt, B; Van den Bergh, R; Hung, WY; Bird, T; Deng, G; Mulder, DW; Smyth, C; Laing, NG; Soriano, E; Pericak-Vance, MA; Haines, J; Rouleau, GA; Gusella, JS; Horvitz, HR; Brown, RH. Mutations in Cu/Zn superoxide dismutase gene are associated with familial amyotrophic lateral sclerosis. *Nature*, 1993 Mar 4, 362, 59-62.

Ruiz-Pesini, E; Lapeña, AC; Díez-Sánchez, C; Pérez-Martos, A; Mantoya, J; Alvarez, E; Díaz, M; Urriés, A; Montoro, L; López-Pérez, MJ; Enríquez, JA. Human mtDNA haplogroups associated with high or reduced spermatozoa motility. *Am J Hum Genet*, 2000 Sep, 67, 682-696.

Samuels, DC; Carothers, AD; Horton, R; Chinnery, PF. The power to detect disease associations with mitochondrial DNA haplogroups. *Am J Hum Genet*, 2006 Apr, 78, 713-720.

Selkoe, DJ. Alzheimer's disease: genes, proteins, and therapy. *Physiol Rev*, 2001 Apr, 81, 741-766.

Simonian, S.A. & Hyman, B.T. (1994). Functional alterations in Alzheimer's disease: selective loss of mitochondrial-encoded cytochrome oxidase mRNA in the hippocampal formation. *J Neuropathol Exp Neurol*, 53, 508-512.

Sorbi, S; Bird, ED; Blass, JP. Decreased pyruvate dehydrogenase complex activity in Huntington and Alzheimer brain. *Ann Neurol*, 1983 Jan, 13, 72-78.

Stewart, PA; Hayakawa, K; Akers, MA; Vinters, HV. A morphometric study of the blood-brain barrier in Alzheimer's disease. *Lab Invest*, 1992 Dec, 67, 734-742.

Sun C, Kong QP, Palanichamy MG, Agrawal S, Bandelt HJ, Yao YG, Khan F, Zhu CL, Chaudhuri TK, Zhang YP. The dazzling array of basal branches in the mtDNA macrohaplogroup M from India as inferred from complete genomes. *Mol Biol Evol*, 2006 Mar, 23, 683–690.

Swerdlow, RH; Parks, JK; Cassarino, DS; Trimmer, PA; Miller, SW; Maguire, DJ; Sheehan, JP; Maguire, RS; Pattee, G; Juel, VC; Phillips, LH; Tuttle, JB; Bennett, JPJr; Davis, RE; Parker, WDJr. Mitochondria in sporadic amyotrophic lateral sclerosis. *Exp Neurol*, 1998 Sep, 153, 135-142.

Takasaki, S. Mitochondrial SNPs associated with Japanese centenarians, Alzheimer's patients, and Parkinson's patients. *Comput Biol Chem*, 2008 Oct, 32, 332-337.

Tanaka, M; Gong, JS; Zhang, J; Yoneda, M; Yagi, K. Mitochondrial genotype associated with longevity. *Lancet*, 1998 Jan 17, 351, 185-186.

Tanzi, R.E. & Bertram, L. (2001). New frontiers in Alzheimer's disease genetics. *Neuron*, 3, 181-184.

The Huntington's Disease Collaborative Research Group. A novel gene containing a trinucleotide repeat that is expanded and unstable on Huntington's disease chromosomes. *Cell*, 1993, 72, 971-983.

Thyagarajan, D; Bressman, S; Bruno, C; Przedborski, S; Shanske, S; Lynch, T; Fahn, S; DiMauro, S. A novel mitochondrial 12SrRNA point mutation in parkinsonism, deafness, and neuropathy. *Ann Neurol*, 2000 Nov, 48, 730-736.

Torroni, A; Achilli, A; Macaulay, V; Richards, M; Bandelt, HJ. Harvesting the fruit of the human mtDNA tree. *Trends Genet,* 2006 Jun, 22, 339-345.

Trimmer, PA; Keeney, PM; Borland, MK; Simon, FA; Almeida, J; Swerdlow, RH; Parks, JP; Parker, WDJr; Bennett, JPJr. Mitochondrial abnormalities in cybrid cell models of sporadic Alzheimer's disease worsen with passage in culture. *Neurobiol Dis*, 2004 Feb, 15, 29-39.

Ugalde, C; Janssen, RJ; van den Heuvel, LP; Smeitink, JA; Nijtmans, LG. Differences in assembly or stability of complex I and other mitochondrial OXPHOS complexes in inherited complex I deficiency. *Hum Mol Genet*, 2004 Mar 15, 13, 659-667.

Van der Walt, JM; Dementieva, YA; Martin, ER; Scott, WK; Nicodemus, KK; Kroner, CC; Welsh-Bohmer, KA; Saunders, AM; Roses, AD; Small, GW; Schmechel, DE; Murali Doraiswamy, P; Gilbert, JR; Haines, JL; Vance, JM; Pericak-Vance, MA. Analysis of European mitochondrial haplogroups with Alzheimer disease risk. *Neurosci Lett*, 2004 Jul 15, 365, 28-32.

Van der Walt, JM; Nicodemus, KK; Martin, ER; Scott, WK; Nance, MA; Watts, RL; Hubble, JP; Haines, JL; Koller, WC; Lyons, K; Pahwa, R; Stern, MB; Colcher, A; Hiner, BC; Jankovic, J; Ondo, WG; Allen, FHJr; Goetz, CG; Small, GW; Mastaglia, F; Stajich, JM; McLaurin, AC; Middleton, LT; Scott, BL; Schmechel, DE; Pericak-Vance, MA; Vance, JM. Mitochondrial polymorphisms significantly reduce the risk of Parkinson's disease. *Am J Hum Genet*, 2003 Apr, 72, 804-811.

Vogler, S; Goedde, R; Miterski, B; Gold, R; Kroner, A; Koczan, D; Zettl, UK; Rieckmann, P; Epplen, JT; Ibrahim, SM. Association of a common polymorphism in the promoter of UCP2 with susceptibility to multiple sclerosis. *J Mol Med*, 2005 Oct;83,806-811.

Walker, FO. Huntington's disease. *Lancet*, 2007 Jan 20, 369, 218-228.

Yoon, T. & Cowan, J.A. (2003). Iron-sulfur cluster biosynthesis. Characterization of frataxin as an iron donor for assembly of [2Fe-2S] clusters in ISU-type proteins. *J Am Chem Soc*, 125, 6078-6084.

Yoshida, S; Mulder, DW; Kurland, LT; Chu, CP; Okazaki, H. Follow-up study on amyotrophic lateral sclerosis in Rochester, Minn., 1925 through 1984. *Neuroepidemiology*, 1986, 5, 61-70.

Yu, X; Koczan, D; Sulonen, AM; Akkad, DA; Kroner, A; Comabella, M; Costa, G; Corongiu, D; Goertsches, R; Camina-Tato, M; Thiesen, HJ; Nyland, HI; Mørk, SJ; Montalban, X; Rieckmann, P; Marrosu, MG; Myhr, KM; Epplen, JT; Saarela, J; Ibrahim, SM. mtDNA nt13708A variant increases the risk of multiple sclerosis. *PLoS ONE*, 2008 Feb 13, 3, e1530.

Zarow, C; Lyness, SA; Mortimer, JA; Chui, HC. Neuronal loss is greater in the locus coeruleus than nucleus basalis and substantia nigra in Alzheimer and Parkinson diseases. *Arch Neurol*, 2003 Mar, 60, 337-341.

In: Horizons in DNA Research, Volume 1
Editor: Jason R. Chesterton, pp. 129-149

ISBN: 978-1-60876-968-1

Chapter 5

THE CONTINUING EVOLUTION OF FORENSIC DNA DATABASES

***Simon J. Walsh*[*], *John S. Buckleton*[**] *and Olivier Ribaux*[***]**
Forensic & Data Centres, Australian Federal Police, Canberra, ACT, Australia
ESR Ltd., Auckland, New Zealand.
Institut de police scientifique, Universite de Lausanne, Switzerland.

ABSTRACT

In terms of forensic-specific applications of DNA technology, few have had a more profound effect on the field than the launching of databases around the world. The expansion of this phenomenon has not only required an appropriate level of scientific sophistication, but also unprecedented legislative, political and financial backing. In practical terms DNA database implementation and operation has brought massive increases in case submissions and a demonstrable change in case and evidence profiles. Forensic DNA databases also remain high profile and retain some element of controversy, typically in relation to socio-legal issues. Despite this impressive impact, it is important to acknowledge that forensic DNA database applications are still young, still only partially developed and understood and still requiring of our attention, monitoring and effort to ensure an ongoing positive contribution.

Fresh complexities are emerging in the management and development of DNA databases. Some of these problems have been examined in detail in this chapter, and, new solutions that allow cross-comparison of database performance are discussed. These methods should assist, not only in identifying present strengths and weaknesses but also in enhancing future database performance and effectiveness. Effective and creative database management has to be seen as a key priority for the forensic community. Achieving it will require a concerted effort of oversight that scientists themselves must be prepared to participate in as it is crucial to maximising the benefits of their work.

---

[*] Corresponding authors: E-mail: simonjoseph.walsh@afp.gov.au
[**] E-mail: john.buckleton@esr.cri.nz
[***] E-mail: Olivier.Ribaux@unil.ch

Forensic science has traditionally focussed on the production of evidentiary outcomes – retrospectively, and with a view to confirming or refuting an already-held belief. DNA databases also offer, however, the potential for proactive forensic contributions to investigations. Delivering these requires functional and philosophical adjustments to the process by which forensic outcomes are produced. This chapter also outlines specific steps that need to be taken to progress beyond the traditional operational framework to one that is also conducive to effective policing and crime resolution.

INTRODUCTION

Typically forensic DNA databases consist of two separate collections of profiles. A database of the profiles of individuals who have either volunteered or been compelled to submit samples, and a database of profiles obtained from samples from crime scenes, or exhibits associated with an alleged offence (Figure 1). The administrator of a database typically has capacity to compare profiles from; (a) individuals to individuals, (b) crime samples to individuals (and individuals to crime samples) and (c) crime samples to other crime samples.

The overall database is comprised of (at least) two separate indices. One contains the DNA profiles from individuals whilst the other stores DNA profiles from crimes. The separate databases are matched either internally (as depicted by arrows (a) and (c) in Figure 1). These matches seek to locate duplicate entries on the "offender database" or crimes with a common DNA profile, respectively. The databases are also matched against each other (as depicted by arrow (b) in Figure 1). This is often regarded as the most informative match process as it links individuals with profiles associated with crimes.

In a relatively short period, the growth of DNA databases internationally has been rapid with tens of millions of short tandem repeat (STR) profiles now held from convicted offenders, suspects and unsolved crimes. Links provided through DNA database searches have contributed valuable intelligence to literally millions of criminal investigations. Often links are provided for crimes which are notoriously difficult to resolve, such as property crime (for example burglary and vehicle theft) and historic unsolved crimes (commonly referred to as "cold cases"). The global scale of DNA database use, its relative infancy, the complexity of addressing requisite socio-legal concerns and the expanding capability of forensic DNA profiling combine to create a challenging law enforcement tool that demands careful assessment and management. In this chapter we introduce the emergence of forensic DNA databases in more detail and isolate aspects of their recent, and future development for more detailed consideration.

A BRIEF SUMMARY OF NATIONAL DNA DATABASE PROGRAMS

Over the past 10-15 years the establishment of a forensic DNA database has been a focal point of development for police and forensic agencies. There has been a wide uptake of the concept and large-scale DNA database operations now exist in most developed countries.

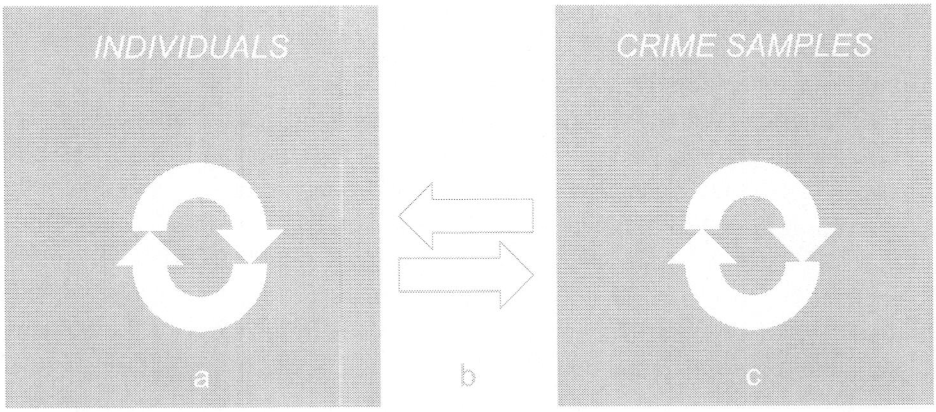

Figure 1. Summary of functionality of a standard DNA database system.

As the initial National DNA Database, the UK has benefited from a broad legislative regime and consistent funding and has grown remarkably since 1995. The UK National DNA Database (UK NDNAD) contains over 4,400,000 person profiles, over 330,000 crime profiles and has contributed investigative links in over 900,000 cases. National DNA databases also exist in at least 24 EU countries Austria, Belgium, Bulgaria, Croatia, Czech Republic, Denmark, Estonia, Finland, France, Germany, Hungary, Lithuania, Luxembourg, The Netherlands, Norway, Poland, Portugal, Romania, Scotland, Slovakia, Spain, Sweden, Switzerland and Ukraine (data provided by ENFSI, see www.enfsi.eu). By late 2008 there were over 2,400,000 person profiles and 385,000 crime profiles on European databases. These had led to over 288,000 links involving previously unsolved crimes.

In 1989 the FBI launched the Combined Offender DNA Index System (or CODIS) as a pilot program. This was fortified in 1994 with the creation of the *DNA Identification Act*. The technology was standardised to a panel of 13 STR loci (known colloquially as the CODIS loci). All 50 states have enacted legislation to establish a State index and once uploaded at the State level the data are combined at the National level through CODIS. At March 2009 there were over 6,830,000 individual profiles and 259,000 crime profiles on CODIS. There have been more than 86,000 investigations aided through CODIS over its history. The information management system that operates CODIS is made freely available by the FBI and has been adopted in at least 27 countries.

The Canadian Government committed to DNA databasing by introducing the *DNA Identification Act in 1998*, and implementing the national database in June 2000. The National DNA Database of Canada has shown consistent growth over its 9-year history. Currently (May 2009) the National DNA Database holds over 162,000 person profiles and 49,000 crime profiles and has contributed links to over 11,700 previously unsolved crimes.

China recently established a National DNA Database program that grew from 28,000 in 2005 to over 1,600,000 by 2008. Profiles are contributed by over 200 laboratories including Hong Kong which itself has over 20,000 profiles on a database that was established in 2001. Japan began DNA DB operations in December 2004 and each province is equipped for standardised DNA analysis. Singapore has over 100,000 profiles on a successful National system. Smaller more restrictive national DB programs exists in Korea and Taiwan but other major Asian countries such as Malaysia, Thailand and Indonesia currently have no National

DNA database system. There is no National DNA database in India although mechanisms to develop one are underway with a draft Bill released in 2006 advocating the sharing of State data at the national level.

Australia is a federation of six States and two Territories. Each has implemented a DNA database with the first (Victoria) beginning in 1997. Due to difficulties harmonising laws data were only combined onto a National system in 2007. The National Criminal identification DNA Database (NCIDD) is managed by the Federal agency CrimTrac and now holds over 315,000 person profiles and 110,000 crime profiles. Since linking the databases there have been over 9,000 interstate links. The New Zealand (NZ) National DNA DB began in 1995 as the 2nd national DB in the world. In global terms it remains small with over 95,900 person samples and 22,000 crime profiles but it contains profiles from 2. 2 % of the NZ population which is a higher proportion than countries such as the USA (2.0%) and Canada (0.5%) but lower than the UK (7.0%). The NZ DNA database has a high crime-to-person hit rate (63 % of crimes loaded) and has contributed over 13,900 links to unsolved investigations.

An appealing aspect of forensic DNA technology has always been the potential for the standardised STR outputs to be shared widely by police and/or forensic agencies. Whilst this potential has not yet reached the point of widespread international exchange, the recent heightened awareness around the threat of trans-national crime and terrorism has precipitated significant early steps towards the formation of international DNA exchange capabilities. In Europe, the Council of the European Union released the first Resolution covering the exchange of DNA analysis results in 1997. The Resolution called on Member States to consider establishing their own national DNA databases and to agree common standards for DNA profiling to facilitate the exchange of data. This approach was acknowledged as an important tool for investigating and combating cross border crime. Work on agreeing common standards for DNA profiling was being advanced under the umbrella of the European Network of Forensic Science Institutes (ENFSI). The Prüm Treaty, signed by Belgium, Germany, France, Luxembourg, the Netherlands, Austria and Spain in May 2005 provides for enhanced cross-border cooperation of the police and judicial authorities. The signatories agreed to give one another access to their DNA and fingerprint files using a *"hit/no hit"* system.[1] This approach creates a necessary distinction between databasing and speculative searching of DNA profiles between states and the preferred process of exchanging specific profiles during serious international investigations.

An example of *bona fide* international databasing is the Interpol DNA Database, set up by the Interpol General Secretariat in 2003 following a resolution by the 67th General Secretariat in 1998. The database (or Gateway) provides a resource through which Interpol's member countries can exchange and compare DNA profile data. Access to the Interpol database (by what are termed "beneficiaries" or "users") is allowable only following a written undertaking. Existing users can also object to any new beneficiary being granted access. The submitting countries retain ownership of the profiles and have direct control of submission, access and deletion, in accordance with (their own) national legislation. Once a match occurs and the submitting country has been notified, then that country can communicate or request additional material to or from another country, subject to restrictions imposed by that country.

---

[1] The "hit/no hit" system describes a process by which a country can query the database of another signatory. The result of the query is an advice that the search has resulted in a "hit", or "no hit". In the event of a hit contact information is provided to allow official follow-up to reveal the particulars of the hit.

This framework is aimed at meeting understandable concerns about privacy of, and control of, profiles once they leave national borders. By 2009 Interpol's DNA database held just over 83,500 profiles from 50 countries and had produced over 180 hits. Several of the successes have involved multiple crimes, some of them most serious, over multiple countries. Given in all cases the act of searching involved only adding a profile already collected and analysed, each of these instances of success must be seen as adding quite significant value towards achieving an otherwise improbable identification.

The National DNA database systems of the selected countries mentioned herein hold over 16,000,000 profiles and collectively they have contributed links to over 1,310,000 unsolved crimes. This represents the product of considerable investment from national governments and the police and forensic agencies responsible for law enforcement. Although, whilst this progress is impressive, DNA database operations are still relatively recent additions to the criminal justice system (CJS) and the principal focus to date has been on their establishment and growth. It is important to acknowledge that forensic DNA database applications are still requiring of our attention, monitoring and effort to ensure an ongoing positive contribution is not only maintained, but maximised.

Operational Impact of Forensic DNA Databases

As the technology that forms the basis for DNA intelligence databases is specialised, the operational components have remained the responsibility of forensic biology laboratories. In general, the database and its products are the property of law enforcement agencies with the analytical and matching processes administered on their behalf by forensic institutions. All aspects of the process whether handled by police or scientists are subject to governing legislation. Often this legislation contains clauses that facilitate external review of operations by delegated parliamentary authorities. From a forensic scientist's perspective the legal basis for the administration of DNA databases represents an additional level of governance over their work and that prescribe the appropriate conditions under which a DNA sample can be collected, analysed and stored, and the criminal sanctions that are enforceable for individuals in breach of these requirements.

In a practical context the impact of the operational management of forensic DNA databases has had a much more profound effect on police and forensic organisations than the need to adjust processes to adhere to the governing legislation. The snapshot of global database models provided above illustrates that a unifying trend in major jurisdictions has been an increase in the scale of database operations. Whilst this has occurred in concert with increasingly broad legislative regimes it is more likely to have been a critical impetus for iterative legislative expansion. The workload generated through DNA database operations has been a major burden on forensic DNA providers and has compelled forensic laboratories to find efficiency gains in the analytical process. Many have simply not coped with the volume of submissions and have experienced considerable backlogs and case processing delays.

This circumstance has resulted in a vastly different model of case management. Historically, police and forensic scientists have had a focus on clearing one crime at a time. Cases were submitted to the forensic laboratory after considerable investigation had occurred and in most cases, a suspect had been identified. The role of the scientist was to process these cases and determine whether there was evidence that could assist either the prosecution or defence of that specific crime. Occasionally there were cases where the forensic analysis contributed vital investigative information that assisted the location and arrest of a suspect. The broad utilisation of DNA databasing has seen this paradigm change. Essentially the analysis of evidence items occurs earlier in the investigative process with the hope that the scientist can produce intelligence information (such as a database link) in crimes for which no suspect has been identified through other means. The increased volume of submissions has also seen a trend away from individual case management and towards batch processing. It has also led to a greater prevalence of automated laboratory techniques that seek to achieve high-throughput analysis without extensive human involvement at all. The drastic increase in the size of databases and volume of information provided from them, has, in turn, imposed a dramatic change on the capacity of the CJS to absorb and utilise this information. For instance, the volume of hits now cannot be fully handled by the police due to sheer lack of resources. This then leads to the event where police work is driven by hits coming from manifold databases, and not by a response to or analysis of the situation (intelligence).

Another trend associated with the emergence of DNA database operations has been the broadening range of crimes submitted to forensic biology laboratories for analysis. In the 1980's DNA profiling was primarily used to solve serious crimes. It now contributes to the investigation of a broad spectrum of crimes, including property offences such as burglary. The variation in jurisdictional legislative and operational frameworks can impact the case submission profile, for example, certain countries in Europe initially focused on the investigation of serious violent and sexual offences (Schneider and Martin, 2001). In general there has been a clear pattern of increase in the proportion of cases submitted from volume crime categories (Walsh, 2007). A unifying factor is that the profile of casework received by the laboratory has changed since the beginning of DNA database operations.

The changing nature of case submissions, represented above, has also been accompanied by an associated change in the type of exhibit located, and hence the type of sample(s) presented for DNA analysis. For example, serious violent crimes would be expected to result in injuries and bleeding from those associated and therefore a higher likelihood of receiving blood as the principle evidence type. Likewise crimes of a sexual nature predominantly involve a male offender and hence will typically be associated with semen evidence. Property crimes and drug crimes are less simple to classify in this way and are not thought to be strongly associated with a characteristic evidence (or sample) type. A general trend however is that these types of incidents often result in the submission of more discrete evidence types, such as cigarette butts, drinking containers, food remnants, tools or swabs from surfaces or objects that the offenders are believed to have touched or handled (commonly referred to as "trace DNA") (Raymond et al., 2004).

As the contribution of forensic DNA evidence has increased in the past 5-10 years, so too has the profile of the field and its effect and impact in terms of positive outcomes of the CJS. However, the increased volume of casework and the alteration in the nature of core evidence types can each be seen as associated with a series of accompanying pressures. In summary, forensic biology practitioners must cope with greater personal expectations in terms of their casework output. This is in the context of more complex technical and interpretative challenges that are associated with the analysis of discrete, low template evidence types (such as touched surfaces, bottles and cans). Frequently, laboratories have been simply unable to cope with the rising volume of casework. Many jurisdictional centres in Australia and overseas have accrued casework backlogs that have in turn brought delays in the presentation of DNA evidence in court, or in some circumstances, trials having to proceed without any DNA evidence at all. At the same time, there is a recurrent demand from the CJS to obtain rapid results in relation to solving specific crimes or so as to decide how a suspect should be handled (detention, surveillance, arrest, etc). Backlogs are in conflict with these new demands and expectations of the CJS and can act as an additional source of pressure on forensic professionals.

From a scientist's perspective this operational environment can obviously cause pressure. This in turn can heighten the potential for error and exacerbate feelings of disempowerment, a lack of ability to adequately prepare personally for court testimony, insufficient time for research, training and professional development, or a diminished emphasis on best-practice models that may incur unmanageable commitments of staff time. In a review of laboratory backlogs in US State Crime Laboratories staff training was ranked 6th out of the top 10 spending priorities behind salaries, equipment, construction, consumables and outsourcing (Lovrich, *et al.*, 2003). This emphasises the importance placed on meeting current demand ahead of future development of staff or capabilities. This problem translates to fewer options for staff progression and monetary reward and these have been documented as affecting staff retention rates in forensic laboratories (Dale and Becker, 2004).

The changing nature of the submitted DNA evidence from traditional sources such as blood and semen to more discrete evidence types such as trace DNA and discarded items has an associated effect on the probative value of such evidence. This can indirectly place the scientist in a difficult position, particularly in interpreting questions focused on the link between any recovered DNA evidence and the actions or activities alleged to have taken place in a criminal event. These are of course the areas that are the primary concern of the courts – as they often contribute directly to resolving pivotal legal points such as culpability or intent. In evaluating these sorts of ambiguities, a forensic witness would routinely turn to a body of experimental data and interpret the current findings in the context of this empirical information. The dual challenge of working with discrete evidence sources is that 1) ambiguities relating to the source of the deposit, the duration it has been in the location it was recovered from and the mechanism by which it was deposited remain unanswerable; and 2) where experimental data exists it is as yet inconclusive and may never adequately address the myriad of potential propositions that could be advanced to explain the origins of trace DNA evidence. When these limitations of the DNA evidence are brought before a court they are often met with concern and/or alarm by members of the judiciary or the jury.

The increasing use of forensic DNA databases also has the effect of incorporating DNA evidence in cases where there is a lack of other supporting evidence. This is particularly true for historic cases which are submitted for DNA analysis as *"cold cases"* in the hope of re-invigorating unresolved investigations (*Using DNA to Solve Cold Cases*, 2002). In many circumstances the probative value of the DNA link may be unequivocal. However in others it will be more tenuous (*R v Hoey*, 2008). This also may present a source of pressure for a forensic scientist. It is important, given the often high profile nature of this kind of testing, to reinforce once more that DNA remains an item of physical evidence only and is not in itself proof of guilt or innocence. In this regard it is scientists who must ensure that the wrong impression of the potential strength of the evidence is not propagated by those wishing to inflate or denigrate its potential. A failure to engender this understanding will only increase pressure on witnesses and foster unreasonable expectations on the part of police, prosecutors and the public.

A final point that forensic scientists must consider in relation to the use of forensic DNA databases is the increased need for awareness of relevant socio-legal issues brought about by an increased public awareness of forensic DNA as a field. This increases the pressure on practitioners but has also increased the level of legislative and policy influence over the field. In a practical context, DNA-based legislation represents an additional level of governance for forensic professionals and one of the first pieces of law that places direct requirements on the manner in which they undertake their professional work. In addition, it prescribes sanctions for individuals or institutions who contravene the administrative processes detailed in these laws. Fervent debate has continued on many of the issues associated with the use of forensic DNA profiling in the CJS, and this debate has now expanded to encompass applications of the scientific process that are primarily the responsibility of the forensic community. Unfortunately, the forensic community has remained largely mute in this discussion. As such it has been dominated by a generalist tone that is abstracted from the practical context. Notwithstanding this, forensic professionals (and particularly the administrators of forensic institutions) would do well to acquaint themselves with these issues and enter the existing debate. Failure to do so could mean that the direction for the application of our scientific tools will become the responsibility of people well removed from the forensic community itself.

There are different DNA database models in different jurisdictions. All can be seen as successful and based on a cursory analysis there is little ability to isolate obvious examples of good or poor practices. The variation in the operating models occurs both between and within countries (Australia and the USA are examples of the latter). There appears from a first-level analysis of database composition to be a distinction in whether a sampling regime focuses on collecting samples from people, or from crimes. Occasionally – but not commonly – there appears to be an equivalent focus on each. The growth of DNA databases seems to have been premised on a belief that *"more is better"*. An analysis of DNA database growth demonstrates that there is a common trend that legal changes occur in favour of increased sampling powers to authorities leading to greater population of the DNA database (Walsh *et al.*, 2008).

Undeniably the area of forensic DNA databasing remains reasonably high profile, and this appears to be a unifying characteristic of various national systems. They also seem to retain some element of controversy. Importantly, we must acknowledge that forensic DNA database applications are still young, still only partially developed and understood and still (in fact more so than ever, it could be argued) requiring of our attention, monitoring and effort to ensure a continued positive contribution.

Assessing the Effectiveness of Forensic DNA Databases

Despite the rather spectacular effects of DNA databases and the considerable degree of notoriety and adulation they have generated, the truth is that constructing a forensic database to provide some links between crimes and suspects is relatively simple. Some truisms of the CJS make this so. DNA evidence is frequently found at crime scenes. Many criminals are recidivist – particularly in property crime offences where it is estimated that a large proportion of the total number of crimes (some say as high as 90%; Smith, 1999) are committed by a small proportion of the criminal population (some say as low as 10%; Smith, 1999). There is considerable evidence to suggest that most serious offenders will have offended previously (Broadhurst and Maller, 1991; Simon, 1997; Zamble and Quinsey, 1997; Heide, *et al.*, 2001; Hunt, 2002) and convicted serious offenders are ubiquitously sampled for DNA database inclusion. The DNA profiles generated in all cases are in a form that greatly assists computerised comparison. The combination of these routine features of the CJS leads to a guarantee of success when the composite parts are united through a database search.

The first crucial question at the macro level is what defines success from a forensic DNA database (solving crimes, impact on certain criminal phenomena, deterrence) and how much success should we expect? In contrast with other forensic disciplines, the social, political, legal and fiscal investment in State or National DNA databases has been extreme. Some reward has clearly flowed from that investment, but does it justify the commitment, or are there ways that we could better ensure optimal efficiency and maximal return? It seems vital that the performance of forensic DNA databases be monitored and optimised. Tracking performance is bound to lead to improvements (Asplen and Lane, 2003).

The second crucial question is: how do we measure the success of forensic DNA databases in order to answer the first question? To do this effectively is a complex undertaking that requires the coalescence of a range of experimental methodologies across numerous spheres of society. Most jurisdictions do not monitor the performance of their databases beyond reporting a one-dimensional index of output relating to the number or proportion of hits. This is a major omission.

To date, there has been little attention to either question – due mainly to the fact that over the history of databases up to now, there has been minimal demand for the answers. Forensic DNA databases have always provided outcomes, many of which involve spectacular and unprecedented contributions to the most serious of cases. These results have taken little strategic thought to achieve, and to some degree this will always be the case. However, it is our view that the days of database models based on overly simplistic expectations and guaranteed success are diminishing. An era is arriving where more profound performance management and consideration of database effectiveness will be essential to ensure an ongoing contribution and to manage future challenges.

As a database ages a number of changes occur, some of which are detrimental. Consider the crime sample database. These are samples from crime scenes and may be from the true offender or may be from an irrelevant source – such as the home owner of a house that has been burgled. Scenes of crime officers (SOCOs) seek to keep the fraction of relevant samples high but certain types of sample, such as cigarette butts have a risk of irrelevancy as they may be discarded or transported randomly, and with little relevance to the scene or offence. Let us say that the fraction of relevant samples when submitted is ω. As time passes crimes are solved which can result in the removal of relevant samples from the database. However, case resolution does nothing for irrelevant samples, and so those samples largely remain on the database. Therefore the fraction of irrelevant samples on the unsolved part of the crime database may slowly rise.

Even relevant samples that generate hits also gradually diminish in their overall value over time. This is based on an expectation that resolving old crimes is of lesser value (and is often harder) than resolving recent crimes. Whilst this is a general statement it is certainly true of property crimes, which make up the vast bulk of database cases and links. There is less justification in investigating old property crimes as there is a reduced chance of recovering property and/or finding associated probative evidence.

The database comprised of samples from people also ages. In a very coarse model this database is comprised of active offenders and non-offenders. In general, there is a movement of some active offenders towards non-offending. This occurs as active offenders retire from crime, die, or become imprisoned. Let us denote the fraction of active offenders on the suspect database α. This fraction has a tendency to decline over time. This decline can be exaggerated if selection criteria for inclusion on the database increases the fraction of non-offenders sampled. We acknowledge that this model is very simplistic and that the division of people into active offenders and non-offenders overlooks the large range of behaviours exhibited by people. However this thinking is very useful when considering the value of database sampling strategies and how they interrelate with performance.

An additional factor that further compromises the efficiency of the person database is that there is a small but almost unavoidable accrual of duplicate samples as people are sampled twice due to aliases, subterfuge (by deliberate repeat submission under similar names or aliases), or due to the existence of numerous legislative models feeding into the one national database (each requiring samples to be collected from locally apprehended suspects or offenders). This latter example exists in Australia.

Presently, assessment of the effectiveness of forensic DNA databases is almost exclusively limited to primary indexes such as profile inclusions and crime-to-person ($C \rightarrow P$) *"match rates"* or *"hit rates"*. These terms are interchangeable and generally are calculated by the total number of hits (or matches, H) divided by the total number of crime samples loaded to the Crime Sample Database (C) at the point in time at which the loading or comparison occurs (t).

$$HR_{C \rightarrow P} = \frac{H_{C \rightarrow P}}{C_t}$$

These indexes are limited in their informativeness, particularly in relation to using them to determine optimal strategies for database operation. They measure output rather than outcomes (Bieber, 2006) and are not corrected for influencing factors.

An inferential model for database performance has been developed and tested (Walsh *et al.*, 2009). The model itself is not data driven but is logic-based, however, the predictions of the model hold particularly well when assessed on the training set (in this case the SDIS data from the FBI's CODIS system). The model provides a mechanism to cross-compare the performance of database systems operating under different legislative and operational frameworks because it is data independent. This comparison was undertaken on publicly available data from four major National DNA Database programs (UK, NZ, Canada and USA).

The modelling of Walsh *et al.* (2009) suggests several key observations related to database performance management:

- Database performance can be assessed by a metric described as the return index (RI) which is estimated by $\frac{H}{NC}$ where H is the number of crime-to-person hits, N is the number of person samples on the database and C is the number of crime samples on the crime sample database.

- A high *RI* indicates a database with a high return; that is maximum hits per person and crime sample tested. In our opinion, given the financial and ethical costs of operating a forensic DNA database this is the objective that database operators, legislators and policy makers must strive for.
- The primary output from a database, hits (H) is modeled by $H = \frac{\alpha N}{M} \times \omega C$. This model suggests that $H \approx NC$ but is conditioned by two parameters α and ω, quality factors for the selection of person and crime samples, respectively and M is the *active criminal population*. High α and ω maximise H and in turn *RI* and are therefore of critical importance in optimising database performance. This has important implications for continuing research into drivers of success across all stages of police, forensic and database operations.

Operational Framework for the Use of Forensic DNA Databases

As mentioned earlier, the establishment of forensic DNA databases has been a contributor to a change in thinking within forensic science. Historically each case was processed as a unit or perhaps as a small series. The analyst adopted a *"cradle-to-grave"* approach and usually sought to refute or corroborate the version(s) of events suggested during either the investigation or trial. Typically a suspect had been identified through other means and the forensic analysis was comparative rather than suggestive.

This model has changed significantly. Forensic DNA databases have catalysed profound changes in the volume and profile of forensic case submissions meaning that with the exception of certain serious crimes, the focus has shifted to a batch-processing model within which the DNA database is an embedded component and where all cases and individual samples are viewed as potentially linked. At an organisational level the crucial objective is efficiency; maximising the processing capacity in the minimal time and with the minimal associated expense. These objectives have typically been set and driven by police users as they align best with their operational priorities.

The ability to link cases and identify a crime series has been an effective strategy in law enforcement for some time (Gotlieb, 1998). Incorporating DNA outcomes into the existing array of investigative data provides an additional mechanism to link cases committed by the same individual or organisation, adding also a highly discriminating mechanism for identification. Through the ability to provide technical information capable of directing police investigations (such as by identifying a suspect or a crime-to-crime link), DNA databases are able to operate as an investigative tool. Another important investigative feature of DNA databases is their ability to transcend jurisdictional boundaries which may have hampered abilities of law enforcement agencies to link crimes and offenders which transcend such boundaries. This is particularly true in large countries, for example the US, Canada and Australia, where states or provinces have responsibility for law enforcement within particular physical and legislative boundaries. This more proactive use of forensic DNA outcomes also presents opportunities for more integrated and intuitive use of the technology. Achieving this relies upon a complete understanding of the nature and scope of the investigative contribution that forensic DNA profiling can make; presently and in the future.

It is the process of databasing which has provided the impetus for regarding forensic outputs as a potential source of investigative leads, rather than its traditional role as corroborative information, provided at the conclusion of an investigation. Champod & Ribaux (2000) correctly point out that *"databases must be seen as tools used in a process which begins with the crime scene investigation and ends with the criminal trial"*. To that end, although they deliver information, databases should not be mis-represented as a stand-alone intelligence cycle or as the sole and final arbiter. The benefits in terms of the CJS will be maximised if databases themselves are part of a broader framework of crime investigation that is capable of integrating forensic outcomes with other systems for qualitative linking based on crime analysis and investigative trends. This section will present a hierarchy of models for forensic database systems to work either independently in an intelligence-providing role or, ideally, as a component of a genuinely integrated investigative framework.

As illustrated in Figure 2 there are a number of examples across the forensic sciences where databases exist within disciplines. Despite the conceptual similarities of these systems there are significant operational barriers to effective integration within police systems or to exchange among disciplines or jurisdictions (notwithstanding legal impediments to such possibilities). A diagram describing the typical structure of a discipline database model (using DNA as the example discipline) is provided in Figure 3.

Under the DNA *Discipline Database Model* there are two repositories of information; the National DNA Database (NDD) and the Crime Sample Database (CSD). The NDD is populated by samples from persons who for some reason have been encountered by law enforcement agencies. As earlier discussed, the practicalities of who (and how many) is (are) sampled is constrained by a combination of policy, legislative and resource factors.

From the perspective of crime sample selection and inclusion onto the CSD there are a number of limiting factors. The list of factors and their key influences include:

1. Whether the crime scene is attended at all

   This is typically based on an assessment of the incident report to the police. A decision whether or not to attend is largely effected by resources, policy, and conflicting priorities at the time.

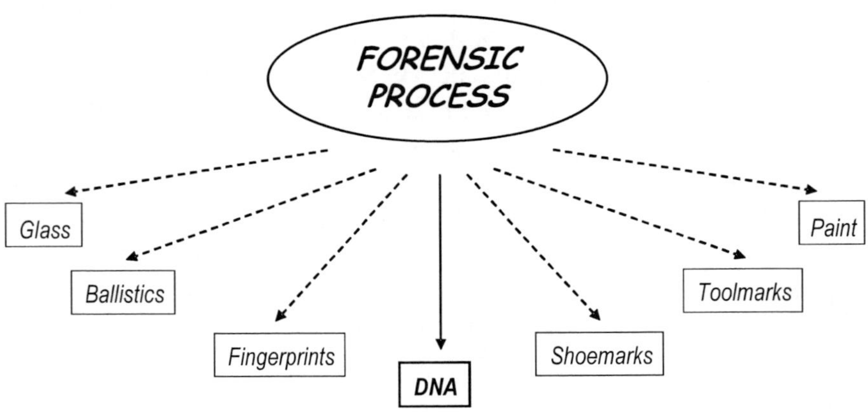

Figure 2. Common examples of discipline forensic databases.

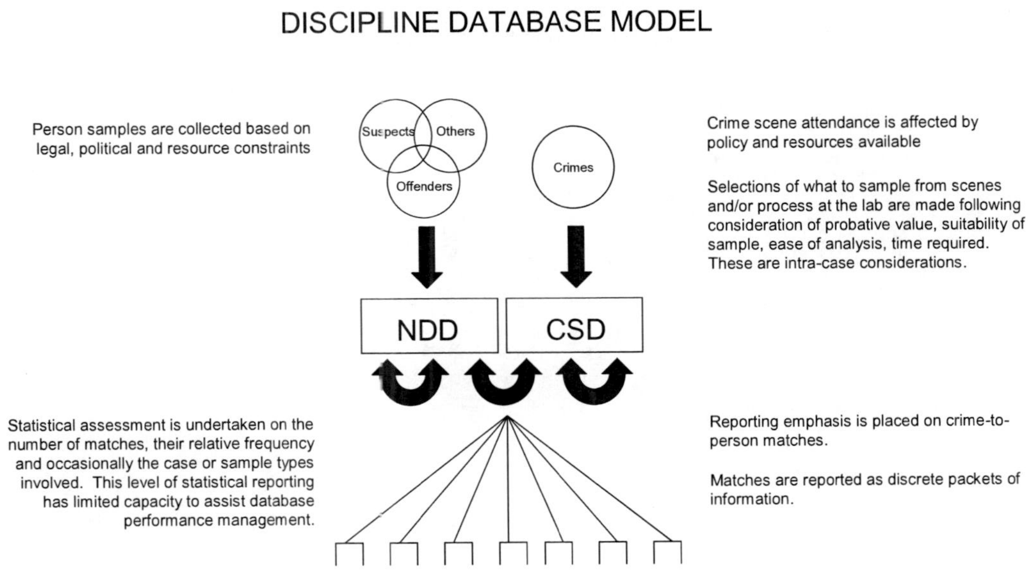

Figure 3. An example of a typical structure of a *Discipline Database Model*. This has been based on forensic DNA databases as this discipline is the focus of this chapter.

2. Whether the crime scene is attended by forensically trained personnel
   This is part of the initial police assessment of the incident report however it will be affected considerably by the jurisdictional framework and policy regarding crime scene response capability.
3. What evidence (if any) is recovered from the crime scene
   This will be affected by the decision of the forensic member attending, which in turn is a product of their training, professionalism and awareness and the timeliness with which the scene is attended. There are also practical limitations of resourcing, time and policy that will constrain what is recovered from the crime scene.
4. Whether the recovered evidence is submitted for laboratory analysis
   This is essentially a decision affected by the probative or investigative value of what is recovered, resources and likelihood of successful analysis. In a volume crime context, resources and likelihood of success are principal limiters.
5. Whether "sub-selection" occurs as part of the laboratory process
   As above, this is essentially a decision affected by the probative or investigative value of what is recovered, resources and likelihood of successful analysis. In a volume crime context, resources and likelihood of success are principal drivers. Importantly, the laboratory may or may be not aware of the specific imperatives of the case, particularly, when the laboratory is remote from the police. This exacerbates the difficulty of making this assessment.
6. Whether analysis is successful.
   This is affected by the laboratory capability, technology, the evidence type submitted for analysis. Internal policy (such as reporting or matching thresholds) may have a minor impact on this.

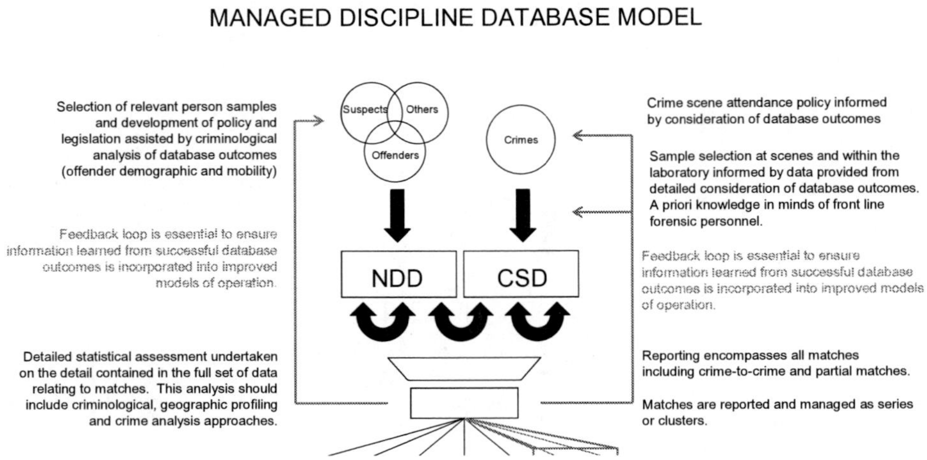

Figure 4. An example of a proposed structure of a *Managed Discipline Database Model*. Once again this model is based on forensic DNA databases.

Matching occurs within each index of both the NDD and the CSD and between the NDD and CSD (refer to Figure 1). Typically the focus of result reporting is in relation to links between entries on the CSD and an entry on the NDD (crime-to-person links). Matches tend to be reported as discrete packets of information (i.e. links generated on a particular day or following the inclusion of a particular sample).

As earlier discussed, there is typically limited statistical analysis of the database holdings and database outputs. When it occurs it is focused only on the number of database entries and the number and frequency of links made. Occasionally the case or sample types involved in database entries and links is reported. This level of statistical reporting limits capacity to assist database performance management.

The contribution of this form of discipline database is mostly confined to low-end tactical level intelligence with outcomes provided in relation to particular cases or series. Occasionally in serial matters there is a potential for this simple discipline database model to contribute operational level intelligence and assist decisions regarding resourcing and investigative strategies in specific inquiries.

The next level of discipline database operation is what we have referred to as the *Managed Discipline Database Model* (Figure 4).

There are subtle but noteworthy changes at all levels of the *Managed Discipline Database Model*. Firstly, all outputs from the database comparison process are collated into a separate holding that is managed and overseen at a global level. Result reporting from this information is multi-layered and goes beyond the automated dissemination of links. The primary purpose of this additional level of analysis is to provide a more efficient capacity to review database outcomes holistically. The vast numbers of cases and crime information residing on forensic DNA databases (and indeed other forensic holdings) is not typically analysed holistically. This is an oversight and is dismissive of unique characteristics of these datasets. That is that they already contain numerous records of links made through highly informative scientific analyses and that they are an amalgamation of crime types and localities that often remain separate in police investigative and intelligence structures.

This additional level of analysis has a number of benefits. Firstly, the outcomes of database comparison are reported as clusters of information. Secondly, trend analysis is more easily undertaken to continuously monitor performance. And thirdly, it establishes an effective feedback mechanism to inform decision-making at the early stages of the forensic process to ensure the information entered into the NDD and CSD indices is of high relevance. Walsh *et al.* (2009) have shown that this is a crucial driver of maximised DNA database performance and one that is best led from within the forensic environment as police agencies notoriously have a limited understanding of the range and scope of forensic capabilities (Lambert, *et al.*, 2007) and cannot be expected to drive more effective uptake of relevant forensic evidence.

This level of operation has been achieved by a number of jurisdictions and should be an accepted minimum operating standard for forensic DNA (or other discipline) databases. There is value in considering forensic DNA database outputs holisitically to identify tactical level intelligence that ultimately can be redirected back into the system to improve its operability. This model advances the application of forensic databases from a consideration of singular outputs (of cases or series) to the analysis of general trends from the totality of the data. Notwithstanding these examples, there has been limited exploration of practical mechanisms whereby the intelligence contribution of forensic DNA science or forensic DNA databases could be enhanced through purposeful analysis of database outcomes. Some examples of potential sources of information that could further enhance the *Managed Discipline Database Model* are provided below:

1. Criminological value of DNA database outcomes
   The expanding use of DNA databases offers significant value as a source of criminological awareness through greater analysis of trends in criminal behaviour. Such analysis could contribute to proactive policing strategies that are based on a contemporary understanding of criminality. It is foreseeable that by analysing outcomes (links) from a DNA Database from a criminological standpoint the operational success of the DNA Database can be capitalised on as a contemporary subset of data on crime, crime investigation and criminals. Analysis of this data could inform current understanding of the extent and nature of high-volume crime and the demographic and criminological profile of those who commit it. The net outcome would be a greater understanding of criminality and recommendations for strategic responses to it that could be implemented across the justice sector.
2. Criminal mobility
   There are limited studies that specifically investigate offender travel. Those that have tend to report a limited extent of travel by offenders, usually within the radius of neighbouring law enforcement jurisdictions (Leitner, *et al.*, 2007). Analysis of the geographic distribution of offences linked through the DNA database and the evaluation of relationships between localities using geographical information systems could be highly profitable. Another way of "measuring" mobility is to globally study crime to person hits, mapping the collection point of the person specimen against the crime scene location (Walsh *et al.*, 2002).
3. Systematic management of crime-to-crime (forensic) links
   Importantly, the identifications of suspects through DNA Database (crime-to-person) links follow a process which is generally well formalised. By contrast, the results of

comparisons between criminal events (crime-to-crime links) are not the subject of the same level of global or systematic management. These links however are capable of being promptly used in specific investigations or intra-jurisdictional series which can be pursued by a single police force. The extensive number of crime-to-crime links (for most database systems they are at least as common as crime-to-person links) shows the extent of this under-utilised potential. The practical value that could be realised through more comprehensive use of crime-to-crime link information includes a contribution at the tactical, operational and strategic levels. It is a serious oversight to not be analysing this information in closer detail.

4. Identifying the existence of a crime series
   Good quality, timely knowledge of a crime series is likely to influence operational decisions by proactive tasking of investigatory or uniform police resources towards the series. Of course DNA outcomes are not the sole factor contributing to these decisions, but are often a fundamental contributor.
5. Assist decisions regarding sample submission
   The analysis of DNA from biological traces is expensive and, in most jurisdictions, decisions as to which samples to submit to the database are important. The decision-making process surrounding database submissions requires the interpretation of several parameters including the available budget, the probative value of the sample, the quality of the sample (as it affects the likelihood of obtaining a complete profile), the probability of obtaining a hit (as much as this is predictable) and the importance of the case. All of these criteria are assisted by a more comprehensive awareness of the criminal situation in one's jurisdiction.

The next level of database operation involves the combination of a range of forensic disciplines in what I have referred to as the *Managed Multi-Discipline Database Model* (Figure 5).

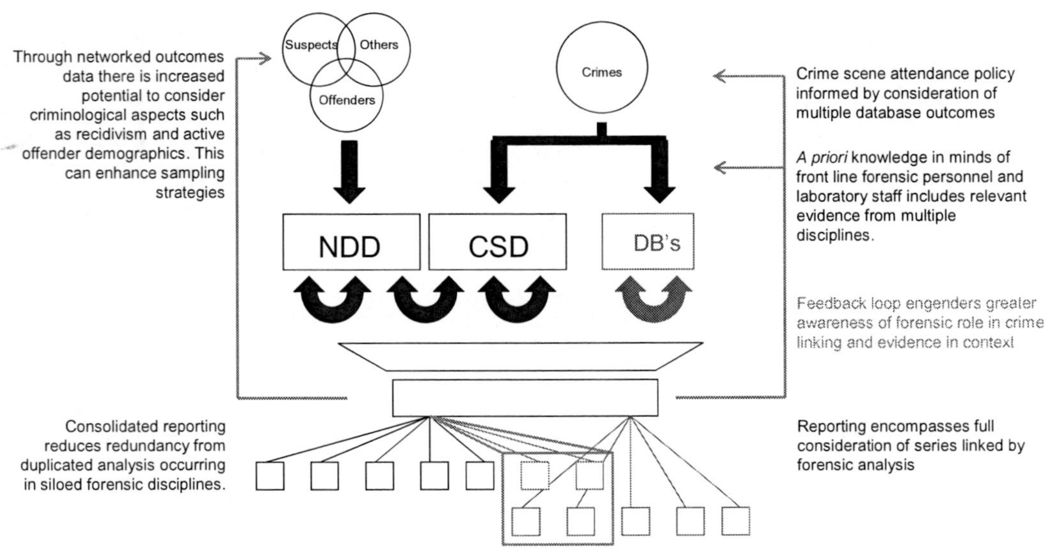

Figure 5. An example of a proposed structure of a *Managed Multi-Discipline Database Model*. Once again this model is based on forensic DNA databases.

Extending the *Managed Discipline Database Model* to incorporate multiple disciplines relies essentially on extending the capability of the results analysis unit to include other relevant forensic disciplines. The value of a multi-discipline approach is based on the assumption that the activity of serial offenders is specific and that this affects the evidence available and how and when it is deposited. As a consequence of this, some criminal signatures will be more easily visualised through certain data (for example modus operandi, shoemarks etc), whereas others will be visible through an entirely different set of data (for example DNA). The principal objective and advantage of this model is that it overcomes the silo-effect of the forensic discipline structure. The benefit of networking forensic outcomes is realised both in terms of efficiency gains by avoiding the duplication of analytical effort and in the value of outcomes and reports. As an example, fingerprints and DNA are both commonly sought in volume crime investigation and examination of the fingerprints at a scene that already has DNA link information has been useful at identifying associates of donor of the DNA link, who are potential co-offenders. By combining the outcomes of various discipline databases additional links will be observed that were not obvious previously. This augments the criminological analyses recommended under the *Managed Discipline Database Model*. In addition, reporting the full extent of these outcomes through the feedback mechanisms develops a high level of situational awareness amongst forensic crime scene and laboratory personnel. This should have a flow-on effect of improving the capacity to select relevant samples and further improve the effectiveness of the entire system. The contribution of this system extends across tactical, operational and strategic intelligence. This should be seen as a desirable end-point for forensic organisations responsible for multiple disciplines.

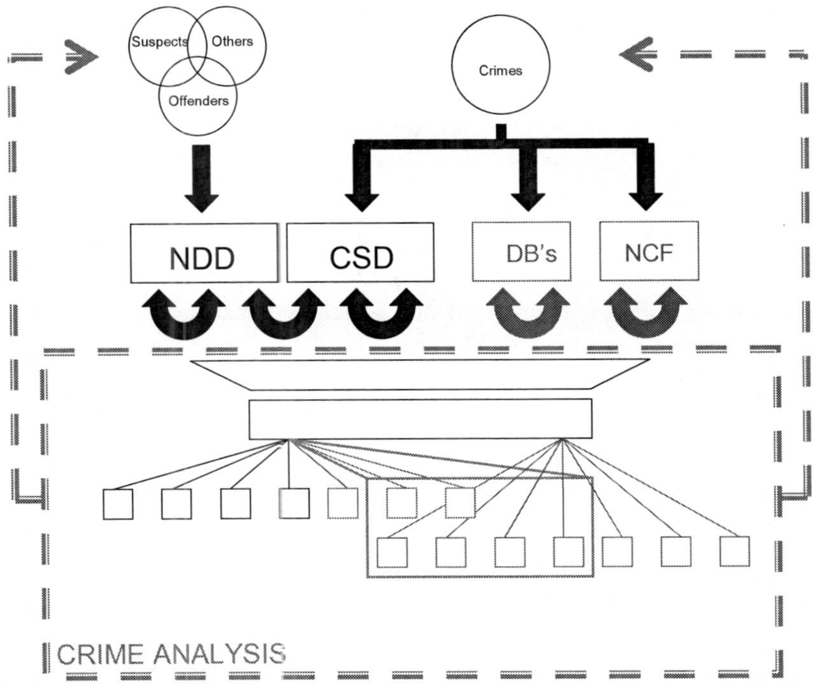

Figure 6. An example of an *Integrated Model*. This model embeds the Managed Multi-Discipline Model within existing crime analysis structures.

To date progressing through the levels of models proposed could be achieved within the walls of a forensic agency and are more examples of effective information management rather than *"intelligence"*. The models above focused on forensic case data and its use and management in the context of an intelligence-based contribution. The ultimate value of this information will be realised when the comprehensive forensic outcomes are maximised and then incorporated into well-developed law enforcement crime analysis and intelligence systems.

The ultimate level of operation, however, proposes full integration of the *Managed Multi-Discipline Database Model* within a broader context of law enforcement crime analysis and intelligence systems (Figure 6). This is based on the assumption that any single evidence type (discipline) is only able to explain part of the story regarding criminal activity. Intelligence definitely requires the logical processing of all available information, whatever its type. The separation of forensic disciplines may be an organisational necessity, but compromises forensic sciences ability to explain singular criminal activities on the basis of the evidence types available. In this sense, it should not be expected that criminality will conform to these organizational constraints, but, in fact, the other way around!

An additional component included in the *Integrated Model* is the use of Non-Comparative Forensics (NCF, in Figure 6). By this term we refer to forensic science tools that have the capacity to suggest intelligence links by virtue of their outcomes and do not rely on direct comparison with material from a known source. Their value in this context should not be underestimated. We have mentioned them as part of the *Integrated Model* as we believe that it is necessary to ensure there is a capacity to utilise the information they generate effectively in an intelligence context, prior to considering them for routine use. NCF techniques included in this category are familial searching, the use of partial (degraded) DNA and mixture and DNA phenotyping.

SUMMARY

Through their ability to provide technical information capable of directing police investigations (such as by identifying a suspect or a crime-to-crime link), DNA databases can operate as a source of tactical and operational intelligence. As mentioned this proactive role is somewhat novel for forensic DNA evidence which is usually employed to corroborate or refute a previously held belief or version of events. In some respects the remnants of this retrospective, identification-driven use of DNA evidence provide a conceptual barrier to its exploitation under a truly intelligence-based model. These include:

- A lack of timeliness such as that identified above through the work of Blakey (2000, 2002) and in the discussion regarding the evidential versus investigative models;
- The lack of a framework to integrate DNA links with other forensic or investigative information. This can be accentuated by attempts to replicate the identification-based approach of DNA and AFIS databases with other forms of forensic evidence. This has proven to be complex and computationally laborious due to the fragmentary and visual nature of other trace evidence types and the complexity of the comparison process;

- An under-utilisation of the value of crime-to-crime DNA links due to a myopic emphasis on identifications. Scene-linking is a fundamental goal of crime analysis and DNA databases can complement and reinforce these efforts;
- An under-utilisation of the potential union between DNA links/outcomes and qualitative links generated through crime analysis and criminal and technical intelligence gathering. Some obvious advantages of this union include a lateral increase in the number of *"detections"* related to each DNA link and the recognition of serial-crimes and crime *"hot-spots"*;
- A tendency for the *"aura of certainty"* associated with DNA matches to supplant rather than complement other investigative strategies;
- A failure to systematically exploit information associated with DNA-based cases and links, such as characteristics of the crime and the evidence, criminal mobility or the offender demographic profile.

As yet, the forensic community has not determined the most effective manner in which to coordinate and exploit the intelligence potential of DNA (and other) evidence and have largely failed to develop systems capable of contributing higher-level operational and strategic intelligence. At the basic level forensic science needs to consider the intelligence cycle and whether we consider ourselves to be a contributor of information into an intelligence process that is abstract from our domain. This is a limited view of the role of forensic science in this area but it is realistic and possibly an accurate reflection of our role in jurisdictions operating under a *Discipline Database Model*. If this is the contribution that we seek then the onus is on us to produce the highest quality information for the purpose. That is, the best outcomes possible from the forensic process that in turn are best suited to use in this way.

At the next level it is possible for forensic science to operate our own (internal) intelligence cycle. If this is our desire then we must question how well we collect, collate, analyse and disseminate our information; and, whether or not the products of our work aide our understanding of the crime problem that we seek to resolve. At present it is not apparent that the forensic DNA field operates systematically at this level, however, this chapter has attempted to provide a blueprint via which forensic DNA database administrators and forensic administrators more generally, can progress towards achieving this objective. At the heart of this framework is effective information management and communication. These are implemented with the objective of improving the situational awareness of forensic and police personnel in order to have a positive influence on their ability to examine crimes and evidence holistically, selecting and processing relevant forensic samples, with the most appropriate forensic mechanism available. The models proposed also engender a greater understanding of criminality and the contribution forensic science can make to its management.

It is important that the forensic science community acknowledges that our focus with regard to forensic DNA database operations has not progressed markedly from the priorities established to support the implementation and growth phases. Undeniably there has been profound success in this area, but the future holds different challenges that will require from us further inventiveness and initiative to address. The true potential of this impressive forensic advance is still to be realised.

Competing Interests

The authors declare that they have no competing interests. The authors alone are responsible for the content and writing of the paper.

References

Asplen, C. H. & Lane, S. A. (2003). Considerations in the development of DNA databases. *Interfaces, 36*, 1-2.

Bieber, F. R. (2006). Turning base hits into earned runs: improving the effectiveness of forensic DNA data bank programs. *Journal of Law and Medical Ethics 34*, 222-233.

Blakey, D. (2000). *Under the Microscope: Thematic Inspection Report on Scientific and Technical Support.* Report of the Her Majesty's Inspectorate of Constabulary, Ditching.

Blakey, D. (2002). *Under the Microscope - Refocused.* Report of the Her Majesty's Inspectorate of Constabulary, Ditching.

Broadhurst, R. G. & Maller, R. A. (1991). Estimating the numbers of prison terms in criminal careers from one-step probabilities of recidivism. *Journal of Quantitative Criminology, 7,* 275-290.

Champod, C. & Ribaux, O. (2000). Forensic identification databases in criminal investigations and trials. In: Nijboer, J. F. and Spranger, W. J. J. M. (Ed). *Harmonisation in Forensic Expertise*. Thela Thesis: Amsterdam, 463-484.

Dale, W. M. & Becker, W. S. (2004) A Case Study of Forensic Scientist Turnover. *Forensic Science Communications, volume 6(3),* available from http://www.fbi.gov/hq/lab/fsc/backissu/july2004/index.htm (accessed 12 January 2007).

Gotlieb, S. (1998). *Crime Analysis - From First Report to Final Arrest.* Alpha Publishing: Montclair, Ca.

Heide, K. M., Spencer, E., Thompson, A. & Solomon, E. P. (2001). Who's in, who's out, and who's back: follow-up data on 59 juveniles incarcerated in adult prison for murder or attempted murder in the early 1980's. *Behavioral Sciences and the Law,* 19, 97-108.

Hunt, N. Crime and crime again: The offenders who will never learn. *The Advertiser* (Adelaide), 5 October 2002, 1.

Lambert, E. G., Hogan, N. L., Nerbonne, T., Barton, S. M., Watson, P. L., Buss, J. & Lambert, J. (2007). Differences in forensic science views and needs in law enforcement: A survey of Michigan law enforcement agencies. *Police Practice and Research,* 8, 415-430.

Leitner, M., Kent, J., Oldfield, I. & Swoope, E. (2007) Geoforensic analysis revisited - The application of Newton's geographic profiling method to serial burglaries in London, UK. *Police Practice and Research,* 8, 359-370.

Lovrich, N. P., Gaffney, M. J., Pratt, T. C., Johnson, C. L., Asplen, C. H., Hurst, L. H. & Schellberg, T. M. (2003). *National Forensic DNA Study Report.* Report of the Division of Governmental Studies and Services, Washington State University and Smith Alling Lane, Professional Services, Available from http://www.dnaresource.com (accessed 13 September 2004).

Raymond, J. J., Walsh, S. J., van Oorschot, R. A., Gunn, P. R. & Roux, C. (2004). Trace DNA: an underutilised resource or Pandora's Box? *Journal of Forensic Identification, 56(4),* 668-686.

Schneider, P. M. & Martin, P. D. (2001). Criminal DNA databases: the European situation. *Forensic Science International, 119*, 232-238.

Simon, L. M. J. (1997). Do criminal offenders specialize in crime types? *Applied and Preventative Psychology,* 6, 35-53.

Smith, D. Cops and swabbers. *The Sydney Morning Herald* (Sydney), 1 December 1999, *19.*

The Queen v Sean Hoey, Crown Court Northern Ireland, Bill No: 341/05, 21 December 2007.

Walsh, S. J., Buckleton, J. S., Ribaux, O., Roux, C. & Raymond, T. (2008). Comparing the Growth and Effectiveness of Forensic DNA Databases. *Forensic Science International: Genetics Supplement Series, 1,* 667-668.

Walsh, S. J., Curran, J. M. & Buckleton, J. S. Modelling forensic DNA database performance. In press *J. Forensic Sci.* (accepted May 2009).

Walsh, S. J. (2007). Current and future trends in forensic molecular biology. In R. Rapley, & D. Whitehouse (Eds.), *Molecular Forensics*. John Wiley & Sons: London, 1-20.

Zamble, F. E. & Quinsey, G. (1997). *The Criminal Recidivism Process*. Chicago University Press: Chicago.

In: Horizons in DNA Research, Volume 1
Editor: Jason R. Chesterton, pp. 151-170
ISBN: 978-1-60876-968-1

Chapter 6

TRANSMITOCHONDRIAL MICE IN STUDYING MTDNA DISTRIBUTION AND TRANSMISSION

Mikhail G. Bass, Vassilina A. Sokolova, Maria E. Kustova, Oksana V. Kidgotko, Faina M. Zakharova and Vadim B. Vasilyev
Research Institute of Experimental Medicine,
Russian Academy of Medical Sciences, St. Petersburg, Russia.

ABSTRACT

Obtaining reliable animal models of human OXPHOS (oxidative phosphorylation) diseases caused by accumulation of mutant mtDNA in cells of certain tissues might be a noticeable step towards understanding their pathogenesis. Despite some attempts, so far no successful approach to the targeted delivery of mutant mtDNA to certain tissues/organs has been elaborated in experiments on animals. We have elaborated a model that allows studying some features of foreign mtDNA distribution at different stages of embryonic development and in organs of transmitochondrial mice developed from zygotes injected with human mitochondria. All pre-implantation embryos and some neonate animals were carrying copies of human mitochondrial genome along with murine mtDNA, which is an indication of artificial heteroplasmy. This feature persisted in the progeny of transmitochondrial animals obtained from injected zygotes i.e. among the mice in generations F_1-F_3 at least 14% were human mtDNA-carriers inheriting foreign genome along the maternal lineage. Only some transmitochondrial embryos had human mtDNA in all the blastomeres. Likewise it was found only in some organs of transmitochondrial mice. Statistical elaboration and Monte Carlo simulation taking into account Markovian process showed, firstly, that human mtDNA was unevenly distributed in zygotic cytoplasm within the time between the injection of human mitochondria and the first cleavage; secondly, that the number of mtDNA segregation units in the entire amount of mitochondria injected (ca. $5 \cdot 10^2$) is about ten. Analysis of human mtDNA distribution among the organs of transmitochondrial animals showed that skeletal muscles, liver and gonads are the targets frequently hit.

INTRODUCTION

The problem of mutant mitochondrial DNA (mtDNA) distribution among tissues and organs in course of embryonic development is of key importance for understanding the mechanisms of occurrence and progress of the so-called 'OXPHOS diseases' (from 'oxidative phosphorylation') developing due to the energy metabolism failure in some 'critical' tissues [Zeviani and Antozzi, 1997; Wallace,1999; DiMauro and Schon, 2003]. A very large part of such diseases are directly connected with altered structure of mtDNA [Holt *et al.*, 1988] and, hence, are inherited along the maternal lineage, though sometimes occur sporadically [Fu *et al.*, 1996]. In view that pathogenesis of the majority, if not of all, diseases caused by mutations in mtDNA is a moot question it has been understood for the last 15 years or more that animal models might be really helpful. Unfortunately, only recently a naturally occurring animal disease resulting from a mutation in mtDNA was found in dogs [Li *et al.*, 2006], neither the underlying nucleotide substitution, nor the phenotype having a direct parallel with humans.

Naturally occurring mtDNA mutations in non-human mammalian species seldom correspond to those causing a clinical phenotype in humans and *vice versa* [de Magalhaes, 2005; Khan *et al.*, 2007], hence, so far the attempts to obtain a genetically modified laboratory animal developing some features of an OXPHOS disease in response to an intervention in the molecular mechanisms of cellular energy production are considered worthwhile. However, obvious discrepancies exist between phenotypic manifestations of the same mutations in human and in murine mtDNA [Nakada *et al.*, 2004]. At the same time attempts to obtain transmitochondrial mice with substantially impaired activity of mitochondrial respiratory chain due to the substitution of murine mtDNA with that from rat species had no success [Kasahara *et al.* 2006], which was explained by the interference in the nuclear-mitochondrial cooperation. The same and other groups presented results indirectly pointing at the expediency of obtaining heteroplasmic mice in contrast to homoplasmic carriers of mutant murine mtDNA as the latter are not likely to clarify the mechanism(s) of dissemination of mutant copies of mitochondrial genome among tissues [Nakada *et al.*, 2001; McKenzie *et al.*, 2004; Kasahara *et al.* 2006]. The difficulties of obtaining an animal phenotype resembling any of nosologies associated with mutations in mtDNA are redoubled by the lack of knowledge about the principles of distribution of mitochondrial genome copies (either wild-type or mutant) among the tissues of an organism in its pre- and postnatal development. No wonder that in the absence of the data about any possible algorithm (if such exists) of that distribution many authors regard it as random.

Our previous studies showed the possibility to obtain laboratory animals that develop from zygotes into which human mitochondria had been injected [Sokolova *et al.*, 2004; Bass *et al.*, 2006]. The foreign for murine organism human mitochondrial genome in our experiments performed as an analog of a mutant mtDNA. It was revealed only in some tissues, this phenomenon of mosaicism being noticed also in patients with energy metabolism failure. Human mtDNA co-existed in cells with the mouse mitochondrial genome, which simulated the heteroplasmy in the tissues of patients provided by a mixture of mutant and wild-type mtDNA copies. At the same time foreign mtDNA did not cause anything that might resemble a clinical phenotype observed in patients with OXPHOS diseases. We suggested that such an experimental approach might be useful for studying the distribution of foreign (and mutant in sight) mtDNA among the tissues of a mouse embryo, and some calculations were made in search of a possible regularity in that distribution [Bass *et al.*, 2006]. However, in that preliminary study no unambiguous algorithm of the foreign mtDNA distribution among the tissues of transmitochondrial mice was revealed. This might be due to the fact that the study was made mainly on neonate and adult animals of generations F_0 and F_1, and on a relatively small sampling of mouse embryos in the second half of pregnancy.

Distribution of different copies of mitochondrial genome in cytokinesis directly depends on segregation of mtDNA molecules. An unclear mechanism functioning as a 'bottleneck' that lets through only some copies of the entire pool of mtDNA molecules underlies segregation, the process virtually unexplored [Howell, 1999]. Apart from the fundamental biological importance of studying the mtDNA segregation, revealing a possibly existing algorithm of its distribution among tissues/organs might be a prominent advance in modeling human mitochondrial diseases in animals. The research of OXPHOS diseases' pathogenesis is seriously hampered as no chances exist to obtain purposeful alterations in predicted animal tissues.

On account of all written above we decided to study the foreign mtDNA distribution at the very early stages of murine embryo development, after which to revert to more mature organisms. This paper describes our study of transmitochondrial embryos at the stage of 2 and 4 blastomeres and of several pedigrees of transmitochondrial mice in generations F_0-F_3.

MATERIALS AND METHODS

Isolation of mitochondria. Human mitochondria were isolated from cultured HepG2 cells as described [Sokolova *et al.*, 2004]. Their functional activity evidencing the overall integrity of organelles was assayed by treatment with MitoTracker Red (Molecular Probes, USA) for 30 min at +4°C dissolved in PBS (phosphate buffer saline) to the final concentration 100 nM. Red fluorescence was viewed under Leica DMRXA (Germany) fluorescent microscope evidencing the integrity of the mitochondrial membranes.

Fertilized murine zygotes were prepared as described in our previous paper [Vasilyev *et al.*, 1999].

Human mitochondria (3-10 pl of suspension) were microinjected using micromanipulator into zygotes after the latter had been immersed for 20-40 min at 37 °C into HT6 medium containing cytochalazin B (5 μg/ml). Microinjections were performed at room temperature using glass injection pipettes with internal diameter 4-6 μm [Vasilyev *et al.*, 1999]. The subsequent procedure is described in [Sokolova *et al.*, 2004]. About 500 human mitochondria were delivered into a zygote by a microinjection.

To evaluate the effect of manipulation on subsequent development of an embryo some zygotes were microinjected at room temperature with buffer solution containing no mitochondria and left for culturing at 37 °C along with the others.

Transplantation of microinjected zygotes to the pseudopregnant mice. Murine zygotes carrying human mitochondria were transplanted to the hybrids F_1 CBAxC57Bl pseudopregnant mice soon after microinjection of mitochondria. Each female received 10-15 zygotes transplanted into the oviduct under ether anaesthesia.

Species-specific primers were constructed previously [Vasilyev *et al.*, 1999]. We selected human mtDNA site 13959-13976 [Anderson *et al.*, 1981] and murine mtDNA site 13437-13455 [Bibb *et al.*, 1981] as non-homologous sequences suitable for discriminative analysis. The corresponding human (A) and murine (B) 5'-primers looked as:

A 5' - CCTTCTTACGAGCCAAAA - 3'

B 5' - ACTCTTCACACAAACATAA - 3'

To construct 3'-primers human mtDNA site 14190-14207 and murine mtDNA site 13584-13601 were selected. These two sequences are highly homologous (15 b.p. of 18 coincide). The corresponding human (A) and murine (B) 3'-primers looked as:

A 5' - CTGGTTGACCATTGTTTG - 3'

B 5' - CTGGGTGATCTTTGTTTG - 3'

Details of PCR performance with these primers are described in [Sokolova *et al.*, 2004].

DNA isolation for control. Human nuclear [Bell *et al.*, 1981] and mitochondrial [Torroni *et al.*, 1992] DNA specimens were purified from HepG2 cells and used as control. Mouse total DNA to be used as control was isolated and purified from the liver of animals [Sambrook *et al.*, 1989].

Isolation of DNA from tissues of embryos and neonate animals. Total DNA was isolated after the lysis of pre-implantation embryos at 2- and 4-blastomere stages [Vasilyev *et al.*, 1999]. Embryos of post-implantation stages were taken out on the 7th, 8th, 9th, 10th, 11th and 13th days of intrauterine development after the females had been sacrificed by dislocation of cervical vertebrae. Lysis of entire 7-11-day embryos and isolation of their DNA or isolation of the latter from organs of 13-day embryos and neonate animals were accomplished as we described earlier [Sokolova *et al.*, 2004].

Statistical evaluation. We calculated a sampling fraction of the total number of animals or embryos in each experimental group. By 'sampling fraction' we understand the portion of embryos or animals that have a characteristic studied (e.g. the embryos that developed to a certain stage, transgenic organisms, etc.). For each sampling fraction an error was estimated. Values obtained were presented as percentage in order to make results comparable. To assay the difference between various sampling fractions we used ζ criterion that is the analogue of Student's test for binominal distribution. The data obtained were also analyzed using χ^2 criterion and the *p* value obtained in Fischer's precise two-sample test [Glantz, 1994].

Criteria ζ and χ^2 were used to test the validity of the difference between a sampling fraction and the general fraction. The latter term means a portion of embryos or of animals that have a feature studied. Those can be apportioned in all the experiments of such kind (i.e. at infinitely large sampling).

In various experimental and control groups the "survival curves" were compared using the Mantel/Haenzel method [Dvoirin and Klimenkov, 1985].

We used both $\chi 2$ and $\chi 2$ with Yates's correction tests providing the comparison of an experimental distribution with a theoretical one to analyze the difference between a modeled case and the real findings in each animal organ.

To probe the differences among the groups we used $\chi 2$ and $\chi 2$ with Yates's correction tests and Fischer's precise one-sided and two-sample tests.

Cluster analysis of organs' similarity according to the frequencies of occurrence of human mtDNA was accomplished. For this analysis the average-link method was used.

Calculation was accomplished using computer programs StatXact-4.0, Statistic 6.0, SPSS 11.0.

RESULTS AND DISCUSSION

When cultured *in vitro*, 21 zygote of 73 microinjected with human mitochondria developed to blastocysts, which is 28.8 ± 5.3%. In parallel intact zygotes were cultured and 19 embryos of 34 reached the blastocyst stage, which is 55.9 ± 8.5%. Proportions of blastocysts in the control and experimental groups are significantly different at 1% level (ζ>2.58, χ^2=7.4, p=0.099). The third group contained 27 zygotes injected with PBS (sham injections) and 8 of those developed to blastocysts, which is 29.6 ± .8%. This proportion differs significantly at 5% level from that found among intact zygotes (ζ>1.96, χ^2=4.5, p=0.04).

We analyzed proportions of mouse embryos that developed in due time with no malformations in comparison to the respective values in the group where zygotes were injected with human mitochondria and among those having got sham injections. Already on the third day of culturing *in vitro* the shares of normally developing embryos were lower both in the group of mitochondria-injected and in that of PBS-injected zygotes (ca. 80%) as compared to intact zygotes (ca. 90%). By the blastocyst stage this difference became highly significant.

Our previous findings [Sokolova *et al.*, 2004; Bass *et al.*, 2006] showed that ca. 4% of both mitochondria- and PBS-injected embryos reached the blastocyst stage about 24 hours later than the rest 96% in each group and than all the intact embryos of the control group. When analyzed by Mantel-Haenzel method, the 'survival curves' reflecting the pace with which the number of normally developing embryos decreased at every developmental stage looked somewhat different [Bass *et al.*, 2006].

Since no significant difference between the two groups with injected embryos was revealed, the data of those two were unified and compared to the pattern of development of control organisms. Indeed, the pace of diminution of the number of normally developing embryos was significantly different between the experimental and the control groups (χ^2=5.15). This difference was confirmed also in Fisher's precise two-sample test (p=0.022).

These results support the conclusion made in our preliminary study [Sokolova *et al.*, 2004] that it is the traumatic effect of microinjection, but not the content of the needle that lessens the number of duly developing embryos. Similar evidence was earlier obtained by another group [Nagao *et al.*, 1998].

In all early embryos that developed *in vitro* from zygotes microinjected with human mitochondria PCR revealed both murine and human mtDNA. This is an evidence of obtaining a simple model that meets the requirements of heteroplasmy, even though the latter was inter-specific i.e. 'human/mouse'.

Examination of separate blastomeres showed that foreign mtDNA cannot be found in all of those. For example, of 156 blastomeres tested at the 2-cell stage (i.e. after the first division of an embryo) only 102 carried human mtDNA in both cells, while the rest 54 had it in one blastomere. Murine mtDNA was detected in all blastomeres analyzed.

The absence of human mtDNA from some blastomeres already after the first division is indicative both of the possible occurrence of mosaicism in the very early embryonic development and at nonrandom distribution of foreign mitochondria during cell division in an embryo.

The amount of mitochondria introduced by a microinjection was somewhat substantial (ca. $5x10^2$), and under conditions of stochastic distribution of foreign mitochondria the probability of their appearance in both daughter-cells after the first division would be very high. The absence of human mtDNA in the second blastomere of 34.6 ± 3.8% embryos (54 of 156) may be caused by several possible mechanisms.

First, the phenomenon observed can be explained by aggregation of mitochondria in clusters during the division of a zygote or shortly before [Van Blerkom *et al.*,, 1998]. This mechanism is often considered as having a pronounced impact on mtDNA segregation [Howell, 1999].

Segregation of mtDNA in the sequence of division blastomeres is occurring most likely according to the 'bottleneck' principle, like it happens in maturing oocytes. Then a noticeable part of mitochondrial genome copies does not leave for the daughter cell in course of cytokinesis [Poulton and Marchington, 1996].The crucial notion characterizing this phenomenon is the 'segregation unit'. According to the data of various groups the size of a segregation unit varies from 5-10 copies of mitochondrial genome to dozens of mtDNA molecules united in nucleoprotein clusters called 'nucleoids' [Howell, 1999; Bogenhagen *et al.*, 2003; 2008; Chen and Butow, 2005; Malka *et al.*, 2006]. The proteins in these clusters are believed to exert the mtDNA-packaging [Brewer *et al.*, 2003; Friddle *et al.*, 2004; Kucej *et al.*, 2008]. The bottleneck mechanism of mtDNA segregation has been described not exclusively for oocytes, but was observed also in several generations of cultured fibroblasts [Matthews *et al.*, 1995].

Let us suppose that segregation through a 'bottleneck' is the only mechanism underlying the observed distribution of mtDNA among blastomeres. If the probability for each segregation unit to appear in a blastomere is regarded as a probability of an independent event, then with two segregation units the chances that only one of two blastomeres gets human mtDNA is 50%. With three units these chances are 25%. In our experiments 34.6±3.8% of embryos had mtDNA in one blastomere. This allows suggesting 2-3 segregation units in the amount of mitochondria introduced into an oocyte. Respectively, more than 10^2 mitochondria donated their mtDNA to each of the segregation units.

Besides, it cannot be excluded that, apart from the aggregation of mitochondria and/or clustering of mtDNA in nucleoids, the minimum threshold concentration of a foreign mtDNA exists that is prerequisite for the latter fixation in the daughter cell.

Another possible explanation of the distribution observed in our study might be that the time between the injection of human mitochondria into a mouse zygote and the first division of the latter is not enough for the introduced mitochondria to spread all over the cell. Such a possibility should not be neglected on account that all foreign mitochondria were purposely injected into the perinuclear space and for some time might remain mostly within that part of the cytoplasm. It should be noted that by spreading of mitochondria we do not mean a stochastic thermodynamic drift of particles in a cell, but a complex sequence of processes that include binding of mitochondria to the cytoskeleton [Tourte *et al.*, 1991; Penman, 1995; Hermann *et al.*, 1997; Griparic *et al.*, 2007]. One should also take into account the tendency of mitochondria (perhaps, of nucleoids as well) to merge [Howell, 1999; Nogawa *et al.*, 1988; Smith and Alcivar, 1993; Kaufman *et al.*, 2003; De Brito and Scorrano, 2008] and frequently occurring joint segregation of neighboring organelles in cytokinesis [Howell, 1996; Howell *et al.*, 1999; Chen and Butow, 2005]. Taking into account the possible effect of mixing host mtDNA with that introduced in experiments does not mean to deny the existence of segregation units. More likely in our case their number was higher than 2 or 3, which implies, respectively, a lower number of mitochondria in each unit.

Finally the third possible explanation of the absence of foreign mtDNA in one of two blastomeres in 34.6±3.8% embryos is a methodic phenomenon. It cannot be excluded that in such embryos the injected mitochondria remained encircled by a vacuole of plasma membrane resulting from only partly penetrating puncture. Though highly improbable, this assumption needs separate consideration.

To choose among three suggestions one needs to analyze the distribution of foreign mtDNA during the next division of embryo into four blastomeres. However, before that one has to view possible options offered by each of the hypotheses put forth.

If the third guess is correct and the absence of foreign mtDNA in one of two blastomeres in 34.6±3.8% embryos occurs as the human mitochondria stay inside vacuoles, then the same share of embryos with human mtDNA only in one blastomere should be observed in a 4-cell embryo (we showed in special experiments that once formed, such vacuoles persist for at least two divisions).

To find possible difference in distribution of foreign mtDNA among four blastomeres if either the first or the second hypothesis is correct, we suggested the following model. Let us consider a 2-cell embryo with foreign mtDNA in each blastomere. The probability to transmit foreign mtDNA to both daughter cells in this case is 65.4%, while transmission to a single daughter cell is probable in 34.6% cases, which is the replica of the first division. The distribution of foreign mtDNA at the 4-cell stage will be as follows. In 12.0% of all cases human mtDNA will remain in one blastomere. It will be found in two cells in 30.5%, in three of four in 29.6%, and in four blastomeres in 28.0% of all embryos.

One should note that such a distribution reflects the "mildest" evaluation of the first hypothesis. The latter is based on an assumption that the size of a segregation unit is the only factor stipulating for an algorithm of foreign mtDNA distribution. Consequently, such an assumption presumes the formation of 2-3 segregation units within the entire amount of human mitochondria that were introduced into a murine zygote. This assertion is based, on the one hand, upon the known fact of the absence of mtDNA replication in the course of first divisions of an embryo, on the other hand, upon the distribution observed in our experiments when at least 65.4±3.8% of 2-cell embryos shared the introduced human mtDNA between the two blastomeres.

Thus, if the first hypothesis is correct one can expect either a coincidence of a distribution observed in experiments with that deduced theoretically, or a slight deviation towards a higher percentage of embryos with small amount of blastomeres carrying foreign mtDNA. Such a deviation seems possible due to the described above distribution of several segregation units between the two daughter cells in course of the first division.

In contrast, if the second hypothesis is correct, the percentage of embryos with larger amount of blastomeres carrying foreign mtDNA is expected. Such an effect may be previewed, firstly, if we suggest that the time interval between the first and the second division is long enough for 'mixing' the segregation units (in the described above meaning of the term 'mixing'). Secondly, it is understood that if an amount of segregation units within the portion of foreign mitochondria is higher than 2-3, the probability of foreign mtDNA transmission to both daughter blastomeres in course of the second division (hence, its occurrence in all four cells) is higher than 65.4%.

Of 102 embryos analyzed at the 4-cell stage, ten carried human mtDNA in a single blastomere (9.8±2.9%). Twenty embryos had it in two blastomeres (19.6±3.9%), twenty five in three cells (24.5±4.3%), and forty seven in all four blastomeres (46.1±4.9%). Murine mtDNA was detected in all blastomeres examined.

Since an error of a fraction depends exclusively on the fraction itself and the size of sampling, an expected 'ideal' distribution of foreign mtDNA among blastomeres of 102 embryos should look as follows. In 12.0±3.2% embryos human mtDNA should be found in a single blastomere. In 30.4±4.6% it should be present in two of four, and in 29.6±4.5% in three of four cells. Finally, 28.0±4.5% embryos should carry foreign mtDNA in all four blastomeres.

The percentage of embryos 9.8±2.9% in which human mtDNA was detected only in one of four blastomeres is significantly lower than the one expected if the third hypothesis is correct, i.e. 34.6±3.8%. The significance of this difference is very high (ζ=5.17, p<0,001). Thus, the third hypothesis can be rejected with a very high probability. However, this result does not reject the probability for human mitochondria to remain within a vacuole in 9.8 ± 2.9 % or a smaller percentage of experiments.

We used χ^2 criterion to compare the predicted distribution of foreign mtDNA among four blastomeres with that observed in our experiments [Figure 1]. The obtained value χ^2=16.62 is indicative of a very high significance (p<0,001) of the difference between the expected and experimental distributions, while a significantly high portion of embryos containing foreign mtDNA in all four blastomeres (46.1±4.9% in experiments *versus* 28.0±4.5% in the theoretical model) favors the second hypothesis.

Besides, an additional indirect support of this hypothesis comes from the results of experiments in which we analyzed the 4-cell transmitochondrial embryos obtained by injecting human mitochondria in one of the two blastomeres of a 2-cell embryo. In that case only 44 of 81 embryos analyzed (54.3±5.5%) contained foreign mtDNA in two of four blastomeres, while in 37 embryos it was found only in one cell (45.7±5.5%). It seems very likely that no 'mixing' of mtDNA happened between the injection of human mitochondria and the subsequent division of the embryo.

So one can make two most important conclusions based on the study accomplished.

First, in accord with our calculations the daughter cells in course of a murine embryo's division get mtDNA as the segregation units, the number of which surely exceeds 2-3. This is in agreement with the data of other researchers who studied different organisms [Ashley *et al.*, 1989; Jenuth *et al.*, 1996]. It should be understood that this number has been calculated with reference to the amount of human mitochondria introduced into a mouse zygote. The conditions of our experiments do not allow evaluating the total number of segregation units, i.e. taking into account also those comprised essentially of murine mtDNA.

Second, it can be concluded that distribution of mtDNA within the entire cytoplasm takes noticeable time, proportionate to the interval between an embryo's divisions. This is correct at least for all cases when foreign mitochondria were introduced into a zygote by microinjection. Such a notion is in line with the proposed mechanisms of mitochondrial mobility within a cell, i.e. the participation of cytoskeleton [Tourte *et al.*, 1991; Penman, 1995; Hermann *et al.*, 1997] and of peculiar locomotor proteins, dynamins [Van der Bliek, 1999; Hinshaw, 2000; Olichon *et al.*, 2002], in that process.

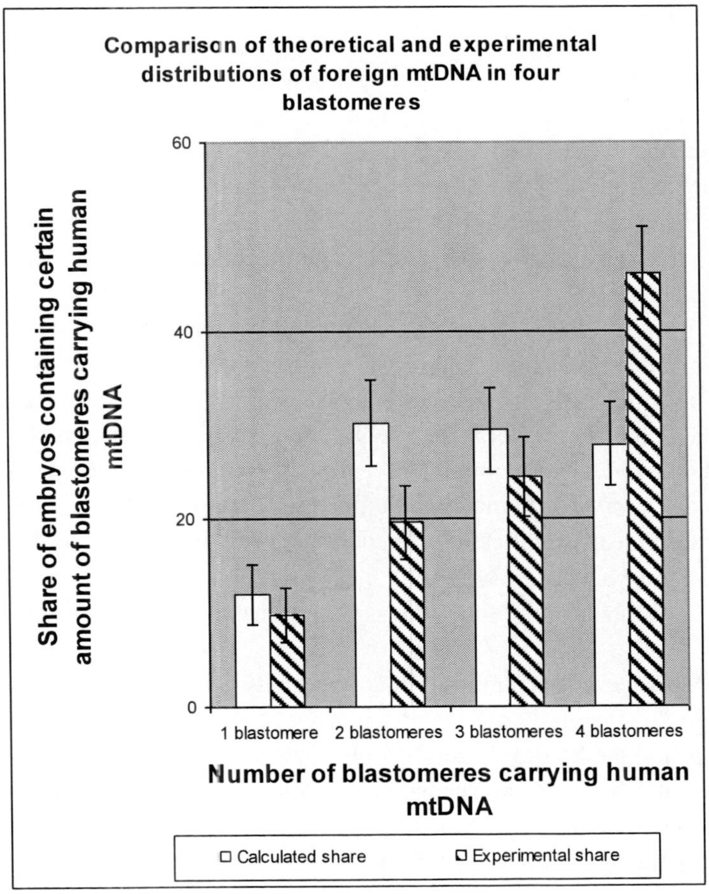

Figure1. Comparison of theoretical and experimental distributions of foreign mtDNA in four blastomeres.

To achieve a more precise evaluation of the number of mtDNA segregation units in the entire bulk of human mitochondria injected into a mouse embryo we used mathematical modeling according to Monte Carlo method accounting for Markovian processes [Borovkov, 1986].

The Monte Carlo model implies characterizing a distribution of objects (mtDNA segregation units in our case) with respect to their possible states (in our case becoming a "daughter" blastomere means acquisition of a state clearly different from the "parental" blastomere). The distribution is based on random and independent transition of an object to any of the two possible states (i.e. each mtDNA cluster can find itself in one of the two "daughter" blastomeres originating from the division "parental" cell). This statistical scenario is repeated at every sequential division of an embryo.

Using Markov's chains allows introducing a correction for the reciprocal linkage of the objects studied. In our case it reflects the situation when mtDNA clusters do not undergo exhaustive mixing within the time between the injection of mitochondria into a zygote and the first division or between the two sequential divisions. In such a case two Markov's coefficients are introduced describing the state of an object *n* depending on the state of an *n-1* object. In our case such objects are mtDNA clusters. Let us term the daughter cell originating from the mtDNA cluster-containing moiety of a zygote as the "old" blastomere and the other one as the "new" blastomere. Then $P_{1\text{-}1}$ coefficient means the probability that the cluster *n* will be transferred from the cytoplasm area giving origin to the "old" blastomere to what will become the "new" blastomere if the *n-1* cluster has made the same way. Respectively, $P_{2\text{-}2}$ means the probability that the cluster n will leave the future "new" blastomere-to-be cytoplasm to the portion of cytoplasm from which the "old" blastomere will be formed if the *n-1* cluster has made the same way. In our case the values are limited to $P_{1\text{-}1}=P_{2\text{-}2}\geq 0.5$. One should notice that when $P_{1\text{-}1}=P_{2\text{-}2}=0.5$ the linkage of clusters is absent, hence, we have the pure Monte Carlo model.

Besides, the model allows describing a situation when in some part of murine zygotes injected with human mitochondria the organelles remain within vacuoles formed of plasma membrane. In this latter case this share of embryos would be by default included in the quota of embryos where human mtDNA was found in a single blastomere as a predetermined statistical percentage. The rest amount of embryos (without mtNDA-containing vacuoles) were subject to the Monte Carlo model including the Markov's chains.

To test the absence of significant differences between the model distribution obtained and the experimental data we used χ^2 criterion.

Our model accounted for the following assumptions.

1. No exhaustive cluster "mixing" occurred between an injection and the subsequent cleavage i.e. the grade of clusters' linkage at the first division is higher than at the next one and the Markov's coefficients are $P_{1\text{-}1}=P_{2\text{-}2} > 0.5$.
2. Between the first and the second division the "mixing" of segregation units occurs, which means that the clusters' linkage is absent at the second and the following divisions and the Markov's coefficients are $P_{1\text{-}1}=P_{2\text{-}2}=0.5$.
3. Human mitochondria remain within the vacuoles of plasma membrane in 10% operated embryos or less.

When the modeled distributions were compared with the experimental data we obtained a range of permissible values for amount of clusters (segregation units). The latter was in the range from 8-9 to 12-13 clusters depending on the value of $P_{1\text{-}1}=P_{2\text{-}2}$ and on account of the mentioned above assumptions. The minimum χ^2 that means the highest compatibility of the model and the experimental data was obtained when the number of clusters was 10-11.

Thus we have shown that about 10 segregation units can be formed in the entire bulk of human mitochondria (ca. $5\text{x}10^2$) injected into a zygote. It should be repeated that such a number was calculated exclusively for the human mitochondria administered in a mouse zygote. The segregation units carrying solely murine mtDNA were not accounted for.

The data obtained concern the very first stages of realization of a complicated mechanism known as cytoplasmic heredity that is inheritance of traits coded by mtDNA. Obviously a thorough investigation of the peculiarities of mitochondrial genome distribution at further stages of embryonic development is needed. It seems likely that the approach we used may be an efficient tool in elucidating an algorithm of mtDNA distribution if such exists.

We observed noticeable differences among the frequencies of detecting human mtDNA in various murine tissues. To illustrate these differences it should be noted that of 111 animals examined not a single tail sampled contained foreign mtDNA, while 15 mice carried human mitochondrial genome in liver. We compared the frequencies of occurrence of human mtDNA in various murine organs with average frequency that should be expected if the chances of foreign mtDNA to get into all organs of transmitochondrial mice were equal. In the latter case the expected frequency of human mtDNA detection would be ca. 8 cases of 111 for each organ examined.

We used both $\chi2$ and $\chi2$ with Yates's correction tests providing the comparison of an experimental distribution with a theoretical one to analyze the difference between a modeled case and the real findings in each animal organ.

Both criteria showed that for such organs as liver where human mtDNA was found in 15 mice of 111, gonads (14 of 111) and skeletal muscles (also 14 of 111) the frequency of foreign mtDNA detection was significantly higher than the mean value. For liver $\chi2$ was 6.60, $p=0.010$, while $\chi2$ with Yates's correction had the value 5.69, $p<0.05$. The same criteria provided values 4.85, $p=0.028$ and 4.08, $p<0.05$ for gonads and skeletal muscles.

In contrast, for such organs as tail where human mtDNA was never revealed, for both spleen and bladder where it was spotted in 1 of 111 animals examined, and for uterus (2 of 111 cases) the frequency of foreign mtDNA detection was significantly lower than the mean value. For tail $\chi2$ was 8.62, $p<0.01$, while $\chi2$ with Yates's correction had the value 7.58, $p<0.01$. For both spleen and bladder the same criteria provided values 6.60, $p=0.010$ and 5.69, $p<0.05$. The respective values calculated for uterus were 4.85, $p=0.028$ and 4.08, $p<0.05$.

One should notice that at least two organs among those in which foreign mtDNA was revealed with significantly higher frequency i.e. liver and skeletal muscles are the target organs/tissues often carrying excessive amounts of mutant mtDNA copies in a number of mitochondrial diseases [Wallace, 1999]. Hence, our results can be considered as obtaining a model somehow mimicking the process of such diseases' development. Our results are in line with those obtained in a study of foreign mitochondria distribution in cloned mice [Inoue *et al.*, 2004].

The significantly higher frequency of foreign mtDNA occurrence in gonads, on the one hand, explains the stably high quote of transgenic animals in several generations and, on the other hand, gives an indication that the model we propose will be securely reproduced in the future experiments.

We analyzed the frequencies of human mtDNA detection in various organs of mice belonging to different comparison groups. Each time the grouping was different i.e. according to the lineage of animals, their sex, the generation to which they belonged. We also formed a group of mice who carried human mtDNA in a single organ and another group of animals who had it in several organs.

To probe the differences among the groups we used $\chi2$ and $\chi2$ with Yates's correction tests and Fischer's precise one-sided and two-sample tests. A difference was proven if its significance was shown by $\chi2$ with Yates's correction test and by Fischer's precise two-sample test. If the results of the two tests gave a discrepancy, those provided by Fischer's precise two-sample test were considered reliable as the test is more adequate in our case (i.e. small groups and less than five objects in a portion compared).

The frequencies of human mtDNA detection in various organs of mice from different pedigrees were analyzed. To simplify further explanations we term mouse pedigrees along generations F_0-F_3 as 'lines'. Pairwise comparing of "ample" lines (those in which nine and more animals were examined) showed that the quota of mice carrying foreign mtDNA in skeletal muscles was significantly higher in Line 61 where it was PCR-detectable in 5 of 19 animals examined (26.32±10.10%) as compared to Line 63 where human mtDNA was not revealed in any of 36 mice analyzed. The significance of the difference between Lines 61 and 63 was confirmed by both $\chi 2$ with Yates's correction test ($\chi 2$=7.48, p=0.006) and Fischer's precise two-sample test (*p*=0.003).

Besides, it should be noted that the distribution among the animals in Line 63 where human mtDNA was not found in skeletal muscles is also significantly different from the total pattern of foreign mtDNA occurrence in muscles of mice from all other lines analyzed. The significance of the difference was demonstrated by both $\chi 2$ with Yates's correction test ($\chi 2$=6.09, p=0.014) and Fischer's precise two-sample test (*p*=0.004).

Similar pattern was observed in Line 63 with respect to the occurrence of human mtDNA in the heart of transgenic mice where it was found in 1 of 36 animals (2.78±2.74%). When this share was compared with the total distribution among all other murine lines the obtained value of Fischer's precise two-sample test (p=0.099) was above the 95% threshold of significance, which does not make the difference reliable. However, this value is sufficiently small to regard the pattern obtained as a tendency with 90% probability.

On the contrary, in gonads of mice from Line 63 human mtDNA was revealed more frequently than its average occurrence in all other lines i.e. we found it in 8 of 36 animals studied (22.22±6.93%). When this share was compared with the mean quota for all other lines analyzed the value of Fischer's precise two-sample test (p=0.063) also did not confirm the difference at the 95% level of significance. However, since the latter was above 93%, this tendency should not be neglected.

It can be concluded that at least one organ (possibly two-three) in mice belonging to different pedigrees/lines becomes the target for foreign mtDNA with different frequencies. Therefore, we evidenced some heterogeneity of murine population that might be indicative either of the "founder effect" or of certain peculiarities of foreign mtDNA segregation in sequential generations of transgenic animals. The first hypothesis is supported by the fact that the only organ/tissue containing human mtDNA in the F_0 female that gave origin to Line 61 were skeletal muscles. It is in this pedigree the percentage of animals carrying human mtDNA in muscles was significantly higher than in Line 63 the progenitress of which had no foreign mtDNA in that tissue.

When frequencies of human mtDNA occurrence in various organs of transmitochondrial mice in sequential generations were compared it appeared that the quota of animals carrying foreign mitochondrial genome in skeletal muscles goes down from F_0 to F_3. In generations F_2 and F_3 the difference with F_0 is highly significant. In F_0 of 24 mice examined, 7 animals carried human mtDNA (29.17±9.28%). Among the animals of F_2 only 1 of 30 (3.33±3.28%) had human mtDNA in muscles, while in F_3 it was not found in the same tissue of any of 14 mice studied. The significance of the difference between F_0 and F_2 was shown both by $\chi 2$ with Yates's correction test ($\chi 2$=5.15, p=0.023) and Fischer's precise two-sample test (p=0.016). In contrast, upon comparing F_0 with F_3 the significant difference was found only by Fischer's precise two-sample test (p=0.033). However, it is properly this test that should be considered as the most adequate in the latter case (with small samplings and a zero share in one of those).

A very similar pattern was observed with respect to foreign mtDNA occurrence in the heart of transgenic mice. In F_0 it was detected in 6 of 24 animals studied (25.00±8.84%), while of 30 mice in F_2 it was found only in 1 (3.33±3.28%). Not a single mouse of 14 analyzed in F_3 carried human mtDNA.

Statistical evaluation showed that the frequencies of foreign mtDNA occurrence in the heart differed significantly between F_0 and F_2 as shown by Fischer's precise two-sample test (p=0.036) that is the most appropriate method for comparing our samplings, as mentioned above. When F_0 and F_3 were compared this test gave p=0.067 which is above the threshold providing 95% significance. The value obtained does not allow regarding the difference as reliable, though it is low enough to speak of a tendency towards that difference with 93% significance. Besides, the lack of significance may be due to a relatively small amount of mice examined in F_3. In such a case the tendency observed will be confirmed when sampling is increased.

It is interesting that the percentage of mice carrying foreign mtDNA in the liver grows from F_1 to F_3. Only 2 of 42 animals analyzed in F_1 had human mtDNA (4.76±3.29%), while the respective figures in F_3 are 4 of 14 (28.57±12.07%). The significance of the difference in this case was shown both by $\chi 2$ test with Yates's correction ($\chi 2$=3.98, p=0.046) and Fischer's precise two-sample test (p=0.029).

It should be noted that we observed not just a trend to an increase or a decrease of the share of organisms carrying foreign mtDNA in some organ, but the oppositely directed processes in different organs. These processes developed whereas the same share of transgenic animals remained in several generations (comparing the quotas of transmitochondrial animals carrying human mtDNA at least in one organ revealed no significant differences among generations).

When the frequencies of spotting human mtDNA in various murine organs depending on a mouse sex were analyzed it appeared that foreign mitochondrial genome was significantly more frequently found in males' intestine (11 of 61 animals, which is 18.3±4.92%) than in the same organ of females (1 of 38 mice, which is 2.63±2.60%). The significance of this difference was shown both by $\chi 2$ test with Yates's correction ($\chi 2$=3.87, p=0.049) and Fischer's precise two-sample test (p=0.026). However, it is hard to find any biological explanation to the phenomenon observed.

It was noticed that of 68 transgenic mice obtained 46 had human mtDNA in one organ, while the rest 22 animals carried it in several (2-6) organs. We compared the frequencies of revealing human mtDNA in various murine organs in two groups of animals. The first group was composed of mice that had human mtDNA in one organ. Mice of the second group carried human mitochondrial genome in several organs. Of 22 mice in the latter group human mitochondrial genome was significantly more frequent in skeletal muscles (9 animals, which is 40.91±10.48%), heart (7 mice, which is 31.82±9.93%), lungs and brain (6 mice each time, which is 27.27±9.50%). Among 46 mice carrying human mitochondrial genome in a single organ that was found equally often in muscles and heart (5 animals, which is 10.87±4.59%), less frequently in lungs (2 mice i.e. 4.35±3.01%) and in brain (1 mouse, which makes 2.17±2.15%).

It is interesting that carrying foreign mtDNA in a single organ happens relatively seldom. Of 7 mice that had it in brain 6 were carrying it also in other organs. Similarly, of 8 mice with human mtDNA in lungs and of 14 animals that had it in skeletal muscles human mitochondrial genome was revealed in other organs of 6 and 9 mice, respectively. Among 12 mice with foreign mtDNA in heart 7 animals carried it in other organs as well.

The difference between the two groups of animals (i.e. with foreign mtDNA in one and in several organs) was significant with respect to every organ taken for comparison. This means that for the brain, lungs and muscles $\chi2$ test with Yates's correction gave the respective figures: $\chi2=7.62$, $p=0.006$; $\chi2=5.49$, $p=0.019$; and $\chi2=6.48$, $p=0.011$. In Fischer's precise two-sample test those were $p=0.004$, $p=0.012$ and $p=0.008$, respectively. With respect to the heart as a target organ only the latter test showed significant difference between the two groups of animals ($p=0.046$).

Thus we revealed the organs that significantly more often become targets for foreign mtDNA when the latter is distributed among several organs. Such are lungs, brain, skeletal muscles and heart. This conclusion is indirectly supported by the fact that these four organs were quickly united within the same group when cluster analysis was carried out.

We accomplished the cluster analysis of organs' similarity according to the frequencies of occurrence of human mtDNA. Average-link method was used for this analysis.

Mean arithmetic distances between an object of one cluster and every object of another cluster were calculated at every step. An object was added to the cluster if the calculated arithmetic mean was smaller than a mean distance to any other cluster.

The starting distance between the objects (i.e. mouse organs) was the inverse value of the frequency of simultaneous occurrence of foreign mtDNA in those. The algorithm was as follows.

1. foreign mtDNA was simultaneously detected in certain organs of five animals;
3. foreign mtDNA was simultaneously detected in certain organs of three animals;
4. foreign mtDNA was simultaneously detected in certain organs of two animals;
5. foreign mtDNA was once detected in those organs;
6. foreign mtDNA was never simultaneously detected in those organs;

Cluster analysis allowed uniting first the heart and the skeletal muscles where foreign mtDNA was simultaneously detected five times (Figure 2). This result is likely to be explained by the similarity of these organs consisting mainly of muscular tissue.

Next step resulted in joining gonads and kidneys in the same cluster, while lungs and brain joined another one. In each pair foreign mtDNA was detected three times. Joining within the same cluster gonads and kidneys with subsequent adding to them intestine and stomach is in line with traditional view on embryological development of these organs [Carlson, 1981].

Uniting lungs and brain is less unequivocal. Yet, one should take into account the results of statistical analysis showing that these organs are among such where foreign mtDNA is detected significantly more often when it is simultaneously found in other organ(s). Similarly we can explain the uniting in one cluster of the "lungs/brain" group and the "heart/skeletal muscles" group in the final dendrogram. The reason why uterus joined the same cluster might lie in the marked share of muscular tissue in this organ.

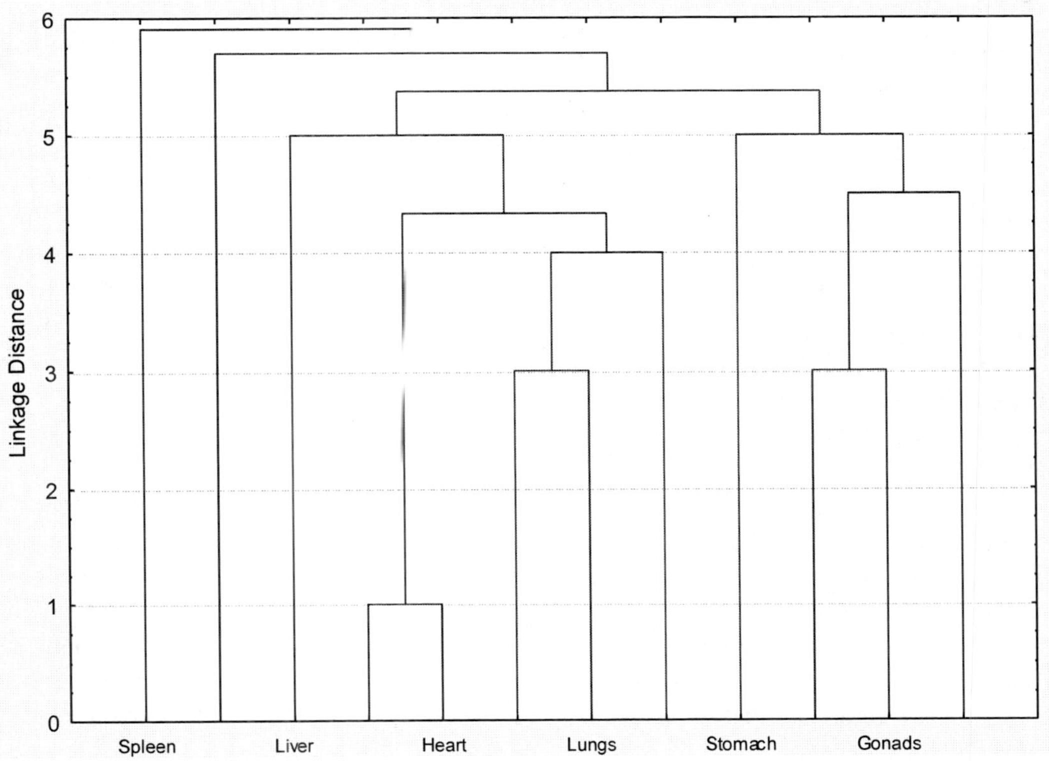

Figure 2. Dendrogram illustrating the grouping of organs into clusters by calculating mean arithmetic distances deduced from frequencies of foreign mtDNA simultaneous occurrence.

CONCLUSION

Speculating on the conceptual model obtained in our experiments one should notice that it has at least one feature uncommon to other models developed so far i.e. human mtDNA introduced in relatively small amounts does not cause any noticeable changes of phenotype, but at the same time is subject to segregation. It seems probable that no negative selection is directed against this foreign mitochondrial genome in murine tissues in contrast to what has been recently shown for mtDNA carrying severe mutations.

From all written above we can make at least three major conclusions. Firstly, it is clearly seen that with high probability foreign mtDNA is unevenly distributed among the cells of a division mouse embryo. The phenomenon of its restricted penetration into each cell of a developing organism cannot be explained by the peculiarities of the method used i.e. of microinjection. In fact, this might have a possible effect on the proportion of mtDNA transferred to a daughter cell after the first cleavage of a zygote, though as shown by our results, such a mechanism can underlie a noticeably smaller share of all cases theoretically possible. The next division into four cells should not have any 'manipulative' aftermath of e.g. non-uniform spreading of administered human organelles in the cytoplasm of the one-cell embryo. We did not obtain unequivocal evidence of the destiny of foreign mitochondria during the first two cleavages, except for indirect evidence of their likely fusion with the host (i.e. mouse) organelles as the MitoTracker fluorescence of the injected mitochondria rapidly decreased in course of culturing an embryo. Therefore, only the distribution of mtDNA itself is subject for discussion.

Despite recent assumptions of mtDNA uniform distribution among mammalian blastomeres [Poulton and Marchington, 2002] and some data supporting this notion [Jenuth *et al.*, 1996], there are well-documented arguments in favor of non-uniform distribution of mitochondria administered into an oocyte by injection [Van Blerkom *et al.*, 1998; 2000]. Similar results were published in our preliminary paper [Kustova *et al.*, 2008]. Moreover, some data were obtained evidencing disproportional distribution of mitochondria among daughter blastomeres of a human embryo [Van Blerkom *et al.*, 2000; Barrit *et al.*, 2001]. In line with these data the results of our study show that no matter how uniformly the injected organelles were distributed in the cytoplasm of zygotes foreign mtDNA not infrequently does not penetrate 25% or even half of the cells constituting an embryo at the earliest stages of its development.

On account that in our experiments human mtDNA mimicked some kind of mutant species one can suggest that similar scenario is applicable to the development of an organism that is destined to manifest the symptoms of an OXPHOS disease upon accumulation of mutant mtDNA copies in certain tissues/organs. In other words, segregation of mtDNA probably occurs also in embryogenesis, which is the second substantial conclusion. The likelihood that mtDNA segregation occurs while an embryo undergoes sequential cleavages means that a future OXPHOS disease may be stipulated by the early events taking place in prenatal development.

This hypothesis provides an explanation for the asymmetric distribution of foreign (and possibly mutant) mtDNA among tissues of an adult mammalian organism. However, it leaves untouched a vast moiety of unexplored mechanisms that ensure a targeted delivery of such mtDNA into every separate cell, tissue and organ.

The third important conclusion from our results is as follows. It seems likely that the preferable target tissues/organs exist; at least this is true for mice. By now their list is limited to the skeletal muscles, liver and gonads. In a previous pilot study liver also was among the most frequent targets [Sokolova *et al.*, 2004]. The same effect was shown in experiments on cloned mice where donor mtDNA was most frequently found in the liver, as mentioned above [Inoue *et al.*, 2004]. Together with that important observation our results pave the way to elaboration of an approach to the targeted delivery of a desired mtDNA species to selected tissues of an experimental animal. This, in turn, might allow obtaining various phenotypes, some of which are likely to resemble the manifestations of OXPHOS diseases in humans.

REFERENCES

[1] Anderson, S., Bankier, A. T., Barrell, B. G., de Bruijn, M. H., Coulson, A. R., Drouin, J., Eperon, I. C., Nierlich, D. P., Roe, B. A., Sanger, F., Schreier, P. H., Smith, A. J., Staden, R. & Young, I. G. (1981). Sequence and organization of the human mitochondrial genome. *Nature, 290,* 457-465.

[2] Ashley, M. V., Laipis, P. J. & Hauswirth, W. W. (1989). Rapid segregation of heteroplasmic bovine mitochondria. *Nucleic Acids Res., 17,* 7325-7331.

[3] Bass, M. G., Sokolova, V. A., Kustova, M. E., Grachyova, E. V., Kidgotko, O. V., Sorokin, A. V. & Vasilyev, V. B. (2006). Assaying the probabilities of obtaining maternally inherited heteroplasmy as the basis for modeling OXPHOS diseases in animals. *Biochim. Biophis. Acta, 1757,* 679-685.

[4] Barrit, J. A., Brenner, C. A , Malter, H. E. & Cohen, J. (2001). Mitochondria in human offspring derived from ooplasmic transplantation. *Human Reprod., 16,* 513–516,

[5] Bell, G. I., Caram, J. H. & Rutter, W. J. (1981). Polymorphic DNA region adjacent to the 5'-end of the human insulin gene, *Proc Natl Acad Sci USA, 78,* 5759-5763.

[6] Bibb, M. J., van Etten, R. A., Wright, C. T., Walberg, M. W. & Clayton, D. A. (1981). Sequence and gene organization of mouse mitochondrial DNA. *Cell, 26,* 167-180.

[7] Bogenhagen, D. F., Wang, Y., Shen, E. L. & Kobayashi, R. (2003). Protein components of mitochondrial DNA nucleoids in higher eukaryotes. *Mol. Cell. Proteomics, 2,* 1205-1216.

[8] Bogenhagen, D. F., Rousseau, D. & Burke, S. (2008). The layered structure of human mtDNA nucleoids. *J. Biol. Chem.* 283, 3665-3675, 2008.

[9] Borovkov, A. A., (1986) *The probability theory*. Moscow, USSR: Nauka. (in Russian).

[10] Brewer, L. R., Friddle, R., Noy, A., Baldwin, E., Martin, S. S., Corzett, M., Balhorn, R. & Baskin, R. J. (2003). Packaging of single DNA molecules by the yeast mitochondrial protein Abf2p. *Biophys. J. 85,* 2519-2524.

[11] Carlson, B. M. (1981). *Patten's foundations of embryology*. New York, USA: McGraw-Hill.

[12] Chen, X. J. & Butow, R. A. (2005). The organization and inheritance of the mitochondrial genome. *Nature Rev. Genet., 6,* 815-825.

[13] De Brito, O. M. & Scorrano, L. (2007). Mitofusin-2, a mitochondria-shaping protein with signaling roles beyond fusion. *Antioxidants and Redox Signaling, 10,* 2008; DOI: 10.1089/ars.1934.

[14] de Magalhaes, J. P. (2005). Human disease-associated mitochondrial mutations fixed in nonhuman primates. *J. Mol. Evol.*, 61, 491–497.

[15] Matthews, P. M., Brown, R. M., Morten, K., Marchington, D., Poulton, J. & Brown, G. (1995). Intracellular heteroplasmy for disease-associated point mutations in mtDNA: Implications for disease expression and evidence for mitotic segregation of heteroplasmic units of mtDNA. *Hum. Genet., 96,* 261-268.

[16] DiMauro, S. & Schon, E. A. (2003). Mitochondrial Respiratory-Chain Diseases. *N. Engl. J. Med., 348,* 2656-2668.

[17] Dvoirin, V.V. & Klimenkov, A.A. (1985). *Methods of controlled clinical trials*. Moscow, USSR: Meditsina. (in Russian).

[18] Friddle, R. W., Klare, J. E., Martin, S. S., Corzett, M., Balhorn, R., Baldwin, E. P., Baskin, R. J. & Noy, A. (2004). Mechanism of DNA compaction by yeast mitochondrial protein Abf2p. *Biophys. J., 86,* 1632-1639.

[19] Fu, K., Hartlen, R., Johns, T., Genge, A., Karpati, G. & Shoubridge, E. A. (1996) A novel heteroplasmic $tRNA^{Leu(CUN)}$ mtDNA point mutation in a sporadic patient with mitochondrial encephalomyopathy segregates in skeletal muscle and suggests an approach to therapy. *Hum. Mol. Genet., 5,* 1835-1840.

[20] Glantz, S.A. (1994). *Primer of Biostatistics*. New York – St. Louis – San Francisco – Auckland, USA: McGraw-Hill Inc.

[21] Griparic, L., Kanazawa, T., & Van der Bliek, A. M. (2007). Regulation of the mitochondrial dynamin-like protein Opa1 by proteolytic cleavage. *J. Cell Biol., 178,* 757–764, 2007.

[22] Hermann, G. J., King, E. J. & Shaw, J. M. (1997). The yeast gene, MDM20, is necessary for mitochondrial inheritance and organization of the actin cytoskeleton. *J. Cell Biol., 137,* 141-153.

[23] Hinshaw, J. E. (2000). Dynamin & its role in membrane fission. *Annu. Rev. Cell Dev. Biol., 16,* 483 – 519.

[24] Holt, U., Harding, A. E. & Morgan-Hughes, J. A. (1988) Deletions of muscle mitochondrial DNA in patients with mitochondrial myopathies. *Nature, 331,* 717–719.

[25] Howell, N., Kubacka, I. & Mackey, D. A. (1996) How rapidly does the human mitochondrial genome evolve? *Am. J. Hum. Genet., 59,* 501-509.

[26] Howell, N. (1999) Human mitochondrial diseases: answering questions and questioning answers. *Intern. Rev. Cytol., 186,* 49-116.

[27] Inoue, K., Ogonuki, N., Yamamoto, Y., Takano, K., Miki, H., Mochida, K. & Ogura, A. (2004). Tissue-specific distribution of donor mitochondrial DNA in cloned mice produced by somatic cell nuclear transfer. *Genesis, 39,* 79–83.

[28] Jenuth, J., Peterson, A., Fu, K. & Shoubridge, E. (1996). Random genetic drift in the female germline explains the rapid segregation of mammalian mitochondrial DNA. *Nature Genet., 14,* 146–151.

[29] Khan, S. M., Smigrodzki, R. M., & Swerdlow, R. H. (2007). Cell and animal models of mtDNA biology: progress and prospects. *Am J. Physiol. Cell Physiol., 292,* 658–669.

[30] Kasahara, A., Ishikawa, K., Yamaoka, M., Ito, M., Watanabe, N., Akimoto, M., Sato, A., Nakada, K., Endo, H., Suda, Y., Aizawa, S. & Hayashi, J. I. (2006). Generation of trans-mitochondrial mice carrying homoplasmic mtDNAs with a missense mutation in a structural gene using ES cells. *Hum. Mol. Genet., 15,* 871–881.

[31] Kaufman, B. A., Kolesar, J. E., Perlman, P. S. & Butow, R. A. (2003). A function for the mitochondrial chaperonin Hsp60 in the structure and transmission of mitochondrial DNA nucleoids in *Saccharomyces cerevisiae*. *J. Cell Biol., 163,* 457-461,.

[32] Kucej, M., Kucejova, B., Subramanian, R., Chen, X. J. & Butow, R. A. (2008). Mitochondrial nucleoids undergo remodeling in response to metabolic cues. *Journal of Cell Science*, 121, 1861-1868.

[33] Kustova, M. E., Sokolova, V. A., Bass, M. G., Zakharova, F. M., Sorokin, A. V., & Vasilyev, V. B. (2008). Distribution of foreign mitochondrial DNA during the first splittings of transmitochondrial mouse embryos. *Cell and Tissue Biology, v 2. No 6,* 609-613.

[34] Li, F. Y., Cuddon, P. A., Song, J., Wood, S. L., Patterson, J. S., Shelton, G. D. & Duncan, I. D. (2006). Canine spongiform leukoencephalomyelopathy is associated with a missense mutation in cytochrome *b*. *Neurobiol. Dis., 21,* 35–42.

[35] Malka, F., Lombes, A. & Rojo, M. (2006). Organization, dynamics and transmission of mitochondrial DNA: focus on vertebrate nucleoids. *Biochim. Biophys. Acta, 1763,* 463-472.

[36] McKenzie, M., Trounce, I.A., Cassar, C.A. & Pinkert, C.A. (2004). Production of homoplasmic xenomitochondrial mice. *Proc. Natl Acad. Sci. USA, 101,* 1685–1690.

[37] Nakada, K., Inoue, K., Ono, T., Isobe, K., Ogura, A., Goto, Y. I., Nonaka, I. & Hayashi, J. I. (2001) Inter-mitochondrial complementation: mitochondria-specific system preventing mice from expression of disease phenotypes by mutant mtDNA. *Nature Med., 7,* 934–940.

[38] Nakada, K., Sato, A., Sone, H., Kasahara, A., Ikeda, K., Kagawa, Y., Yonekawa, H. & Hayashi, J. I. (2004).Accumulation of pathogenic DmtDNA induced deafness but not diabetic phenotypes in mito-mice. *Biochem. Biophys. Res. Commun., 323,* 175–184.

[39] Nagao, Y., Totsuka, Y., Atomi, Y., Yonekawa, H. & Imai, H. (1998). Effect of different type of mitochondrial DNA on preimplantation embryonic development in the mouse, *J. Reprod. Dev., 44,* 129-134.

[40] Nogawa, T., Sung, W. K., Jagiello, G. M. & Bowne, W. (1988). A quantitative analysis of mitochondria during fetal mouse oogenesis. *J. Morphol., 195,* 225-234.

[41] Olichon, A., Emorine, L. J., Descoins, E., Pelloquin, L., Brichese, L., Gas, N., Guillou, E., Delettre, C., Valette, A., Hamel, C. P., Ducommun, B., Lenaers, G. & Belenguer, P. (2002). The human dynamin-related protein OPA1 is anchored to the mitochondrial inner membrane facing the inter-membrane space. *FEBS Lett., 523,* 171-176.

[42] Penman, S. (1995). Rethinking cell structure. *Proc. Natl. Acad. Sci. USA, 92,* 5251-5257.

[43] Poulton, J. & Marchington, D. R. (1996). Prospects for DNA-based prenatal diagnosis of mitochondrial disorders. *Prenatal Diag., 16,* 1247-1256.

[44] Poulton, J. & Marchington, D. R. (2002) Segregation of mitochondrial DNA (mtDNA) in human oocytes and in animal models of mtDNA disease: clinical implications. *Reproduction, 123,* 751–755.

[45] Sambrook, J., Fritsch, E. F. & Maniatis, T. (1989). *Molecular Cloning: a Laboratory Manual.* New York, USA: Cold Spring Harbor Lab Press, v 2, pp. 9.16-9.19.

[46] Smith, L. C. & Alcivar, A. A. (1993). Cytoplasmic inheritance and its effects on development & performance. *J. Reprod. Fertil. (Suppl.), 48,* 31-43.

[47] Sokolova, V. A., Kustova, M. E., Arbuzova, N. I., Sorokin, A. V., Moskaliova, O. S., Bass, M. G. & Vasilyev, V. B. (2004). Obtaining mice that carry human mitochondrial DNA transmitted to the progeny. *Mol. Reprod. Dev., 68,* 299-307.

[48] Torroni, A., Schurr, T. G., Yang, C. C., Szathmany, E. J. E., Williams, R. C., Schanfield, M. S. & Troup, G. A. (1992). Native American mitochondrial DNA analysis indicates that the Amerind of the Nadene populations were founded by two independent migrations. *Genetics, 130,* 153-162.

[49] Tourte, M., Besse, C. & Mounolou, J. C. (1991). Cytochemical evidence of an organized microtubular cytoskeleton in *Xaenopus laevis* oocytes: Involvement in the segregation of mitochondrial populations. *Mol. Reprod. Dev., 30,* 353-359.

[50] Van Blerkom, J., Sinclair, J. & Davis, P. (1998). Mitochondrial transfer between oocytes: potential applications of mitochondrial donation and the issue of heteroplasmy. *Human Reprod., 13,* 2857–2868, 1998

[51] Van Blerkom, J., Davis, P. & Alexander, S. (2000). Differential mitochondrial distribution in human pronuclear embryos leads to disproportionate inheritance between blastomeres: relationship to microtubular organization, ATP content and competence. *Human Reprod., 15,* 2621–2633.

[52] Van der Bliek, A. M. (1999). Functional diversity in the dynamin family. *Trends Cell Biol., 9,* 96 – 102.

[53] Vasilyev, V. B., Sokolova, V. A, Sorokin, A. V., Bass, M. G., Arbuzova, N. I., Patkin, E. L., Golubkov, V. I., Dyban, A. P. & Gaitskhoki, V. S. (1999). Persistence of human mitochondrial DNA throughout the development to the blastocyst of mouse zygotes microinjected with human mitochondria. *Zygote, 7,* 279-283.

[54] Wallace, D. C. (1999). Mitochondrial diseases in man and mouse. *Science, 283,* 1482-1488.

[55] Zeviani, M. & Antozzi, C. (1997). Mitochondrial disorders. *Mol. Hum. Reprod., 3,* 133-148.

In: Horizons in DNA Research, Volume 1
Editor: Jason R. Chesterton, pp. 171-186

ISBN: 978-1-60876-968-1

Chapter 7

TRACE DNA ANALYSIS

Kaye N. Ballantyne*
Department of Forensic Molecular Biology, Erasmus University Medical Center, Rotterdam, The Netherlands.

ABSTRACT

Trace DNA has become a large part of the average forensic laboratories' workload. Remarkably low DNA amounts (<100pg) have been successfully analysed to obtain profiles from a wide range of sample types. Touched objects constitute the most common source of trace DNA, but any type of biological material present in low amounts may be considered as trace, including minute blood deposits, saliva residue on partially consumed food, or even epithelial cells from the interior surface of condoms. In addition to supplementing existing analysis techniques in serious crime cases, trace DNA can allow investigation of volume crime cases such as burglary or vehicle theft, where previously DNA evidence was not considered usable. However, despite the widespread use of trace DNA, at present there are very few specific validated methods. This has lead to controversy in the use of trace DNA, and particularly the low copy number amplification technique. It has been established that the use of existing methodology (developed for high-copy number samples) leads to significant levels of artefacts with trace DNA, including allele drop-out and drop-in, stutter, and allelic/locus imbalance. To minimize these, there are numerous modifications that can be made to existed methods to increase the success of trace DNA analysis. These include reduced extraction volumes, increased cycle number, reduced PCR volume, and increased injection time for capillary electrophoresis. In addition, new research shows that the introduction of techniques such as whole genome amplification, molecular crowding, and post-PCR purification can significantly increase success rates with trace DNA. It is clear that each step of the analysis procedure (including collection, extraction, amplification and fragment detection) can, and should, be optimized with regards to trace DNA. However, the use of increasingly sensitive trace DNA analysis techniques must bring an increasing awareness of the potential for contamination, both within the laboratory and at crime scenes. In particular, this has implications for the analysis of trace DNA from cold cases, which

* Corresponding author: E-mail: k.ballantyne@erasmusmc.nl

were not collected or stored with highly sensitive DNA detection techniques in mind. Although it is certain that trace DNA will continue to be used within forensic biology, there may need to be modifications to a wide range of practices to ensure accuracy and reliability.

INTRODUCTION

Trace (n): 1. a sign or evidence of some past thing. 2. A minute and often barely detectable amount or indication. 3. An amount of a chemical constituent not always quantitatively determinable because of minuteness. (Merriam-Webster Dictionary, 2009)

The use of trace amounts of evidentiary material for DNA profiling is now common within forensic science, ever since the first description in 1997 of the production of DNA profiles from fingerprints (van Oorschot and Jones 1997). Following this, it has been demonstrated that trace DNA can be obtained from a wide variety of substrates, including touched objects (van Oorschot and Jones 1997), drinking containers (Abaz et al 2002), bedding (Petricevic et al 2006), shoes (Bright and Petricevic 2004), documents (Sewell et al 2008) and ropes (Wickenheiser 2002), amongst many other items. As profiling techniques become more sensitive, the number and type of cases that can now be considered for DNA profiling has been considerably expanded, particularly when considering volume crime such as auto theft or burglary. Trace DNA has now become both a large research field within forensic genetics, and an important issue for many laboratories and courts. As increasing numbers of trace DNA samples are analysed as part of routine case work, and more challenges are raised to its theory and use, it has become apparent that there is still a considerable amount to be learnt, researched and validated within this field. This chapter will provide an overview of the current state of trace DNA knowledge and usage within forensic biology, and describe some of the newer methods that can be used to increase success rates from these minute samples.

Before beginning, it will be useful to define exactly what is meant by the term trace DNA. There has come to be a large number of terms associated with minute quantities of DNA within the realm of forensic molecular biology. Trace DNA is sometimes used to refer to *any* small quantity of biological material, regardless of cellular origin (such as this chapter, or Wickenheiser 2002). In other instances, trace DNA may refer to only epithelial cells transferred by skin contact with items. This may also be referred to as 'contact DNA', or 'touch DNA'. Further terms often seen are 'Low Copy Number', and also 'Low Template DNA typing' (Gill et al 2001, Caddy et al 2008), which are commonly used to refer to the specialised analysis procedures (see below), rather than the nature of the sample itself. This terminology refers to any DNA sample containing less than 100pg of amplifiable DNA, regardless of its source (epithelial, salvia, blood, semen etc). For the purposes of this chapter, the term 'trace DNA' shall refer to any minute quantity of cellular material from any source, which yields less than 100pg of amplifiable DNA for analysis. The 100pg limit, although somewhat arbitrary and laboratory/analyst dependant, is commonly regarded as the lower threshold to which the current commercial STR amplification kits (such as Identifiler, SGM Plus and PowerPlex) can be reliably used. However, any user familiar with these kits will have experienced that incomplete profiles may occur from 500pg and less, and conversely that 50pg may result in complete profiles. As such, it may not always be a simple matter to interpret trace DNA results within the casework framework, or in researching effective strategies for the analysis of trace DNA.

ORIGIN AND DEPOSITION OF TRACE DNA

While trace DNA can be from any biological material, it is most commonly seen originating from epithelial cells on touched objects. Any skin contact, be it with hands, ears, legs etc, can result in cellular transfer from the donor to the object. The process of deposition of epithelial cells is essentially a transfer process, where sloughed epithelial cells, oils and sweat are transferred to the surface contacted. In addition to the transfer of nucleated and non-nucleated cells, it has been demonstrated that small amounts of fragmented DNA from apoptotic cells are carried within the sweat and sebaceous oils that are transferred (Kita et al 2008), thus suggesting that STR profiles may be retrievable from touched objects even when no cells are transferred. Crucially, the length of contact is of little consequence to the amount of material deposited – a short contact will deposit similar amounts of material to prolonged contact (Wickenheiser 2002).

There are several factors important in the level of DNA transferred. It has been suggested by several authors that there are 'good' and 'bad' shedders – certain individuals deposit more cells/DNA than others (Lowe et al 2002). However, the effect does not appear to be consistent, with variation occurring within the same individual over time, and also between different hands on the same individual at the same time (Phipps and Petricevic 2007). While it may be true that some individuals are consistently good shedders, it would appear that their percentage in the population is considerably lower than first estimated (45%, Rutty et al 2003), and that the same individual may be both a good and bad shedder at different time points. In addition to variation between (and within) individuals, the substrate onto which the cells are deposited plays a large role. Porous substrates such as paper or wood show in considerably higher transfer rates than non-porous glass or plastic (Alessandrini et al 2003). However, the substrate may also have a considerable effect on the recovery of the deposited DNA, rather than the actual amount deposited.

While the presence of trace DNA on an object is usually indicative of the last person to touch an object, this may not always be the case. There is very little knowledge of the persistence of trace DNA – whether it lasts days, weeks, or months, and the extent to which environmental conditions play a role in persistence. It is conceivable that under the correct conditions, cellular deposits may adhere to surfaces for weeks or months, although empirical research testing persistence is still lacking. In preliminary research, Raymond et al (2008) suggested that the retrieval of DNA from deposits declines rapidly, with the majority not detectable after 1 day, and completely gone after 2 weeks outside

The dearth of knowledge regarding primary transfer rates and persistence of DNA leads to an important issue which is currently facing the trace DNA field. The occurrence of secondary transfer has been a much debated issue for around a decade. While primary transfer (person-person or person-object) is accepted without exception, secondary transfers (person-person-object, person-object-person etc), and the rate at which they occur, are yet to be exhaustively examined. Early papers gave conflicting results, with some demonstrating the phenomenon (van Oorschot and Jones 1997, Petricevic et al 2006, and Polley et al 2006), and others failing to find evidence it occurred frequently enough to be a concern (Ladd et al 1999). However, subsequent studies have shown secondary transfer can occur relatively easily, although estimates of frequency differ widely. Phipps and Petricevic (2007) found only 2% of alleles did not match the primary donor (and were theorised to be transferred from another, secondary donor), while Lowe et al (2002) found transfer rates between 11% and 75%, depending on the 'shedder' status of each individual in the transfer. The occurrence of secondary transfer, and the variation in its occurrence, raises considerable issues for forensic biology and criminology, as discussed below.

CURRENT METHODOLOGY FOR TRACE DNA ANALYSIS

At present, there are no globally recognised, validated procedures for the analysis of trace DNA. The vast majority of laboratories use standard methodologies for all samples, regardless of quantity – all of which have been developed, optimised and validated for use on higher quantities (generally 1ng) of DNA. Despite considerable research, it can be argued that there are no completely new methods specifically developed for trace DNA. However, there have been many modifications developed to existing methods, which can increase success rates. Within this section, current methodologies for each stage of the analysis procedure (collection, extraction, quantification, amplification, detection and profile interpretation/ genotyping) will be described briefly, with a summary of potential improvements for trace DNA.

DNA Collection

The detection and collection of trace DNA at the crime scene could be argued to be one of the most significant factors affecting the success rate of subsequent DNA profiling. Unfortunately, at present there are no defined guidelines on the best methods to use. Fingerprint enhancement methodology is commonly used to detect the presence of epithelial deposits, although dusting powders and ninhydrin may negatively affect subsequent PCR amplifications (Sewell et al 2008, Grubweiser et al 2003). However, there do not appear to be any new methods developed specifically for detecting trace epithelial cells at crime scenes. Additionally, there is no clear consensus on the best methods to collect trace DNA deposits. Different collection matrices, such as cotton cloths, nylon, foam and cotton swabs, adhesive tape, and vacuuming have all been trialled with no clear consensus on the best method for trace DNA. All methods leave considerable amounts of material on the surface of the object, or that which is collected is not able to be retrieved from the matrix for extraction. However, the double wet-dry swabbing method has been shown to be the most efficient of the current methods (Sweet et al 1997, Pang and Cheung 2007), while the use of solvents such as dilute ethanol has also been shown to retrieve more material than sterile water, yet the vast majority of crime scene examiners still use the single swab method with water (Franke et al 2008).

Certainly, one of the best methods to increase the collection of trace DNA from crime scenes and exhibits would be improved education of crime scene examiners and laboratory staff. Knowledge is one of the most powerful tools to increase the collection rate. Further research into the best method to collect, and then recover, the maximal amount of sample would be beneficial. However, teaching crime scene personnel to use the double swab method, whilst rotating the swab around the stain/contact trace would increase collection rates above those currently seen (Raymond et al 2008).

DNA Extraction

It has previously been demonstrated that two commonly used extraction methods, Chelex and phenol chloroform, result in the loss of significant amounts of DNA (van Oorschot et al 2003). Chelex extraction commonly loses 20-76% of the DNA present on a substrate, while phenol-chloroform extractions have 37-75% losses. Although the percentage loss decreases as lower amounts of DNA are extracted, it is for these trace amounts where the loss of any material is the most critical for success. There are conflicting reports on the effectiveness of the newer extraction kits, including silica columns (such as Qiagen's QiaAmp) or silica-coated magnetic bead extraction (such as Promega's DNA IQ). Some reports suggest that they are not as effective with trace DNA samples as simpler Chelex or lysis methods (Schiffner et al 2005), while others suggest that they produce better results (Greenspoon et al 2004).

For most extraction methods, there have been modifications to standard methods developed to attempt to increase the success rate, and particularly the final concentration, of trace DNA extracts. Commonly, this involves reducing the volume of extraction reagents used (for example from 200µl to 50µl in the case of Chelex extractions, Walsh et al 1991). While this effectively concentrates the DNA, it may also result in lower yields due to incomplete recovery of cells from the substrate, or only partial elution of the DNA from silica columns. Alternatively, the introduction of carriers such as salmon sperm DNA or Poly A RNA (see Schiffner et al 2005), can increase the yield from silica-based extractions of very low template amounts (Kishore et al 2006). Carriers in DNA extractions are thought to increase efficiency and recovery rates either by preventing the sample binding to the plastic tube or filters, or by increasing the DNA-silica binding rate by binding to free water molecules, thus preventing them binding to the silica. This may aid in recovering the maximum amount of DNA possible.

With the increasing use of laser-microdissection (LMD) techniques in forensic biology, the use of the direct lysis technique is becoming more possible. Direct lysis, using either proteinase K/SDS or alkaline solutions, has the distinct advantage of having no tube transfers during extraction. For this reason, it is commonly used in reproductive genetics, to avoid the loss of precious cells. Previously, lysis has not been considered a viable option for forensic samples, as the sample is not purified in any way, potentially leaving PCR inhibitors in the sample. However, if cells are microdissected away from the substrate they were deposited on, direct lysis methods can then be used without concern of substantial inhibition. For example, LMD has been successfully used to isolate buccal cells from soil mixtures (Anoruo et al 2008). Direct lysis of the sample within the PCR tube has been shown to minimise allelic drop-out, through the removal/reduction of stochastic sampling effects (Thornhill et al 2001, Thornhill & Snow 2002).

Quantitation of Trace DNA

Quantitation of trace DNA samples is not always regarded as necessary, as if the sample if assumed to have very low template amounts it is customary to simply add the maximum amount possible to the amplification reaction. However, in practice, it may be advisable to quantitate all samples, to prevent over-amplification of samples containing more DNA than expected. Real-time PCR multiplexes such as Quantifiler and Quantifiler Duo (Applied Biosystems), now in common use in forensic laboratories, allow accurate quantitation and an indication of potential PCR inhibitors present in the sample. It is often possible to quantitate as little as 6pg/μl with these kits, offering much greater sensitivity than previous methods. As recommended in the Caddy report (2008), quantitation of trace DNA samples prior to amplification offers much greater control over the amplification, improved results, and can prevent the loss of precious samples through the need for re-amplifications.

Amplification of Trace DNA

At present, most laboratories use standard amplification methodology with trace DNA samples. Most commercially available, and widely used, STR multiplex kits are optimised for use with between 0.5 and 2ng of DNA. While these kits all work exceptionally well with 1ng of high quality DNA, when the template amounts are reduced below 100-200pg, a range of artefacts are commonly seen. Allelic and locus dropouts are commonly seen, allele drop-in (contamination) may be seen, and stutter percentages are often increased (Kloosterman and Kersbergen 2003, Gill et al 2000). Drop-out and stutter are both usually the result of stochastic processes, which may occur at any point in the reaction set-up and analysis. At such low template levels, random fluctuations in the representation of each template molecule in the DNA sample added to the PCR will have a large influence on the resultant STR profile. Some alleles/loci may be over-represented within the aliquot of sample taken for amplification, while others will be under-represented. To compound this, imbalanced amplification between alleles or loci within the reaction itself can add to the bias seen in the final profile. Navidi et al (1992) give a comprehensive mathematical simulation of stochastic fluctuations, while Taberlet et al (1996) give a practical demonstration of the effects of low template amounts. Both papers demonstrate the need for replicate reactions to give statistical accuracy of low template amplifications – at least 10 are recommended for very low template amounts (such as trace DNA amounts).

There is also a certain level of stochastic variation in the sampling of the PCR product for electrophoresis. Certainly, different aliquots of the same PCR product will produce slight variations in peak heights and allele balance. Additionally, when the products are near the baseline detection limit of the capillary electrophoresis instrument, this can result in profiles differing between replicate injections of the same aliquot of PCR product (even when separated by only 45 minutes in time, data not shown). Different machines will have different baseline thresholds, and the correct limits for each machine need to be determined empirically.

Unfortunately, there is little that can be done to reduce the stochasticity of trace DNA amplifications. Most strategies have revolved around increasing the generation of PCR product, or subsequently increasing the concentration/purity of the product following amplification. There are several strategies currently in use, and some contention over the best methods to use. The longest standing method is the Low Copy Number (LCN) technique, pioneered by the Forensic Science Service in the UK. This involves increasing the number of PCR cycles used to amplify the STRs from 28 to 34. With the increased levels of amplification, it is possible to obtain STR profiles from as little as 30-60pg, compared to 300-600pg with the standard 28 cycles. This strategy results in large increases in peak height, with higher percentages of alleles observed (from 0% at 28 cycles to 100% at 34, Gill et al 2000). While this technique has proved very effective for profiling trace DNA samples, there are a number of factors to be aware of when applying this protocol. Increases in artefacts are observed, with peak imbalance and stutter commonly seen (Whitaker et al 2001). Additionally, there is increased mixture and contamination detection, due to the greater sensitivity. Due to allele drop-in and drop-out, it is strongly recommended to perform a minimum of 2 independent amplifications for each sample, allowing the production of a consensus profile without the artefacts. Because there are a number of limitations with this technique, it has not been widely adopted. At present, only the Forensic Science Service in the UK, the Netherlands Forensic Institute, the Switzerland and New York Chief Medical Offices and New Zealand's Environmental Science Research (ESR) routinely use the LCN technique.

Several alternatives to the 34 cycle LCN method have been proposed, including the use of 30 instead of 28 or 34 cycles (Roeder et al 2009) and a 28+6 approach, where additional Taq polymerase is added after 28 cycles, and thermal cycling is continued for a further 6 cycles (Kloosterman and Kersbergen 2003). One advantage to this 28+6 approach is the ability to stop the PCR after 28 cycles, analyse the results, and only perform the additional cycles if required. However, like the standard 34 cycle approach, artefacts are often observed.

There are also many other options to increase the amplification success of trace DNA. Reducing the reaction volume from 50µl to 5 or 10µl can substantially increase the amount of product generated, with a four-fold increase in sensitivity (Gaines et al 2002). When the volume is reduced to 10 or 25µl, there is no significant increase in the level of artefacts, and relative peak height information is retained, allowing accurate mixture interpretation (Fregeau et al 2003, Leclair et al 2003). The DNA sample may also be dried directly in the PCR tube, further increasing the amplification through the addition of large amounts of DNA. The development of chemically structured glass chips, such as Advalytix's AmpliGrid slides, allow the amplification of single or few cells in a total volume of 1µl. Successful amplification of 15 STRs was achieved from only 32pg (6 cells), and partial STR profiles from 4pg (Schmidt et al 2006). As well as having increased sensitivity, the use of chip-PCRs dramatically reduces the volume of sample needed; allowing either concentration of extracts, or increased numbers of replicate reactions to be performed.

It is also possible to alter the chemical components of the PCR to maximise amplification efficiency and success. Altering the concentrations of certain components can substantially change the way a PCR assay responds to various sample types or template amounts. Increasing the amount of Taq polymerase (from 1 unit to 3-5 units) can give higher amplification success, while the use of hot-start polymerases (such as Ampli-Taq Gold from Applied Biosystems) increases specificity (Moretti et al 1998). The addition of bovine serum albumin can aid in overcoming inhibition of the polymerase.

New research has shown that the use of molecular crowders can substantially increase PCR efficiency (Ballantyne et al, manuscript in preparation). For many enzymatic reactions, it can be beneficial to 'crowd' the reaction by adding a large inert polymer. This has the effect of reducing the space available to reactants, and increases the probability of diffusion-constrained reactions between template, primers, dNTPs and polymerase occurring. Previously, crowders such as polyethylene glycol (PEG) have been used to increase the amplification success of whole genome amplification methods (see below, and Ballantyne et al 2008). Therefore, three common crowders, PEG, Ficoll and Dextran were all trialled as potential crowders for commonly used STR multiplexes. As molecular crowding has a complex effect on the biochemistry of reactions, different crowders, and levels of crowding, may have very different effects on different amplification systems.

Following extensive empirical testing of each crowder, it was found that a 1% v/v solution of Dextran crowded amplifications most effectively. Multiple STR multiplexes were tested for amplification success of trace amounts of DNA both with and without crowding with Dextran 70. PowerPlex (Promega), Identifiler, SGM Plus, Yfiler (Applied Biosystems) and the miniSTR multiplex NC01 (Coble and Butler 2005) all showed increased numbers of alleles amplified, and increased peak heights, when the reaction was crowded. Across all multiplexes, there was an average increase of ~2 fold in peak height in amplifications containing less than 100pg. The number of alleles and loci called were also substantially increased, with 2.5 fold increases in the number of alleles seen from 6pg amplifications. The greatest effect was seen with PowerPlex, where up to 9-fold increases were seen in the number of alleles. No increases in artefacts were seen – no instances of allele drop-in were observed, peak balance was retained compared to non-crowded samples, and stutter was not significantly increased. A substantial advantage of this technique is that it does not require any additional steps, or sample manipulations – the dextran can simply be added to, and thoroughly mixed with the PCR master mix. The dextran does not affect capillary electrophoresis, and interpretation of the resultant profiles is simplified due to the increased peak height.

Whole Genome Amplification

All of the previous modifications mentioned work by increasing the efficiency of the amplification, or detection of the product. However, in some cases this may not be enough to generate a usable product. Whole genome amplification (WGA) is a useful technique to increase the amount of DNA available prior to the STR genotyping PCR. A subtype of WGA, multiple displacement amplification (MDA) has been shown to be the most effective for small sample amounts (Ballantyne et al 2007a). This method uses a highly processive enzyme, Phi29, to replicate long (up to 70kb) sequences, with random hexamer primers ensuring coverage across the majority of the genome. Commercial MDA kits, such as GenomiPhi (GE Biosciences), are able to amplify trace DNA quantities to produce increased material for STR profiling, although not with complete success. Although more complete profiles (up to 70% increases in numbers of alleles) were produced from the WGA material, compared to standard protocols, there was also a large increase in the level of artefacts observed. Although allele drop-in was not commonly observed, allele drop-outs were still common, and there was greatly increased peak height imbalance seen (Ballantyne et al 2007a).

A series of optimisation tests were undertaken, both to further increase the amplification success, and to decrease the level of amplification bias seen. Initially, to increase the level of amplification, molecular crowding with PEG400 was used during the WGA reaction. Amplification success (in terms of the numbers of STR alleles observed) nearly doubled with crowding the solution, particularly for amplifications containing 50-100pg (Ballantyne et al 2006). As with standard PCR, it is theorised that crowding the amplification reaction increases the chance of diffusion-limited reactions between components occurring. However, the amplification bias between alleles and loci is still frequently observed with crowded WGA products, as it is an inherent feature of the Phi29 polymerase. Therefore, an altered denaturation strategy was applied, where the initial denaturation phase (prior to WGA) was omitted for half of the DNA template, and performed for the other half. By combining the two reactions, it was possible to drastically reduce the amount of bias seen (Ballantyne et al 2007b). With standard WGA, the average peak height ratio was a very low 39%, indicating substantial bias. When the modified denaturation strategy was used, the bias was substantially reduced, with an average PHR of 68%.

Combining both molecular crowding, and the denaturation strategy, resulted in more complete and even WGA amplifications, with an average 34% increase in STR profile success compared to standard PCR alone (Ballantyne et al 2007c). Although the improved MDA method developed does require additional sample manipulations, it can be used to generate substantial amounts of material from limited samples, allowing repeated testing for STRs, or the use of greater numbers of loci (such as testing autosomal, Y-, X- and mitochondrial markers) on precious crime scene or archived samples.

Post-PCR Detection/Genotyping

In addition to manipulating/optimising all the sample collection, extraction and amplification steps, it is also possible to improve trace DNA genotyping success by optimising the detection of the PCR product. Rather than performing increased cycle numbers, it is also possible to use post-PCR purification to maximise trace DNA amplification success. Silica gel membrane columns such as the MinElute kit (Qiagen) are most commonly used, for their ease of use and effectiveness. In addition to removing excess primers, dNTPs and salts, purification can be used to concentrate the sample to 5-10μl, allowing more product to be injected into the capillary. This procedure gives greater success than the standard, non-purified 28 cycle method, increasing peak heights 3-67 fold, depending on the method of purification used (Smith and Ballantyne, 2007, Forster et al 2008). Although artefacts such as allele drop-in, drop-out, peak imbalance and increased stutter are still observed, they are commonly less severe when the product is purified, compared to 34-cycle products.

Increased injection times for capillary electrophoresis may also be used to increase genotyping success. This effectively introduces more product into the capillary, allowing increased detection and thus greater success. However, this may also increase the artefact level, particularly in terms of stutter allele heights and spectral pull-up. It may be advisable to purify the product prior to increasing the injection time, to prevent excess salts and ions interfering with the electrokinetic injection (Forster et al 2008).

The concentration of PCR product can also be increased prior to injection, either by concentrating the product during purification, or by drying the product completely and re-constituting with formamide. While this provides improved peak heights and genotyping success, it must be balanced against the increased levels of artefacts and inhibition of injection by excess salts and ions.

Interpretation of Trace DNA Profiles

When low amounts of DNA are used for amplification, and high levels of artefacts are present, it is necessary to apply strict interpretation guidelines for genotyping and profile use. Peak height imbalance between alleles and loci can reduce genotyping accuracy, and particularly increase the likelihood of false homozygote calls. Inadvertent contamination (either within the laboratory or at the crime-scene) is more likely to be detected when low levels of DNA and methodological enhancements are used. Mixture genotyping can be extremely difficult when very low levels of DNA are used, due to peak imbalance and increased stutter allele heights. As such, replicate reactions for any trace DNA should be performed to enable elimination of spurious alleles. The use of consensus profiles generated from replicates can aid in determining the 'correct' alleles – only when alleles are duplicated can they be reported (Gill et al 2000). In addition, a new statistical method for determining the stochastic threshold for allele drop-out has been described, allowing the calculation of the risk of false inclusions and exclusions when very low template levels are used. This may allow the use of the most conservative likelihood ratio for potential matches, and the determination of a conservative peak height threshold for homozygote calls that reduces the chance of false inclusions (Gill et al 2009).

Past the genotyping and profile interpretation stages, the trace DNA evidence must be placed into the context of the case. There is a level of concern within the forensic biology community that trace DNA evidence may at times be over-stated in terms of its evidentiary value. A crucial feature of trace DNA is that the origin of the biological material cannot be determined. Unlike visible saliva, semen or blood stains, there is no presumptive test for epithelial cells, and at times the level of biological material may be so low that current tissue origin tests will not provide a result. Therefore, any speculation on the origin of the cells is scientifically invalid, and may unfairly bias the DNA results against the defendant.

Coming Issues for Trace DNA

Although there has already been considerable research and discussion within the community regarding trace DNA analysis and interpretation, there are a number of unresolved issues which may impact its future use in criminal cases. The lack of validated techniques for trace DNA analysis was criticised in the recent Omagh bombing trial (the Queen v Sean Hoey, Northern Ireland, 341/05), leading to a review of the science of low template DNA analysis in the United Kingdom (Caddy et al 2008). This report highlighted the need for training of personnel involved, the need for stricter contamination control and interpretation guidelines, and the need for considerably more research and validation into specific areas of trace DNA (low-template DNA) analysis. The 21 recommendations of the investigating committee, and the newly established Forensic Science Regulator's responses, all appear to be a positive step towards adopting a 'best-practice' approach to trace DNA analysis. However, while implementing stricter controls and procedures into the laboratory will assist in resolving some of the current issues with trace DNA, there are several still remaining. These include secondary transfer and persistence, contamination and the over-reliance on trace DNA as probative evidence.

The occurrence of secondary transfer has been a much debated issue for around a decade. While primary transfer (person-person or person-object) is accepted without exception, secondary transfers (person-person-object, person-object-person etc), and the rate at which they occur, are yet to be exhaustively examined. Early papers gave conflicting results, with some demonstrating the phenomenon (van Oorschot and Jones 1997), and others failing to find evidence it occurred frequently enough to be a concern (Ladd et al 1999). However, subsequent studies have shown secondary transfer can occur relatively easily, although estimates of frequency differ widely. Phipps and Petricevic (2007) found only 2% of alleles did not match the donor (and were theorised to be transferred from another donor), while Lowe et al (2002) found transfer rates between 11% and 75%, depending on the 'shedder' status of each individual in the transfer.

This large variation in results obtained creates considerable debate over the potential effects of secondary transfer. If it does occur with high or moderate frequencies, there are large implications for trace DNA use as probative evidence. In some scenarios, it raises the possibility that an individual's DNA may be found at a crime scene, *without that person ever being at the scene*. This scenario, although unlikely, is possible, particularly if the first individual is a 'good' shedder, while the second (transfer) individual is a poor shedder, and the time between contacts is short. It has been reasonably well demonstrated that the last person to touch an object is usually the major profile donor, but with the increasingly sensitive analysis techniques described in this chapter, it is becoming more likely that the minor profile of the 'non-present' individual could be detected in its entirety, despite being present in very small amounts. Whether or not this is demonstrated conclusively, the possibility of it occurring may be enough to raise doubt about the ability to match an individual to a scene using only trace amounts of DNA. At the moment, the presence of an individual's DNA on an object is generally considered enough to imply their use/contact with the object. However, as research into secondary transfer grows, there is a distinct possibility that this will no longer be possible. As such, it will be imperative to have an understanding of the rate of secondary transfer, situations where it is likely or unlikely to occur, and how length of time since contact affects transfer.

The persistence of trace DNA in different environments is also an area where further research is required. No large scale studies have been conducted, particularly with long time periods. While it could be reasonably theorised that trace DNA would degrade at a standard rate, and would therefore become undetectable with relative speed, there has been no research confirming this. Therefore, the possibility that trace DNA could be detected after months, or conversely, not detected after a week, remain open, and sources of uncertainty for interpretation and attribution of profiles.

The rate of contamination of trace DNA samples is also a growing concern. Any addition of DNA to an object after a crime is contamination, whether it occurs at the crime scene by investigators or the public, or in the laboratory. Within the laboratory, it is necessary to take specific precautions when working with trace DNA exhibits. Simply speaking may introduce detectable amounts of DNA to an exhibit as far away as 120cm (Finnebraaten et al 2008). Simulations of crime scene examinations showed that individuals may contaminate exhibits simply by moving or talking near exhibits (Rutty et al 2003). More worryingly, cross-exhibit contamination within the laboratory is entirely possible, with tertiary transfer occurring from exhibits, onto the gloved hands of the examiner, then onto examination equipment (Poy and van Oorschot, 2006). While quaternary transfer back to other exhibits was not detected, it remains a possibility. Thus, stringent cleaning and decontamination procedures must be in place following each examination. Manufacturer contamination of 'sterile' plastic ware has also been demonstrated (Howitt et al 2003). Other forensic scientists must also be aware of the possibility of contamination – if an exhibit is sent for fingerprint enhancement or handwriting analysis prior to swabbing for DNA, the analyst may introduce their DNA inadvertently to the exhibit, or between two exhibits, due to incomplete knowledge of the risks involved, and proper procedure. Thus, it is not only biologists that need to be educated about contamination prevention, but all members of forensic investigation team.

While it is (relatively) easy to control for contamination within laboratories, the events prior to and during sample collection are considerably harder to control. The use of and frequent changing of gloves, cover-alls and masks for crime-scene personnel considerably reduce contamination (Rutty et al 2003). The addition of police and crime-scene personnel to DNA exclusion databases would also aid in detecting contamination at the scene, and has been recommended in the United Kingdom (Caddy et al 2008). The occurrence of contamination at the scene raises considerable issues regarding the analysis of cold cases for DNA. If exhibits were collected by personnel not wearing protective equipment, there is a strong possibility that contamination occurred. Prior to knowledge about trace DNA, there were few precautions taken during scene examinations. Gloves were changed infrequently (if worn at all), thus introducing the possibility for cross-exhibit contamination, or even cross-scene contamination. Exhibits may have been packaged together for transport or storage, resulting in transfer of DNA between exhibits. As such, placing significance on the finding of DNA on particular objects in such cases could be considered doubtful, and, in light of the current knowledge of transfer and trace DNA, unscientific. Such factors were key in the dismissal of the previously mentioned Omagh bombing case (the Queen v Sean Hoey, Northern Ireland 341/05), and should be kept firmly in mind when any cold cases are considered for DNA analysis.

CONCLUSION

Trace DNA is an expanding and important area within forensic molecular biology. The increasing sensitivity of methods allows profiles to be generated from smaller and smaller traces of biological material, and the limit of detection is now approaching single cell level. Although not yet applied in most laboratories, there are numerous improvements to existing methods which can be used to further increase success rates with trace samples.

However, simply because we are capable of obtaining profiles from single cells, it does not mean we necessarily should. At the present time, there are large gaps in our knowledge of many areas of trace DNA. Without having the entire picture, there is a question over relying on trace DNA as probative evidence. At best, trace DNA should be considered as circumstantial evidence, where it *may* imply presence at a crime-scene, or involvement in a crime, rather than definite proof of guilt. Over-reliance on trace evidence can be prejudicial against suspects, and should be avoided until the remaining questions on trace DNA can be answered.

There is no question that trace DNA is extremely useful, and very common, in forensic cases. Considering the evidence with an eye on the limitations of our current knowledge is far more conservative to the defendant, and beneficial to the entire forensic biology community, than blindly applying trace DNA analysis to every case that comes in.

Competing Interests

The authors declare that they have no competing interests. The authors alone are responsible for the content and writing of the paper.

Acknowledgments

Kaye Ballantyne is supported by the Erasmus University Medical Center Rotterdam and the Netherlands Forensic Institute. Research described in this chapter was supported by the Department of Forensic Molecular Biology, Erasmus University, the Victorian Police Forensic Services Department in Melbourne, Australia, and La Trobe University in Melbourne, Australia. Roland van Oorschot and R. John Mitchell are thanked for their considerable assistance with research into trace DNA amplification.

References

Abaz, J; Walsh, SJ; Curran, JM; Moss, DS; Cullen, J; Bright, J; Crowe, GA; Cockerton, SL; Power, EB. Comparison of the variables affecting the recovery of DNA from common drinking containers. *Forensic Science International,* 2002, 126, 233-240

Alessandrini, F; Cecati, M; Pesaresi, M; Turchi, C; Carle, F; Tagliabracci, A. Fingerprints as evidence for a genetic profile: Morphological study on fingerprints and analysis of exogenous and individual factors affecting DNA typing. *Journal of Forensic Science,* 2003, 48, 586-592.

Anoruo, B; van Oorschot, R; Mitchell, J; Howells, D. Isolating cells from non-sperm cellular mixtures using the PALM microlaser micro dissection system. *Forensic Science International,* 2007, 173, 93-96.

Ballantyne, KN; van Oorschot, RA; Mitchell, RJ; Koukoulas, I. Molecular crowding increases the amplification success of multiple displacement amplification and short tandem repeat genotyping. *Analytical Biochemistry,* 2006, 355, 298-303.

Ballantyne, KN; van Oorschot, RA; Mitchell, RJ. Comparison of two whole genome amplification methods for STR genotyping of LCN and degraded DNA samples. *Forensic Science International,* 2007a, 166, 35-41.

Ballantyne, KN; van Oorschot, RA; Muharam, I; van Daal, A; Mitchell, RJ. Decreasing amplification bias associated with multiple displacement amplification and short tandem repeat genotyping. *Analytical Biochemistry,* 2007b, 368, 222-229.

Ballantyne, KN; van Oorschot, RAH; Mitchell, RJ. Increasing amplification success of forensic DNA samples using multiple displacement amplification. *Forensic Science, Medicine and Pathology,* 2007c, 3, 182-187.

Bright, JA; Petricevic, SF. Recovery of trace DNA and its application to DNA profiling of shoe insoles. *Forensic Science International,* 2004, 145, 7-12.

Caddy, B; Taylor, GR; Linacre, AMT. A review of the science of low template DNA analysis; Executive Summary. Available from http://police, 2008.

Coble, MD; Butler, JM. Characterisation of new miniSTR loci to aid analysis of degraded DNA. *Journal of Forensic Science,* 2005, 50, 43-53.

Finnebraaten, M; Graner, T; Hoff-Olsen, P. May a speaking individual contaminate the routine DNA laboratory? *Forensic Science International: Genetics Supplement Series,* 2008, 1, 421-422.

Forster, L; Thomson, J; Kutranov, S. Direct comparison of post-28-cycle PCR purification and modified capillary electrophoresis methods with the 34-cycle "low copy number" (LCN) method for analysis of trace forensic DNA samples. *Forensic Science International: Genetics,* 2008, 2, 318-328

Franke, N; Augustin, C; Puschel, K. Optimisation of DNA-extraction and typing from contact stains. *Forensic Science International: Genetics Supplement Series*, 2008, 1, 423-425

Fregeau, CJ; Bowen, KL; Leclair, B; Trudel, I; Bishop, L; Fourney, RM. AmpFlSTR Profiler Plus short tandem repeat DNA analysis of casework samples, mixture samples, and nonhuman DNA samples amplified under reduced PCR volume conditions (25μl). *Journal of Forensic Science*, 2003, 48, 1014-1034

Gaines, ML; Wojtkiewicz, PW; Valentine, JA; Brown, CL. Reduced volume PCR amplification reactions using the AmpFlSTR Profiler Plus kit. *Journal of Forensic Science,* 2002, 47, 1224-1237

Gill, P; Whitaker, J; Flaxman, C; Brown, N; Buckleton, J. An investigation of the rigor of interpretation rules for STRs derived from less than 100 pg of DNA. *Forensic Science International,* 2000, 112, 17-40.

Gill, P, Puch-Solis, R; Curran, J. The *low-template DNA* (stochastic) threshold-its determination relative to risk analysis for national DNA databases. *Forensic Science International: Genetics,* 2009, 3, 104-111

Greenspoon, SA; Ban, JD; Sykes, K; Ballard, EJ; Edler, SS; Baisden, M; Covington, BL. Application of the BioMek 2000 laboratory automation workstation and the DNA IQ system to the extraction of forensic casework samples. *Journal of Forensic Science,* 2004, 49, 29-39

Grubweiser, P; Thaler, A; Kochl, S; Teissl, R; Rabl, W; Parson, W. Systematic study on STR profiling on blood and saliva traces after visulation on fingerprint marks. *Journal of Forensic Science,* 2003, 48, 733-741

Howitt, T; Johnson, P; Cotton, L; Rowlands, D; Sullivan, K. Ensuring the integrity of results: A continuing challenge in forensic DNA analysis. *Proceedings of the 14th International Symposium on Human Identification.* Madison, Wisconsin. Available at http://www.promega.com/geneticicproc/ussymp14proc/oralpresentations/Howitt.pdf. (2003).

Kita, T; Yamaguchi, H; Yokoyama, M; Tanaka, T; Tanaka, N. Morphological study of fragmented DNA on touched objects. *Forensic Science International: Genetics,* 2008, 3, 32-36

Kishore, R; Reef Hardy, W; Anderson, VJ; Sanchez, NA; Buoncristiani, MR. Optimisation of DNA extraction from low-yield and degraded samples using the BioRobot EZ1 and BioRobotM48. *Journal of Forensic Science,* 2006, 51, 1055-1061

Kloosterman, AD; Kersbergen, P. Efficacy and limits of genotyping low copy number (LCN) DNA samples by multiplex PCR of STR loci. *Journal de la Societe de biologie,* 2003, 197, 351-359.

Ladd, C; Adamowicz, MS; Bourke, MT; Scherczinger, CA; Lee, HC. A systematic analysis of secondary DNA transfer. *Journal of Forensic Science,* 1999, 44, 1270-1272.

Leclair, B; Sgueglia, JB; Wojtowicz, PC; Juston, AC; Fregeau, CJ; Fourney, RM. STR DNA typing: increased sensitivity and efficient sample consumption using reduced PCR reaction volumes. *Journal of Forensic Science,* 2003, 48, 1001-1013.

Lowe, A; Murray, C; Whitaker, J; Tully, G; Gill, P. The propensity of individuals to deposit DNA and secondary transfer of low level DNA from individuals to inert surfaces. *Forensic Science International,* 2002, 129, 25-34.

Moretti, T; Koons, B; Budowle, B. Enhancement of PCR amplification yield and specificity using AmpliTaq Gold DNA polymerase. *BioTechniques,* 1998, 25, 716-722.

Navidi, W; Arnheim, N; Waterman, MS. A multiple tubes approach for accurate genotyping of very small DNA samples by using PCR: statistical considerations. *American Journal of Human Genetics,* 1992, 50, 347-359.

Pang, BC; Cheung, BK. Double swab technique for collecting touched evidence. *Legal Medicine (Tokyo).* 2007, 9, 181-184.

Petricevic, SF; Bright, JA; Cockerton, SL. DNA profiling of trace DNA recovered from bedding. *Forensic Science International.,* 2006, 159, 21-26.

Phipps, M; Petricevic, S. The tendency of individuals to transfer DNA to handled items. *Forensic Science International.* 2007, 168, 162-168.

Polley, D; Mickiewicz, P; Vaughn, M; Miller, T; Warburton, R; Komonski, D; Kantautas, C; Reid, B; Frappier, R; Newman, J. An investigation of DNA recovery from firearms and cartridge cases. *Canadian Society of Forensic Science,* 2006, 39, 217-228.

Poy, A; van Oorschot, RAH. Beware; gloves and equipment used during the examination of exhibits are potential vectors for transfer of DNA-containing material. *International Congress Series,* 2006, 1288. 556-558

Raymond, JJ; van Oorschot, RAH; Walsh, SJ; Roux, C. Trace DNA analysis: Do you know what your neighbour is doing? A multi-jurisdictional survey. *Forensic Science International: Genetics,* 2008, 2, 19-28.

Raymond, JJ; Walsh, SJ; van Oorschot, RAH; Gunn, PR; Evans, L; Roux, C. Assessing trace DNA evidence from a residential burglary: Abundance, transfer and persistence. *Forensic Science International: Genetics Supplement Series,* 2008, 1, 442-423.

Roeder, AD; Elsmore, P; Greenhalgh, M; McDonald, A. Maximising DNA profiling success from sub-optimal quantities of DNA: A staged approach. *Forensic Science International: Genetics,* 2009, 3, 128-137.

Rutty, GN; Hopwood, A; Tucker, V. The effectiveness of protective clothing in the reduction of potential DNA contamination of the scene of the crime. *International Journal of Legal Medicine,* 2003, 117, 170-174.

Schmidt, U; Lutz-Bonengel, S; Weisser, HJ; Sanger, T; Pollak, S; Schon, U; Zacher, T; Mann, W. Low volume amplification on chemically structured chips using the PowerPlex 16 DNA amplification kit. *International Journal of Legal Medicine,* 2006, 120, 42-48

Schniffer, LA; Bajda, EJ; Prinz, M; Sebestyen, J; Shaler, R; Caragine, TA. Optimisation of a simple, automatable extraction method to recover sufficient DNA from low copy number DNA samples for generation of short tandem repeat profiles. *Croatian Medical Journal,* 2005, 46, 578-586.

Sewell, J; Quinones, I; Ames, C; Multaney, B; Curtis, S; Seeboruth, Moore, S; Daniel, B. Recovery of DNA and fingerprints from touched documents. *Forensic Science International:Genetics,* 2008, 2, 281-285

Smith, PJ; Ballantyne, J. Simplified low copy number DNA analysis by post-PCR purification. *Journal of Forensic Science,* 2007, 52, 820-829

Sweet, D; Lorente, M; Lorente, J; Valenzuela, A; Villaneuva, E. An improved method to recover saliva from human skin: the double swab technique. *Journal of Forensic Science.,* 1997, 42, 320-322.

Taberlet, P; Griffin, S; Goossens, B; Questiau, S; Manceau, V; Escaravage, N; Waits, LP; Bouvet, J. Reliable genotyping of samples with very low DNA quantities using PCR. *Nucleic Acids Research,* 1996, 24, 3189-3194.

Thornhill, AR; McGrath, JA; Eady, RA; Braude, PR; Handyside, AH. A comparison of different lysis buffers to assess allele dropout from single cells for preimplantation genetic diagnosis. *Prenatal Diagnostics,* 2001, 21, 490-497

Thornhill, AR; Snow, K. Molecular diagnostics in preimplantation genetic diagnosis. *Journal of Molecular Diagnostics,* 2002, 4, 11-29

van Oorschot, RA; Jones, MK. DNA fingerprints from fingerprints. *Nature,* 1997, 387, 767.

van Oorschot, RAH; Phelan, DG; Furlong, S; Scarfo, GM; Holding, NL; Cummins, MJ. Are you collecting all the available DNA from touched objects? *International Congress Series,* 2003, 1239, 803-807.

Walsh, PS; Metzger, DA; Higuichi, R. Chelex 100 as a medium for simple extraction of DNA for PCR-based typing from forensic material. *BioTechniques,* 1991, 10, 506-513

Wickenheiser, RA. Trace DNA: a review, discussion of theory, and application of the transfer of trace quantities of DNA through skin contact. *Journal of Forensic Science,* 2002, 47, 442-450.

Whitaker, JP; Cotton, EA; Gill, P. A comparison of the characteristics of profiles produced with the AmpFlSTR SGM Plus multiplex system for both standard and low copy number (LCN) STR DNA analysis. *Forensic Science International,* 2001, 123, 215-223.

In: Horizons in DNA Research, Volume 1
Editor: Jason R. Chesterton, pp. 187-202
ISBN: 978-1-60876-968-1

Chapter 8

THE USE OF MITOCHONDRIAL DNA IN FORENSIC IDENTIFICATION

C. D. Matheson
Lakehead University, Thunder Bay, Canada.

ABSTRACT

Mitochondrial DNA has been used for the purposes of forensic identification since the mid 90s. Since then dedicated laboratories have been set-up and guidelines for the analysis and interpretation of mitochondrial DNA have been established. The most common method of analysis has been to amplify and sequence the mitochondrial DNA to generate a mitochondrial profile. This method has been applied to species identification used on forensically important animal species, and human identification whereby the mitochondrial DNA is used for forensic identification of low copy number DNA and degraded remains. The use of mitochondrial DNA does not allow discrimination to an individual like nuclear DNA as it can only identify maternally related individuals. However this has the potential for the exclusion of an individual in forensic identification.

INTRODUCTION

The genetics of mitochondrial DNA (mtDNA) is unique and when understood can be applied to genetic identification in the forensic sciences. Mitochondrial DNA is an extra-chromosomal mini-circle found within the mitochondrial organelle of the cell. Due to the process of fertilisation mitochondria are usually only inherited from the mother (Giles et al. 1980). There are thousands of copies of mtDNA within each cell (Budowle et al. 1990). The reference sequence for the mitochondrial genome is 16569bp in size and was fully sequenced in 1981 (Anderson et al. 1981). The presence of deletions and insertions means that the mitochondrial genome size can vary in different maternal lineages. The mitochondrial genome is now very well characterised with 13 coding genes, 2 ribosomal RNA genes, 22 transfer RNA genes and the control region. Within the control region, D-Loop or non-coding region are the hypervariable regions, these are specific regions that have a higher variability and can be used to identify sub-population groups or lineages due to the increased variation. The genetic variation present in the coding region of the mitochondrial genome can also be useful for species identification (Bataille et al. 1999; Bellis et al. 2003; Wolf et al. 1999). The analyses of these various regions (coding and non-coding regions) have thus found their way into the forensic sciences. Forensic analysts all over the world are starting to use mtDNA, some reluctantly, some with caution and some as a last resort on degraded or low copy number DNA (LCN-DNA) samples or particularly on bone, teeth and hair samples. The use of mtDNA in forensic science requires careful interpretation and a great understanding of the unique features and problems of mtDNA. However, complexity within mtDNA, its damage, associated diseases, its inheritance and its interaction with the nuclear DNA (nDNA) genome requires greater understanding, which provides for much research in the future.

HUMAN IDENTIFICATION

Traditionally mtDNA analysis was performed using the technique of restriction fragment length polymorphisms (RFLP). This technique requires the fully intact mitochondrial genome and then cuts it up into small fragments using enzymes known as restriction endonucleases. The presence or absence of the recognition site of these restriction enzymes allowed these researchers to identify some single nucleotide polymorphisms. If the enzyme was able to cut then the 4-16bp sequence at that specific recognition site would be known and if the restriction enzyme did not cut then a polymorphism or mutation present within the recognition site was preventing it from being cut. In this way researchers separated the human populations into groups based on the similarity of these restriction patterns. These groups became known as a haplogroups. Further studies then began to show finer scale difference using different restriction enzymes and this generated sub-population groups known as haplotypes. It was proposed that all individuals within the same haplotype were maternally related and that those within the same haplogroup were of similar maternal ethnic groups. These haplogroups began to identify continent specific types and ethnic groups began to be studied in a biological way. The choice of restriction enzymes used to identify these groups were chosen by the researchers based on patterns and these variations were called discrete markers they later became known as single nucleotide polymorphisms (SNP).

With the advent of the Polymerase Chain Reaction (PCR), the technology changed from cloning to various methods involving PCR (Mullis et al. 1986a; Mullis and Faloona 1987; Saiki et al. 1988). It became possible to analyze even small amounts of degraded DNA (Chou et al. 1992; Mullis et al. 1986b; Mullis and Faloona 1987; Paabo et al. 1989; Saiki et al. 1988; Saiki et al. 1985). One of the first PCR based method was PCR-RFLP, a combination of PCR and the first method of RFLP analysis. In this technique they would amplify a specific fragment of mtDNA using specifically designed primers and analyze the DNA fragments by gel electrophoresis. If the amplification was successful and a band was produced then a restriction enzyme was used to cut the PCR product and much like the older RFLP technique researchers would be able to identify discrete markers without the need to recover intact mitochondrial genomes.

History of Forensic Cases

The first examination of skeletal remains for forensic identification used the existing ancient or degraded DNA research techniques. In 1991 the mtDNA was recovered from skeletal material (Hagelberg et al. 1991) followed by the analysis of a murder victim from dental remains (Ginther et al. 1992). This work was followed by the identification of civil war soldiers and Vietnam soldiers (Fisher et al. 1993; Holland et al. 1993). Mitochondrial DNA laboratories were being set up in law enforcement agencies like the FBI (completed 1996) and FSS (completed 1994) to begin the process of developing mtDNA analytical procedures and carrying out developmental method validation which was required before forensic casework on mtDNA could begin. The first case that went to court in the USA that involved the use of mtDNA was the case of Paul Ware (1996 Tennessee v. Ware) and Shannon Leonard Murrin in Canada (1999 R. v. Murrin) since then many cases have employed mtDNA when there is not enough sample for STR analysis (LCN-DNA) or when there are just hair shafts, degraded skeletal or dental remaining. The FBIs method validation was completed in 1995 (Wilson et al. 1995). One of the most renowned cases involving mitochondrial DNA was its application to the Romanov remains from Russia which also identified the presence of heteroplasmy as a potential difficulty for mtDNA analysis (Ivanov et al. 1996). The use of mtDNA from forensically important species began to be used like that of domestic dog (*Canis familiaris*) (Savolainen et al. 1997). Once mtDNA analysis become routine international guidelines began to be developed for the analysis and interpretation of mtDNA (Bar et al. 2000; Carracedo et al. 2000). However forensic analysis involving mtDNA is still predominantly used in cases of missing persons due to the highly degraded nature of the usually buried remains (Bender et al. 2000; Melton and Nelson 2001)

Historical Cases

Historical investigations were carried out as forensic cases. The first historical case was the identification of Mengele (Holden 1992; Jeffreys et al. 1992), followed by the Romanovs (Gill 1997; Gill et al. 1994; Ivanov et al. 1996), Duke Christian II (Gerstenberger et al. 1998), Martin Bormann (Anslinger et al. 2001), King Louis XVI (Jehaes et al. 1998; Jehaes et al. 2001), Thomas Jepherson (Abbey 1999; Foster et al. 1998), Francesco Petrarca (Caramelli et al. 2007), Che Guevara's guerrillas from Bolivia (Lleonart R et al. 2000) and Jesse James (Stone et al. 2001). The ability to analyse degraded and LCN-DNA allowed mtDNA to be used to identify the buried remains of victims of war beginning with the civil war (Fisher et al. 1993), Vietnam war (Holland et al. 1993), the Baltic conflict (Primorac et al. 1996) and mass graves in Guatemala (Boles et al. 1995) and Argentina (Corach et al. 1997; Skolnick 1993). While some of this research has generated controversy with for example the Romanov study (Hofreiter et al. 2004; Knight 1998; Stone 2004; Zhivotovsky 1999). Over time most of these concerns have become resolved (Coble et al. 2009).

Species Identification

The mitochondrial genome can be used for the identification of species, populations, sub-populations and lineages. The first application of mtDNA used for species identification in a forensic investigation used PCR-RFLP analysis (Murray et al. 1995). However sequence variation became the preferred method used for species identification using the sequence variation to identify specific species (Savolainen et al. 1997). Numerous report were published on various methods for the analysis of mtDNA for species identification (Angleby and Savolainen 2005; Bellis et al. 2003; Nakaki et al. 2007). By multiplexing these mtDNA detection systems rapid screening of species was made possible in a forensic context (Bataille et al. 1999). All of the species identification studies have focused on genetic coding regions within the mitochondrial genome because they exhibit enough genetic variation for the identification of species. Some studies have included the mtDNA control region but these are to help identify breeds, population groups and sub-population groups.

Methods Apllied to Forensic Mitochondrial DNA Analysis

Presently there are only a few different methods employed for the analysis of mtDNA in forensic laboratories. The tradition method of PCR amplification followed by sequencing is the standard method used in most laboratories. Some laboratories have incorporated PCR amplification followed by cloning but most laboratories have designed specific multiplex PCR methods for the analysis of mtDNA. The genetic targets range from the non-coding control region for population and maternal lineage identification to the coding genes (cytochrome b, 16s rRNA gene, 18s rRNA gene, cytochrome oxidase I and II) for species identification. More recently laboratories have been developing a multiplex single nucleotide extension (SNE) system for the simultaneous detection of the SNP associated with tradition discrete markers for haplotype and haplogroup determination (Nelson et al. 2007; Quintans et al. 2004; Vallone et al. 2004). Using these multiplex SNE systems many discrete markers are able to be analyzed in a single PCR. The haplogroup has become helpful for ethnic identification but caution must still be taken as the ethnic ancestry of an individual does not indicate his physical appearance. Researchers have also begun to look at microarrays but these are yet to be applied to forensic analysis (Erdogan et al. 2001).

Mitochondrial Considerations

The challenges when using mtDNA for the identification of biological remains depend upon the methods being used, the quantity and quality of the DNA being recovered from the biological sample, the type of biological sample and an understanding of mtDNA.

Inheritance

The ability to identify an individual must also take into account the mitochondria's mode of inheritance, which is through the maternal line. All maternal relatives will share the same mtDNA type and siblings will share the same mtDNA they inherited from their mother. This is a distinct advantage in which any maternally related individual (MRI) shares the same mtDNA type and can therefore be used to identify maternal relatedness. This benefit can span generations but the ability to discriminate between close relationships or identify an individual is not possible and that mtDNA type is not necessarily highly unique.

Copy Number

The high copy number provides the greatest benefit for mtDNA analysis, the quantity of mtDNA far exceeds the quantity of nDNA and can be analyzed much more easily. Its genome is a well characterized with between 1 and 10 copies of mtDNA within each mitochondrion (Nass 1969; Satoh and Kuroiwa 1991) and between 1 and 1000 mitochondria per cell (Shoffner and Wallace 1992; Tully and Levin 2000). The reason it has been used in degraded samples and in particular cases like hair shaft samples, is because it is often the most abundant type of DNA available for analysis. However, due to the multiple copies that exist per cell, there is the chance that these copies can differ within the same organism.

Structure

The structure of the mitochondrial genome as a mini-circle affects its preservation. It differs from the nuclear genome that it does not have chromatin, nucleosomes or histones. This structural difference is important as the nucleosome formation of the nDNA has been used to explain the preservation of nDNA into 200-400bp fragments but the mtDNA is exposed. There are some proteins that interact with the mtDNA in the nucleoid but not like the histones in nDNA. The lack of histones to protect the mtDNA and this exposed nature allows more damage to accumulate and to be generated within the mtDNA.

Metabolic Function

The function of the mitochondria *in vivo* also has its affects on the DNA. The reactive oxygen species (ROS) that are formed as a by-product of its metabolism increases the amount of *in vivo* damage to the mtDNA. This damage is usually adequately repaired by endogenous repair mechanisms however gone unchecked this inherent damage can persist and become detrimental to the function of the mitochondria. Once the body dies these ROS may still be around and may still cause damage on the mtDNA post-mortem that will not be repaired by the mitochondria's repair mechanism, as it is not longer functioning. This can cause an increased amount of damage within the mtDNA, which will cause problems when using of the mtDNA sequence.

Sample Type

The source of the biological material is a prime consideration for mitochondrial analysis. In a crime scene especially, almost any biological material can be collected for analysis. These can include blood, semen, saliva, urine, hair, teeth, bone, and tissue samples. All of these samples can be analyzed using mtDNA, especially hair shaft as the hair shaft can be difficult to analyzed using nDNA. Hair is peculiar in that it undergoes different forms of degradation compared to other biological materials, which are less harsh and allow for a greater preservation of mtDNA within the hair shaft in comparison to most others (Gilbert et al. 2006; Gilbert et al. 2004). This has lead to the analysis of mtDNA in hair as being usually very successful.

Genetic Variation

The genetic variation and mode of inheritance limits the power of discrimination with mtDNA profiles. One can gain a genetic profile but it will not be unique to that individual, as all maternally related individuals will share the same profile unless there is an immediate mutation in one individual or generation. In forensic use mtDNA has the power to exclude an individual but definitely does not have the power for inclusions. So that a suspect's profile that does not match the evidence by more than two polymorphisms can be excluded while if it matches the evidentiary samples does not allow exclusion that means that suspect is still a possible suspect. If there is one or two differences between the suspect and the evidentiary sample it is usually considered inconclusive due to the possibility of mutations in one of the samples.

Heteroplasmy

Heteroplasmy is where more than one population of mtDNA existence within the same cell, tissue or organism (Melton 2004). Unlike nDNA, there are up to thousands of copies of mtDNA in a single cell if there is any difference between these multiple copies at a single nucleotide position or in length there will be heteroplasmy. Heteroplasmy is of great concern when attempting to analyze the mtDNA of a degraded samples. There are numerous forms of heteroplasmy (Comas et al. 1995; Melton 2004; Wilson et al. 1997), such as inherited heteroplasmy, where an offspring is passed two mitochondrial copies from their mother or when the father unexpectedly contributes his own mitochondria which has not yet been observed in humans. Inherited heteroplasmy results in all or some of the tissues within the body containing heteroplasmic mtDNA, though the presence and relative proportion or percentage of heteroplasmy in the different tissues can vary (Gocke et al. 1998). Alternatively, there is acquired heteroplasmy, which occurs within the organism itself within its own lifetime, and can occasionally be passed on to offspring if the heteroplasmy occurs in the germ cells. Acquired heteroplasmy can be SNPs that differ between the two copies as point mutations or point heteroplasmy, which occurs in about 3% of humans(Comas et al. 1995). This form of point heteroplasmy where there is a variation at one nucleotide or SNP can be quite common and can differ between tissues (Bendall et al. 1996; Bendall et al. 1997; Comas et al. 1995). Length heteroplasmy is another form of heteroplasmy, which can occur in disease states where there has been a replication error or a deletion induced through a metabolic process or deficient repair mechanisms (Bendall and Sykes 1995; Melton 2004). Additionally, length heteroplasmy can occur in the biological aging of an individual if the mitochondrial biological model is accepted (Osiewacz and Hermanns 1992; Wallace et al. 1995). Ultimately, the accumulation of small deletions within the mitochondrial genome will normally occur over an individual's lifetime. These differences will affect the ability to identify an individual and discriminate between heteroplasmy and contamination.

Pseudogenes

The interpretation of mtDNA results can be further complicated by the presence of nuclear copies of mtDNA sequences. These are called pseudogenes or nuclear-mitochondrial inserts (NUMTs) (Herrnstadt et al. 1999). These sequences are old fragments of mtDNA that have become incorporated into the nuclear genome. Once inserted into the nuclear genome they no longer retain their function and are therefore susceptible to mutations. These increased mutations, if amplified and incorrectly thought to be mtDNA, would show greater variation than is accurate and could be confused for unique endogenous sequences or mislabeled as heteroplasmy within the sample. The frequency of these NUMTs are unknown but if there is a high number of NUMTs then the chances of amplifying one increases or the chance of amplifying contaminant NUMTs is also a possibility (Morgan 1997).

Aged and Degraded Samples

Mitochondrial DNA is usually used on buried or degraded forensic samples. The problems faced when analyzing aged or degraded samples is the degradation of the DNA and the preservation of the sample which can all affect the ability to recover the DNA for analysis. These problems are degradation of DNA, inhibition and contamination.

Degradation of DNA

The ability to recover and amplify degraded DNA is hindered by the forms of damage within the DNA these can be categorized into modification, fragmentation and cross-linkage of the DNA. Many factors play a part in the preservation of DNA within forensic samples.

Modifications are very common in degraded DNA samples, and these processes more often than not serve to inhibit genetic analysis or increase the errors introduced through the analytical process. The sugar-phosphate backbone and the nitrogenous bases of the DNA are attacked by a variety of chemical, biological and enzymatic molecules (Hoss et al. 1996). These will modify the chemical structures of the DNA which can prevent the DNA polymerase from amplifying, or cause misincorporation of nucleotides (Hoss et al. 1996; Poinar and Stankiewicz 1999). Some of the most common forms of modification include oxidation, hydrolysis and alkylation. Hydrolysis can cause deamination, demethylation, depurination and depyrimidation. All of these chemical processes generate numerous modified bases that will form irregular base pairing so that when DNA polymerases replicate the DNA misincorporations will be generated. Some of the modifications to the DNA backbone can increase the fragmentation of the DNA.

Fragmentation is present in the degraded DNA as single stranded breaks (SSB) and double stranded breaks (DSB). Single stranded Breaks can be generated through hydrolysis of the phosphate-sugar bond, ring opening of the sugar, oxidation, radiation and free radical damage of the DNA backbone. The presence of two SSB in close proximity to each other can cause DSB alternatively other physical mechanisms can shear double stranded DNA (dsDNA) like repeated freeze thaw cycles.

Cross-linkage can occur between DNA and carbohydrates, hydrocarbons, lipids, proteins, metals and DNA. All of these will affect the ability to amplify the cross-linked DNA and has been suggested to be the largest cause of non-viable template DNA in some burial environments (Hansen et al. 2006). Metal ions can crosslink to the DNA which can catalyze further degradation or inhibit the analytical process (Schweigert et al. 2000). DNA polymerases usually fail to amplify DNA that is heavily cross-linked.

Taphonomy is the study of the processes of death and decay of biological material. These processes begin with cell death, autolysis and necrosis of the cell through cellular decomposition aided by the release of lytic enzymes from the liver, the acidic release of the stomach and the chemical and bacterial release of the bowels. During the period following death a combination of factors including enzymatic processes and chemical reactions modify the existing DNA (Perry et al. 1988). Taphonomy then continues through deposition of the sample and diagenesis to full decomposition of the hard skeletal and dental remains. It is within the framework of taphonomy that the key factors are examined to distinguish the contribution of physical and biological agents on the degradation process. These can be considered abiotic and biotic respectively. The abiotic agents can include water content, temperature, PH, environmental stability and exposure.

Water content is considered to be the most important factor to the preservation of biological remains; it is this physical attribute that can accelerate the decomposition of DNA. The water content of the body and the burial environment also has a great effect on the fragmentation of DNA. It has been shown to initiate strand breakage by attacking the sugar base bonds. Where the base is lost the chain is weakened and eventually cleaved or broken (Eglinton and Logan 1991; Lindahl 1993). High water content will increase the hydrolytic damage thought to be the predominant cause of DNA fragmentation and damage. Water allows for the activity of biological organisms, lytic enzymes and chemicals, which will quickly degrade and fragment the body's DNA. If the body dries quickly after death or the water is immobilized (as in freezing) these processes are arrested and will aid in preservation. The presence of water can also assist in ionic and molecular reactions that can fragment DNA. In low pH environments this can also cause acid hydrolysis, which has the same effect. There have also been many cases of good molecular preservation of underwater remains, however, in both salt and fresh water (Crainic et al. 2002; Iscan et al. 2005; Sweet and Shutler 1999). Water content in conjunction with temperature may be a better determinant as good preservation of DNA is more likely to be induced from a cool, anaerobic, stable environment than with other agents. By freezing a sample it causes the water content to lose activity as it is most likely not the presence of the water but rather the activity of the water which is the detrimental agent in preservation. The chemical reaction of the water molecule with the nucleic acids causes the DNA to become more damaged. In a frozen sample this water activity is halted by the processes of physical state change of the water molecule into a solid, which prevents it from remaining active. Therefore frozen material preserves very well.

Oxygen free radicals can generate oxidative damage to the DNA molecule, causing fragmentation (Matsuo et al. 1995; Paabo 1989). This form of damage is high, particularly in mtDNA, due to the metabolic processes of the mitochondrion and its metabolic by-products. In living organisms, these free radicals are removed by the generation of 'free radical capture molecules' like caratenoids, but such processes cease to be operational at death and attacks on the DNA can continue unchecked. Metals and metal ions can also induce fragmentation (Okayasu et al. 2003).

Temperature is another taphonomic factor that can affect preservation. In the case of heat or fire, burning of the biological material can even remove the carbon content of the biological material altogether (Nelson and Melton 2007; Pusch et al. 2000; Ye et al. 2004). In contrast, colder temperatures can increase the preservation of DNA, even to the extent of freezing the sample.

The depositional environment of the buried remains can be of great significance, especially the pH and soil chemistry of the location. A neutral pH will preserve biological materials the best. It should be noted, however, that it is often the lack of oxygen in some of these unexpectedly stable environments that allows for the optimal preservation of such biological materials. When a body is lying on the surface of the ground, raw exposure to the elements can contribute to its degradation. This exposure can increase UV irradiation, wind and water damage, and can be detrimental to the preservation of the sample.

The biotic agents of taphonomy can include bacteria, insects, scavengers and plant roots that can affect the preservation of molecules within biological material. The bacteria begin its attack on molecular preservation as soon as the body dies. It is the endogenous bacteria that will find nutrients anywhere they can within the now decaying body of the organism. These endogenous bacteria could include pathogenic bacteria as well depending on the health state of the organism. Then the exogenous bacteria begin to encroach on the molecules of the remains. These can be prevented in a variety of states like a frozen body, anaerobic environment, or in a bactericidal environment like in oils or bitumen. Once death occurs insect will begin to penetrate from the outside and start the process of oviposition, the laying of eggs. These insects will degrade the body along with beetles, flies, moths and worms. Larger scavengers can then pull the body apart and increase the rate of degradation.

Inhibition

Any environment that biological material is found can contain inhibition. There are many forms of inhibition, and these can occur at a number of different steps in the analytical process. Any environment in which degraded biological material is found can contain inhibitors; the molecular techniques that are then applied to analyze the remains can be inhibited through the presence of metal ions, organic acids, biological breakdown products, collagen, haemoglobin or even too much DNA usually excess amounts of exogenous DNA that will inhibit the analysis of limited quantities of endogenous DNA. These inhibitors have to be dealt with prior to analysis so that genetic study can begin. Current techniques in degraded DNA research take some of these inhibitors into account, and attempt to remove the inhibitors from the sample prior to the extraction and analyses of ancient or degraded materials.

There are two types of inhibition, enzymatic inhibition and template inhibition. Enzymatic inhibition affects the enzymes in the analytical methods. These enzymes can include the proteinase K in the enzymatic extraction methods or the DNA polymerase in the PCR amplifications. Some of this enzyme inhibition can be caused by divalent cations, humic acids, heme from hemoglobin, DNA quantity, collagen and other biological breakdown products. Another form of inhibition is template inhibition, where the template cannot be amplified due to a problem within the template itself, such as secondary structures, base dimerisation (pyrimidine dimers) and metal cross-linking, preventing the DNA polymerase from proceeding (Alonso et al. 2004). Many studies have begun to investigate causes of inhibition in degraded DNA (Hall et al. 1992).

Contamination

There are a number of certain types of contamination that should be considered for any analysis. The most obvious form of contamination is modern contamination through contact from crime scene analysts, law enforcement personnel, the public and handlers in the field, as well as analysts and residual DNA from the laboratory at the time of genetic analysis. Also if the remains are found in mass graves where multiple bodies have been placed then the possibility for the remains of one body contaminating another is present. Caution must be taken in attempts to remove this contamination, as this can sometimes prove more detrimental to the analysis. Some techniques can be more detrimental than others in removing contaminants. Hypochlorite washes to oxidatively damage the DNA (Hayatsu et al. 1971) and UV irradiation (Fox et al. 1991; Niederhauser et al. 1994; Ou et al. 1991) to damage the modern contamination are techniques that have been applied to reduce contamination. Many protocols have used these on sample surfaces, but in some cases this has modified the modern DNA to behave as degraded DNA. Contamination is a concern with mtDNA due to the multiple copies per cell. If a single cell contaminates the analytical process then possibly thousands of copies of DNA have contaminated the extract and will out-compete the damaged, fragmented and degraded DNA template in the PCR.

CONCLUSIONS

There are features of mitochondrial genetics that have to be understood to be able to use mtDNA accurately and reliably for identification purposes. All of these unique characteristics have been incorporated into forensic interpretational guidelines and a very conservative approach is taken for the interpretation of the mtDNA results. Mitochondrial DNA is used in many forensic laboratories around the world for human and animal identification. A variety of different methods are used in these laboratories but PCR amplification followed by sequencing remains the most common approach. At the same time forensic science has begun to push the limits of their analytical procedures to examine degraded DNA samples and LCN-DNA samples. The technologies for mtDNA analysis are now moving forward with techniques like microarrays, microfluidic devices and real-time on-site analysis. The next stage is headed in the direction of whole genome analysis where the whole mtDNA genome from each individual would be analyzed for more power of discrimination.

REFERENCES

Abbey, DM. The Thomas Jefferson paternity case. *Nature,* 1999, 397(6714), 32.

Alonso, A; Martin, P; Albarran, C; Garcia, P; Garcia, O; de Simon, LF; Garcia-Hirschfeld, J; Sancho, M; de La Rua, C; Fernandez-Piqueras, J. Real-time PCR designs to estimate nuclear and mitochondrial DNA copy number in forensic and ancient DNA studies. *Forensic Sci Int,* 2004, 139(2-3), 141-149.

Anderson, S; Bankier, AT; Barrell, BG; de Bruijn, MH; Coulson, AR; Drouin, J; Eperon, IC; Nierlich, DP; Roe, BA; Sanger, F; others. Sequence and organization of the human mitochondrial genome. *Nature*, 1981, 290(5806), 457-465.

Angleby, H; Savolainen, P; Forensic informativity of domestic dog mtDNA control region sequences. *Forensic Sci Int,* 2005, 154(2-3), 99-110.

Anslinger, K; Weichhold, G; Keil, W; Bayer, B; Eisenmenger, W. Identification of the skeletal remains of Martin Bormann by mtDNA analysis. Int *J Legal Med,* 2001, 114(3), 194-196.

Bar, W; Brinkmann, B; Budowle, B; Carracedo, A; Gill, P; Holland, M; Lincoln, PJ; Mayr, W; Morling, N; Olaisen, B; others. DNA Commission of the International Society for Forensic Genetics: guidelines for mitochondrial DNA typing. *Int J Legal Med,* 2000, 113(4), 193-196.

Bataille, M; Crainic, K; Leterreux, M; Durigon, M; de Mazancourt, P. Multiplex amplification of mitochondrial DNA for human and species identification in forensic evaluation. *Forensic Sci Int,* 1999, 99(3), 165-170.

Bellis, C; Ashton, KJ; Freney, L; Blair, B; Griffiths, LR; A molecular genetic approach for forensic animal species identification. *Forensic Science International,* 2003, 134(2-3), 99-108.

Bendall, KE; Macaulay, VA; Baker, JR; Sykes, BC. Heteroplasmic point mutation in the human mtDNA control region. *AmJHumGenet,* 1996, 59(6), 1276-1287.

Bendall, KE; Macaulay, VA; Sykes, BC. Variable levels of a heteroplasmic point mutation in individual hair roots. *AmJHumGenet,* 1997, 61(6), 1303-1308.

Bendall, KE; Sykes, BC. Length heteroplasmy in the first hypervariable segment of the human mtDNA control region. *Am J Hum Genet,* 1995, 57(2), 248-256.

Bender, K; Schneider, PM; Rittner, C. Application of mtDNA sequence analysis in forensic casework for the identification of human remains. *Forensic Science International*, 2000, 113(1-3), 103-107.

Boles, TC; Snow, CC; Stover, E. Forensic DNA testing on skeletal remains from mass graves : a pilot project in Guatemala. *JForensic Sci,* 1995, 40(3), 349-355.

Budowle, B; Adams, DE; Comey, CC; Merrill, CR. Mitochondrial DNA: A Possible Genetic Material Suitable for Forensic Analysis. In: Lee HC, and Gaensslen RE, editors. DNA and Other Polymorphisms in Forensic Science. Chicago: Year Book Medical Publishers Inc. 1990, 76-97.

Caramelli, D; Lalueza-Fox, C; Capelli, C; Lari, M; Sampietro, ML; Gigli, E; Milani, L; Pilli, E; Guimaraes, S; Chiarelli, B; others. Genetic analysis of the skeletal remains attributed to Francesco Petrarca. *Forensic Sci Int.,* 2007.

Carracedo, A; Bar, W; Lincoln, P; Mayr, W; Morling, N; Olaisen, B; Schneider, P; Budowle, B; Brinkmann, B; Gill, P; others. DNA commission of the international society for forensic genetics: guidelines for mitochondrial DNA typing. *Forensic Sci Int*, 2000, 110(2), 79-85.

Chou, Q; Russell, M; Birch, DE; Raymond, J; Bloch, W. Prevention of pre-PCR mis-priming and primer dimerization improves low-copy-number amplifications. *Nucl Acids Res,* 1992, 20(7), 1717-1723.

Coble, MD; Loreille, OM; Wadhams, MJ; Edson, SM; Maynard, K; Meyer, CE; Niederstatter, H; Berger, C; Berger, B; Falsetti, AB; others. Mystery solved: the identification of the two missing Romanov children using DNA analysis. PLoS ONE 2009, 4(3), e4838.

Comas, D; Paabo, S; Bertranpetit, J. Heteroplasmy in the control region of human mitochondrial DNA. *Genome Res,* 1995, 5(1), 89-90.

Corach, D; Sala, A; Penacino, G; Iannucci, N; Bernardi, P; Doretti, M; Fondebrider, L; Ginarte, A; Inchaurregui, A; Somigliana, C; others. Additional approaches to DNA typing of skeletal remains: the search for "missing" persons killed during the last dictatorship in Argentina. *Electrophoresis*, 1997, 18(9), 1608-1612.

Crainic, K; Paraire, F; Leterreux, M; Durigon, M; de Mazancourt, P. Skeletal remains presumed submerged in water for three years identified using PCR-STR analysis. *J Forensic Sci*, 2002, 47(5), 1025-1027.

Eglinton, G; Logan, GA. Molecular preservation. *PhilTransRSocLondserB*, 1991, 333, 315-328.

Erdogan, F; Kirchner, R; Mann, W; Ropers, HH; Nuber, UA. Detection of mitochondrial single nucleotide polymorphisms using a primer elongation reaction on oligonucleotide microarrays. *Nucleic Acids Res*, 2001, 29(7), e36.

Fisher, DL; Holland, MM; Mitchell, L; Sledzik, PS; Wilcox, AW; Wadhams, M; Weedn, VW. Extraction, evaluation, and amplification of DNA from decalcified and undecalcified United States Civil War bone. *J Forensic Sci,* 1993, 38(1), 60-68.

Foster, E; Jobling, M; Taylor, P; Donnelly, P; de Knijff, P; Mieremet, R; Zerjal, T; Tyler-Smith, C. Jefferson fathered slave's last child. *Nature,* 1998, 396(November):27-28.

Fox, JC; Ait-Khaled, M; Webster, A; Emery, VC. Eliminating PCR contamination: is UV irradiation the answer? *J Virol Methods,* 1991, 33(3), 375-382.

Gerstenberger, J; Hummel, S; Herrmann, B. Assignment of an Isolated Skeletal Element to the Skeleton of Duke Christian II through STR-typing. *Ancient Biomolecules,* 1998, 2(1), 63-68.

Gilbert, MT; Menez, L; Janaway, RC; Tobin, DJ; Cooper, A; Wilson, AS. Resistance of degraded hair shafts to contaminant DNA. *Forensic Sci Int,* 2006, 156(2-3), 208-212.

Gilbert, MT; Wilson, AS; Bunce, M; Hansen, AJ; Willerslev, E; Shapiro, B; Higham, TF; Richards, MP; O'Connell, TC; Tobin, DJ; others. Ancient mitochondrial DNA from hair. *Curr Biol*, 2004, 14(12), R463-464.

Giles, RE; Blanc, H; Cann, HM; Wallace, DC; Maternal inheritance of human mitochondrial DNA. *Proc Natl Acad Sci U S A*, 1980, 77(11), 6715-6719.

Gill, P. The Romanovs. *Med Leg J,* 1997, 65 (Pt 3), 122-132.

Gill, P; Ivanov, PL; Kimpton, C; Piercy, R; Benson, N; Tully, G; Evett, I; Hagelberg, E; Sullivan, K. Identification of the remains of the Romanov family by DNA analysis. *Nat Genet*, 1994, 6(2), 130-135.

Ginther, C; Issel-Tarver, L; King, MC. Identifying individuals by sequencing mitochondrial DNA from teeth. *Nature Genetics*, 1992, 2(2), 135-138.

Gocke, CD; Benko, FA; Rogan, PK. Transmission of mitochondrial DNA heteroplasmy in normal pedigrees. *HumGenet,* 1998, 102(2), 182-186.

Hagelberg, E; Gray, IC; Jeffreys, AJ. Identification of the skeletal remains of a murder victim by DNA analysis. *Nature,* 1991, 352(August), 427-429.

Hall, LM; Ashworth, C; Bartsiokas, A; Jones, DS. Experiments on inhibition problems in old tissues. *Ancient DNA Newsletter*, 1992, 1(2-December), 9-10.

Hansen, AJ; Mitchell, DL; Wiuf, C, Paniker, L; Brand, TB; Binladen, J; Gilichinsky, DA; Ronn, R; Willerslev, E. Crosslinks rather than Strand Breaks Determine Access to Ancient DNA Sequences from Frozen Sediments. *Genetics.*, 2006.

Hayatsu, H; Pan, S; Ukita, T. Reaction of sodium hypochlorite with nucleic acids and their constituents. *Chem Pharm Bull* (Tokyo) 1971, 19(10), 2189-2192.

Herrnstadt, C; Clevenger, W; Ghosh, SS; Anderson, C; Fahy, E; Miller, S; Howell, N; Davis, RE. A novel mitochondrial DNA-like sequence in the human nuclear genome. *Genomics*, 1999, 60(1), 67-77.

Hofreiter, M; Loreille, O; Ferriola, D; Parsons, TJ. Ongoing controversy over Romanov remains. *Science*, 2004, 306(5695), 407-410.

Holden, C. Mengele's DNA. *Science*, 1992, 255(February.):801.

Holland, MM; Fisher, DL; Mitchell, LG; Rodriquez, WC; Canik, JJ; Merril, CR; Weedn, VW. Mitochondrial DNA sequence analysis of human skeletal remains : identification of remains from the Vietnam war. *JForensicSci,* 1993, 38(3), 542-553.

Hoss, M; Jaruga, P; Zastawny, TH; Dizdaroglu, M; Paabo, S. DNA damage and DNA sequence retrieval from ancient tissues. *Nucleic acids research*, 1996, 24(7), 1304-1307.

Iscan, MY; Solla, HE; McCabe, BQ. Victim of a dictatorial regime: identification of Mr. Roberto Gomensoro Josman. *Forensic Sci Int,* 2005, 151(2-3), 213-220.

Ivanov, PL; Wadhams, MJ; Roby, RK; Holland, MM; Weedn, VW; Parsons, TJ. Mitochondrial DNA sequence heteroplasmy in the Grand Duke of Russia Georgij Romanov establishes the authenticity of the remains of Tsar Nicholas II. *Nat Genet,* 1996, 12(4), 417-420.

Jeffreys, AJ; Allen, MJ; Hagelberg, E; Sonnberg, A. Identification of the skeletal remains of Josef Mengele by DNA analysis. *Forensic Sci Int*, 1992, 56(1), 65-76.

Jehaes, E; Decorte, R; Peneau, A; Petrie, J; Boiry, P; Gilissen, A; Moisan, J;P, Van den Berghe, H; Pascal, O; Cassiman, JJ. Mitochondrial DNA analysis on remains of a putative son of Louis XVI, King of France and Maris-Antoinette. *European Journal of Human Genetics*, 1998, 6, 383-395.

Jehaes, E; Pfeiffer, H; Toprak, K; Decorte, R; Brinkmann, B; Cassiman, JJ. Mitochondrial DNA analysis of the putative heart of Louis XVII, son of Louis XVI and Marie-Antoinette. *Eur J Hum Genet*, 2001, 9(3), 185-190.

Knight, J. Bearing false witness. *New Scientist*(February 28), 1998, 13.

Lindahl, T. Instability and decay of the primary structure of DNA. *Nature*, 1993, 362(April), 709-715.

Lleonart, R; Riego, E; Sainz de la Pena, MV; Bacallao, K; Amaro, F; Santiesteban, M; Blanco, M; Currenti, H; Puentes, A; Rolo, F; others. Forensic identification of skeletal remains from members of Ernesto Che Guevara's guerrillas in Bolivia based on DNA typing. *Int J Legal Med*, 2000, 113(2), 98-101.

Matsuo, S; Toyokuni, S; Osaka, M; Hamazaki, S; Sugiyama, T. Degradation of DNA in dried tissues by atmospheric oxygen. *Biochem Biophys Res Commun,* 1995, 208(3), 1021-1027.

Melton, T. Mitochondrial DNA Heteroplasmy. *Forensic Science Reviews,* 2004, 16, 3-19.

Melton, T; Nelson, K. Forensic mitochondrial DNA analysis: two years of commercial casework experience in the United States. *Croat Med J,* 2001, 42(3), 298-303.

Morgan, MA. Amplification of Human Nuclear Pseudogenes Derived from Mitochondrial DNA : A Problem for Mitochondrial DNA Identity Testing ?; 1997. Promega. p 128.

Mullis, K; Faloona, F; Scharf, S; Saiki, R; Horn, G; Erlich, H. Specific enzymatic amplification of DNA in vitro: the polymerase chain reaction. *Cold Spring Harb Symp Quant Biol*, 1986a, 51 Pt 1, 263-273.

Mullis, K; Faloona, F; Scharf, SJ; Saiki, RK; Horn, GT; Erlich, H. Specific synthesis of DNA in vitro via a polymerase catalysed chain reaction. *Cold Spring Harbor Symp Quant Biol,* 1986b, 51, 263-273.

Mullis, KB; Faloona, FA. Specific synthesis of DNA in vitro via a polymerase-catalyzed chain reaction. *Methods Enzymol*, 1987, 155, 335-350.

Murray, BW; McClymont, RA; Strobeck, C. Forensic Identification of Ungulate Species Using Restriction Digests of PCR-Amplified Mitochondrial DNA. *J Forensic Sci,* 1995, 40(6), 943-951.

Nakaki, S; Hino, D; Miyoshi, M; Nakayama, H; Moriyoshi, H; Morikawa, T; Itohara, K. Study of animal species (human, dog and cat) identification using a multiplex single-base primer extension reaction in the cytochrome b gene. *Forensic Sci Int*, 2007, 173(2-3), 97-102.

Nass, MM. Mitochondrial DNA. I. Intramitochondrial distribution and structural relations of single- and double-length circular DNA. *J Mol Biol,* 1969, 42(3), 521-528.

Nelson, K; Melton, T. Forensic mitochondrial DNA analysis of 116 casework skeletal samples. *J Forensic Sci*, 2007, 52(3), 557-561.

Nelson, TM; Just, RS; Loreille, O; Schanfield, MS; Podini, D. Development of a multiplex single base extension assay for mitochondrial DNA haplogroup typing. *Croat Med J,* 2007, 48(4), 460-472.

Niederhauser, C; Hofelein, C; Wegmuller, B; Luthy, J; Candrian, U. Reliability of PCR decontamination systems. *PCR Methods Appl*, 1994, 4(2), 117-123.

Okayasu, R; Takahashi, S; Sato, H; Kubota, Y; Scolavino, S; Bedford, JS. Induction of DNA double strand breaks by arsenite: comparative studies with DNA breaks induced by X-rays. *DNA Repair (Amst)*, 2003, 2(3), 309-314.

Osiewacz, HD; Hermanns, J. The role of mitochondrial DNA rearrangements in aging and human diseases [see comments]. *Aging Milano*, 1992, 4(4), 273-286.

Ou, CY; Moore, JL; Schochetman, G. Use of UV irradiation to reduce false positivity in polymerase chain reaction. *Biotechniques*, 1991, 10(4), 442, 444, 446.

Paabo, S. Ancient DNA : Extraction,characterization,molecular cloning, and enzymatic amplification. *Proc Natl Acad Sci USA,* 1989, 86(march):1939-1943.

Paabo, S; Higuchi, R; Wilson, AC. Ancient DNA and the polymerase chain reaction. *The Journal of Biological Chemistry*, 1989, 264(17), 9709-9712.

Perry, WLd; Bass, WM; Riggsby, WS; Sirotkin, K. The autodegradation of deoxyribonucleic acid (DNA) in human rib bone and its relationship to the time interval since death. *J Forensic Sci*, 1988, 33(1), 144-153.

Poinar, HN; Stankiewicz, BA. Protein preservation and DNA retrieval from ancient tissues. *Proc Natl Acad Sci U S A,* 1999, 96(15), 8426-8431.

Primorac, D; Andelinovic, S; Definis-Gojanovic, M; Drmic, I; Rezic, B; Baden, MM; Kennedy, MA; Schanfield, MS; Skakel, SB; Lee, HC. Identification of war victims from mass graves in Croatia, Bosnia, and Herzegovina by use of standard forensic methods and DNA typing. *J Forensic Sci,* 1996, 41(5), 891-894.

Pusch, CM; Broghammer, M; Scholz, M. Cremation practices and the survival of ancient DNA: burnt bone analyses via RAPD-mediated PCR. *Anthropol Anz,* 2000, 58(3), 237-251.

Quintans, B; Alvarez-Iglesias, V; Salas, A; Phillips, C; Lareu, MV; Carracedo, A. Typing of mitochondrial DNA coding region SNPs of forensic and anthropological interest using SNaPshot minisequencing. *Forensic Sci Int,* 2004, 140(2-3), 251-257.

Saiki, RK; Gelfand, DH; Stoffel, S; Scharf, SJ; Higuchi, R; Horn, GT; Mullis, KB; Erlich, HA. Primer-directed enzymatic amplification of DNA with a thermostable DNA polymerase. *Science,* 1988, 239(4839), 487-491.

Saiki, RK; Scharf, SJ; Faloona, F; Mullis, KB; Horn, GT; Erlich, HA; Arnheim, N. Enzymatic amplification of β-globulin genomic sequences and restriction site analysis for diagnosis of sickle cell anemia. *Science,* 1985, 230, 1350-1354.

Satoh, M; Kuroiwa, T. Organization of multiple nucleoids and DNA molecules in mitochondria of a human cell. *Exp Cell Res,* 1991, 196(1), 137-140.

Savolainen, P; Rosen, B; Holmberg, A; Leitner, T; Uhlen, M; Lundeberg, J. Sequence analysis of domestic dog mitochondrial DNA for forensic use. *J Forensic Sci,* 1997, 42(4), 593-600.

Schweigert, N; Acero, JL; von Gunten, U; Canonica, S; Zehnder, AJ; Eggen, RI. DNA degradation by the mixture of copper and catechol is caused by DNA-copper-hydroperoxo complexes, probably DNA-Cu(I)OOH. E*nviron Mol Mutagen*, 2000, 36(1), 5-12.

Shoffner, JM; Wallace, DC. Mitochondrial Genetics: Principles and Practice. *American Journal of Human Genetics*, 1992, 51, 1179-1186.

Skolnick, AA. Mitochondrial DNA studies help identify lost victims of human rights abuses. *Jama*, 1993, 269(15), 1911-1913.

Stone, AC; Starrs, JE; Stoneking, M. Mitochondrial DNA analysis of the presumptive remains of Jesse James. *J Forensic Sci,* 2001, 46(1), 173-176.

Stone, R. DNA forensics. Buried, recovered, lost again? The Romanovs may never rest. *Science*, 2004, 303(5659), 753.

Sweet, D; Shutler, GG. Analysis of salivary DNA evidence from a bite mark on a body submerged in water. *J Forensic Sci,* 1999, 44(5), 1069-1072.

Tully, LA; Levin, BC. Human mitochondrial genetics. *Biotechnol Genet Eng Rev*, 2000, 17, 147-177.

Vallone, PM; Just, RS; Coble, MD; Butler, JM; Parsons, TJ. A multiplex allele-specific primer extension assay for forensically informative SNPs distributed throughout the mitochondrial genome. *Int J Legal Med*, 2004, 118(3), 147-157.

Wallace, DC; Shoffner, JM; Trounce, I; Brown, MD; Ballinger, SW; Corral-Debrinski, M; Horton, T; Jun, AS; Lott, MT. Mitochondrial DNA mutations in human degenerative diseases and aging. *Biochim Biophys Acta,* 1995, 1271(1), 141-151.

Wilson, MR; DiZinno, JA; Polanskey, D; Replogle, J; Budowle, B. Validation of mitochondrial DNA sequencing for forensic casework analysis. *IntJLegalMed*, 1995, 108(2), 68-74.

Wilson, MR; Polanskey, D; Replogle, J; DiZinno, JA; Budowle, B. A family exhibiting heteroplasmy in the human mitochondrial DNA control region reveals both somatic mosaicism and pronounced segregation of mitotypes. *Hum Genet*, 1997, 100(2), 167-171.

Wolf, C; Rentsch, J; Hubner, P. PCR-RFLP analysis of mitochondrial DNA: a reliable method for species identification. *J Agric Food Chem,* 1999, 47(4), 1350-1355.

Ye, J; Ji, A; Parra, EJ; Zheng, X; Jiang, C; Zhao, X; Hu, L; Tu, Z. A Simple and Efficient Method for Extracting DNA From Old and Burned Bone. *J Forensic Sci,* 2004, 49(4).

Zhivotovsky, LA. Recognition of the remains of Tsar Nicholas II and his family: a case of premature identification? *Ann Hum Biol,* 1999, 26(6), 569-577.

In: Horizons in DNA Research, Volume 1
Editor: Jason R. Chesterton, pp. 203-223
ISBN: 978-1-60876-968-1

Chapter 9

DNA-WRAPPED CARBON NANOTUBES IN AQUEOUS SOLUTION AND AIR

Masaru Tachibana*
Department of Nanosystem Science, Yokohama City University, 22-2 Seto, Kanazawa-ku, Yokohama 236-0027, Japan.

ABSTRACT

The production of DNA-wrapped single-wall carbon nanotubes (SWNTs) in aqueous solution has led to a bid advance for separation or isolation of SWNTs. The understanding of the effect of DNA-wrapping on physical properties of SWNTs is important for basic understanding of DNA-SWNT interaction as well as engineering applications of DNA-wrapped SWNTs. Raman scattering of SWNTs are significantly influenced by DNA wrapping. The Raman spectra of DNA-wrapped SWNTs in air strongly depend on relative humidity. In addition, the DNA-wrapping mechanism for smaller diameter SWNTs is different from that for larger ones. Such properties of DNA-wrapped SWNTs can provide further applications in air. In this chapter, first, the production of DNA-wrapped SWNTs is described. Next, the Raman spectral features of DNA-wrapped SWNTs in aqueous solution and air are reviewed. Finally, the morphologies of DNA-wrapped SWNTs are discussed.

1. INTRODUCTION

Many unique properties of DNA are very attractive for the functionalization of novel nanoscale materials. One of the most exciting applications of DNA for the materials functionalization is the production of the complex of DNA and single-wall carbon nanotube (SWNT) which is one of the most promising nanoscale materials for next generation [1,2]. The production of SWNTs wrapped by DNA in aqueous solution has led to a bid advance for

* Corresponding author: E-mail: tachiban@yokohama-cu.ac.jp

the separation or isolation of SWNTs [3]. This separation of SWNTs by DNA wrapping has provided new methods to study previously unaccessible processes in single nanotube, such as phonon dynamics [4]. In addition, the unveiled properties of DNA-wrapped SWNTs have provided sensing applications in various environments including a living cell [5-9].

The understanding of the effect of DNA-wrapping on physical properties of SWNTs is important for basis understanding of DNA-SWNTs interaction as well as technological applications of DNA-wrapped SWNTs. On the other hand, almost studies on DNA-wrapped SWNTs have been carried out in aqueous solution. For further applications, the understanding of the properties of DNA-wrapped SWNTs in air or vacuum is required.

Raman scattering of SWNTs are significantly influenced by DNA wrapping [10-12]. The Raman spectra of DNA-wrapped SWNTs in air strongly depend on relative humidity [13]. In addition, the DNA-wrapping mechanism for smaller diameter SWNTs is different from that for larger ones [14]. Such properties can provide further applications in air. In this chapter, first, the production of DNA-wrapped SWNTs is described. Next, the Raman spectral features of DNA-wrapped SWNTs in aqueous solution and air are reviewed. Finally, the morphologies of DNA-wrapped SWNTs are discussed.

2. Production of DNA-wrapped SWNTs

DNA-wrapped SWNTs dispersed in aqueous solution are prepared by below procedure as reported previously [2,3]. As-produced SWNT bundles are used as the starting materials. 1 mg of as-produced SWNTs is mixed in 1 ml aqueous single-stranded DNA (ssDNA) solution (0.3 mg ml^{-1} ssDNA, 20 mM HEPES buffer at pH 7.5). The DNAs are a sequence of repeats of alternating dG (deoxyguanylate) and dT (deoxythymidylate), d(GT)n, with total length of n = 20. The mixture is sonicated on ice for 60 min at a power level of 6.5 W. After sonication, the samples are divided into 0.5 ml aliquots, and centrifuged for 120 min at 25000 g to remove insoluble materials, leaving DNA-wrapped SWNT solution. The DNA-wrapped ones dispersed in aqueous solution are shown in Figure 1(a). For comparing, SWNTs without DNA dispersed in aqueous solution are also show in Figure 1(b). From the pictures in Figure 1, it is obvious that the dispersibility of DNA-wrapped SWNTs is much high compared with that of ones without DNA.

In addition, the dried DNA-wrapped SWNTs sample is prepared by dropping 3 μL of the stock solution onto a Si/SiO_2 substrate (1 μL at a time) in air. The drops are allowed to dry into a thick layer of DNA-wrapped SWNTs. The dried state of DNA-wrapped one on the substrate is also quite different from that of one without DNA as shown in Figure 2. The dried state of DNA-wrapped one exhibits a uniform thickness of film on the substrate, whereas one without DNA gives arise to inhomogeneously condensed one. The formation of the uniform film of DNA-wrapped one will be useful for fundamental studies and various applications in air or vacuum.

Figure 1. Photographs of (a) DNA-wrapped SWNTs and (b) SWNTs without DNA in aqueous solution, dispersed with sonication.

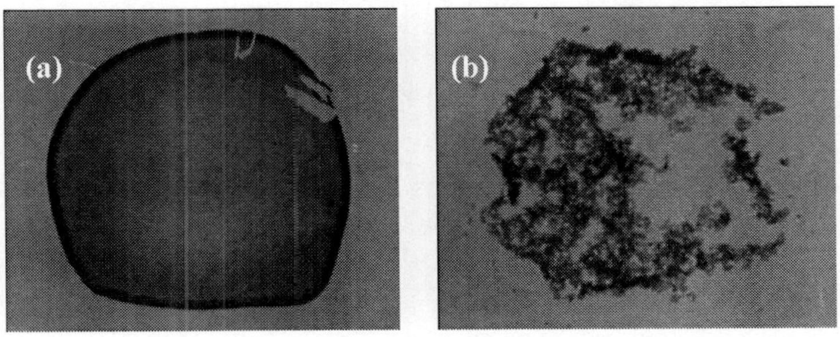

Figure 2. Photographs of dried (a) DNA-wrapped SWNTs and (b) SWNTs without DNA on the quartz substrate, prepared by dropping their aqueous solutions dispersed with sonication.

3. RAMAN SPECTRA OF DNA-WRAPPED SWNTS IN AQUEOUS SOLUTION AND AIR

Raman scattering, that is one of the most important physical properties of SWNTs, is related to not only phonon modes but also structures and electronic states [15,16]. The resonant Raman spectra of SWNTs include two main features: a radial breathing mode (RBM) and a tangential mode (so-called G band). The RBM band can give an easy and quick determination of the tube diameter distribution in SWNTs [17]. The G band is a more complex feature. The G band splits into two main components, so-called G^+ line at high frequency (typically 1590 cm^{-1}) and G^- line at low frequency. Both of G^+ and G^- lines are fitted with Lorentzian lines for semiconducting SWNTs [18]. On the other hand, for metallic SWNTs, G^+ and G^- lines are fitted with a Lorentzian line and an broad and asymmetric Breit-Wigner-Fano (BWF) line, respectively [19,20]. These Raman bands are sensitive to the surrounding environment including defects [21-25]. In this section, Raman spectral features of DNA-wrapped SWNTs in aqueous solution and air are reviewed.

3.1. Semiconducting SWNTs

Figure 3 shows RBMs and G band in Raman spectra for as-grown bundles of SWNTs, DNA-wrapped SWNTs in aqueous solution and dried DNA-wrapped SWNTs, taken with an excitation energy of E_{exc} = 2.33 eV [12]. The diameters of SWNTs resonantly contributing to the Raman spectrum are estimated to be d = 1.4 ± 0.1 nm from the RBM frequencies in Figure 3(a). Note that the RBM frequency ω_{RBM} [cm^{-1}] = 234/d [nm] + 10 for typical SWNT bundles [26]. According to the (revised) Kataura plot [19,27] referring to the excitation energy and SWNT diameters listed above, it is found that semiconducting SWNTs is predominantly resonant in the Raman spectrum of the as-grown bundles of SWNTs in Figure 3(a)

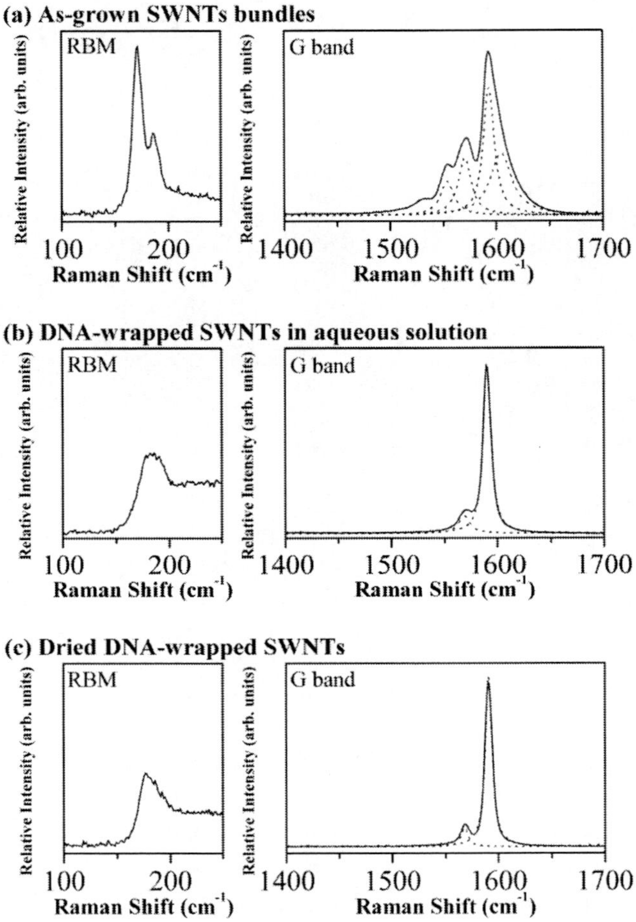

Figure 3. RBMs and G band in typical Raman spectra for (a) as-grown SWNT bundles, (b) DNA-wrapped SWNTs in aqueous solution and (c) dried DNA-wrapped SWNTs, taken at E_{exc} = 2.33 eV (λ = 532 nm) at 296 K. The G band is fitted with Lorentzian (dotted) lines. [12]

The G band for as-grown bundles of SWNTs is best fitted with four Lorentzian lines, as shown in Figure 3(a). The values of the fitting parameters are presented in Table 1. The Lorentzian lines located at 1553, 1570, 1593 and 1605 cm^{-1} can be assigned to the E_2^{LO}, unresolved (A^{TO}+E_1^{LO}), unresolved (A^{LO}+E_1^{TO}) and E_2^{TO} modes, respectively, in the G band associated with semiconducting SWNTs [18,28], as presented Table 1. Note that LO and TO indicate the longitudinal and transverse optical characters with respect to the tube axis, respectively. These spectral features of the G band suggest that semiconducting SWNTs are predominantly resonant in the Raman spectrum of as-grown bundles of SWNTs in Figure 3(a). This is consistent with that predicted from RBM frequencies as mentioned above.

Table 1. Peak values (ω) and FWHM (2Γ) of Lorentzian lines used to fit the G band in Raman spectra [a] [12].

Fitting curve	Component (Mode symmetry)	As-grown SWNT bundles		DNA-wrapped SWNTs in aqueous solution	
		ω (cm^{-1})	2Γ (cm^{-1})	ω (cm^{-1})	2Γ (cm^{-1})
Lorentzian	S (E_2^{LO})	1553	15		
Lorentzian	S (A^{TO}+E_1^{LO})	1570	15	1569	13
Lorentzian	S (A^{LO}+E_1^{TO})	1593	13	1591	9
Lorentzian	S (E_2^{TO})	1605	24		

Fitting curve	Component (Mode symmetry)	Dried DNA-wrapped SWNTs	
		ω (cm^{-1})	2Γ (cm^{-1})
Lorentzian	S (E_2^{LO})		
Lorentzian	S (A^{TO}+E_1^{LO})	1568	8
Lorentzian	S (A^{LO}+E_1^{TO})	1591	8
Lorentzian	S (E_2^{TO})		

[a] Taken at E_{exc} = 2.33 eV (λ = 532 nm) at 296 K. S with mode symmetries denotes the line components associated with semiconducting SWNTs. The fitting parameters are obtained from the smallest root-mean-square error of the fitting procedure.

As shown in Figure 3(b), significant changes in the RBMs and G band associated with semiconducting SWNTs are observed for the DNA-wrapped SWNTs in aqueous solution. For the RBMs, not only mode frequencies but also relative intensities are changed. The frequencies are upshifted by 2-5 cm^{-1} relative to the as-grown bundles of SWNTs. The change in the RBM relative intensity implies a variation in some kinds of SWNTs resonantly contributing to the Raman spectrum. The G band for the DNA-wrapped SWNTs in aqueous solution is best fitted with only two Lorentzian lines located at 1591 and 1569 cm^{-1}, as shown in Figure 3(b). These two peaks can be assigned to unresolved (A^{LO}+E_1^{TO}) and unresolved (A^{TO}+E_1^{LO}) modes, so-called the G^+ and G^- bands, respectively, associated with semiconducting SWNTs, as listed in Table 1. These results show that the E_2 modes of SWNTs are vanished by DNA-wrapping in aqueous solution. This is direct evidence for the vanishing of the E_2 modes, which has been predicted from the narrowing of the G^+ band by DNA-wrapping for CoMoCAT and HiPco samples [10,11] as mentioned below.

In addition, the relative intensity of the G^- band to the G^+ band associated with semiconducting SWNTs changes significantly by DNA-wrapping, as shown in Figure 3(b). The relative intensity of G^- to G^+ band is drastically reduced by 78 %, relative to that of the as-grown bundles of SWNTs. The G^+ and G^- bands can be predominantly ascribed to the vibrations of carbon atoms along the axis and the circumferential direction of the SWNT, respectively [29]. Thus, the reduction of the relative intensity of the G^- band means suppression of the vibrations of carbon atoms along the circumference direction of the

SWNTs. This shows that the DNA-wrapping can suppress carbon atom vibrations along the circumferential direction, not along the axis of the SWNTs.

The frequency and width of the G^+ band associated with semiconducting SWNTs also change by DNA-wrapping. The G^+ band is downshifted by 2 cm^{-1} relative to that of the as-grown bundles of SWNTs, as listed in Table 1. The frequency of the G^+ band is sensitive to charge transfer from dopant additions to SWNTs [30], as in graphite intercalation compounds [31,32]. Thus, the downshift of the G^+ band is probably due to electron transfer from DNA to semiconducting SWNTs. This implies that DNA can play a role as donor for SWNTs in aqueous solution. Such charge transfer can also be supported by the change of RBMs as mentioned above. In addition, the width of the G^+ band for DNA-wrapped SWNTs is reduced to 9 cm^{-1}, which is close to 5 cm^{-1} of isolated SWNTs [29]. This suggests that narrowing of the G^+ band can be ascribed to isolation of SWNTs by DNA-wrapping in aqueous solution. These spectral features for semiconducting SWNTs remain almost unchanged after the DNA-wrapped SWNTs were dried in air, as shown in Figure 3(c).

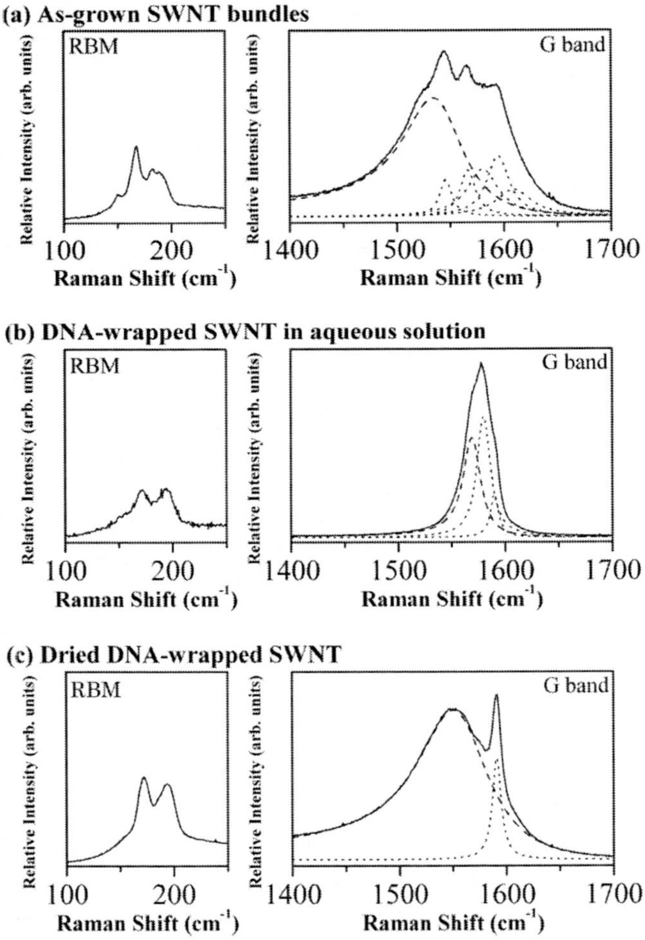

Figure 4. RBMs and G band in typical Raman spectra for (a) as-grown SWNT bundles, (b) DNA-wrapped SWNTs in aqueous solution and (c) dried DNA-wrapped SWNTs, taken at E_{exc} = 1.96 eV (λ = 632.8 nm) at 296 K. The G band is fitted with Lorentzian (dotted) lines, and a broad and asymmetric BWF (dashed) line. [12]

Table 2. Peak values (ω) and FWHM (2Γ) of Lorentzian lines and one BWF line used to fit the G band in Raman spectra [a] [12].

Fitting curve	Component [b] (Mode symmetry)	As-grown SWNT bundles			DNA-wrapped SWNTs in aqueous solution		
		$\omega(cm^{-1})$	$2\Gamma(cm^{-1})$	1/q [c]	$\omega(cm^{-1})$	$2\Gamma(cm^{-1})$	1/q
BWF	M (A^{LO})	1535	66	-0.16	1570	20	-0.06
Lorentzian	S (E_2^{LO})	1546	18				
Lorentzian	S ($A^{TO}+E_1^{LO}$)	1565	19				
Lorentzian	M (A^{TO})	1580	23		1580	15	
Lorentzian	S ($A^{LO}+E_1^{TO}$)	1594	20		1590	9	
Lorentzian	S (E_2^{TO})	1610	36				

Fitting curve	Component [b] (Mode symmetry)	Dried DNA-wrapped SWNTs		
		$\omega(cm^{-1})$	$2\Gamma(cm^{-1})$	1/q
BWF	M (A^{LO})	1556	80	-0.15
Lorentzian	S (E_2^{LO})			
Lorentzian	S ($A^{TO}+E_1^{LO}$)			
Lorentzian	M (A^{TO})			
Lorentzian	S ($A^{LO}+E_1^{TO}$)	1591	10	
Lorentzian	S (E_2^{TO})			

[a] Taken at E_{exc} = 1.96 eV (λ = 632.8 nm) at 296 K. The fitting parameters are obtained from the smallest root-mean-square error of the fitting procedure.

[b] S and M with mode symmetries denote the line components associated with semiconducting and metallic SWNTs, respectively.

[c] Asymmetric parameter (1/*q*) for the BWF line.

3.2 Metallic SWNTs

RBMs and G band for as-grown bundles of SWNTs, DNA-wrapped SWNTs in aqueous solution and dried DNA-wrapped SWNTs, taken with E_{exc} = 1.96 eV, are shown in Figures 4. From the RBM frequencies in Figure 4(a), the diameters of SWNTs resonantly contributing in the Raman spectrum can be estimated to be $d = 1.5 \pm 0.2$ nm by the relation used above. According to the (revised) Kataura plot as mentioned above, it is found that both metallic and semiconducting SWNTs are resonant in the Raman spectrum of as-grown SWNT bundles in Figure 4(a).

The G band shown in Figure 4(a) is best fitted with five Lorentzian lines and one Breit-Wigner-Fano (BWF) line [19,20,33],

$$I(\omega) = I_0 \frac{[1+(\omega-\omega_{\mathrm{BWF}})/q\Gamma]^2}{1+[(\omega-\omega_{\mathrm{BWF}})/\Gamma]^2}, \quad (1)$$

where I_0, ω_{BWF}, Γ, and $1/q$ are the intensity, renormalized frequency, broadening parameter, and asymmetric parameter, respectively. Note that ω_{BWF} and 2Γ are called, respectively, the peak frequency and full width at half maximum (FWHM) of the BWF line. The BWF line results from the coupling of phonons to the electronic continuum of metallic SWNTs [19,20], although the nature is still controversial [33-36]. More recently, it was reported that the broad peak is attributed to electron-phonon coupling with Kohn anomaly [37-39]. As clarified by Eq. (1), $1/q$ determines the departure of the line shape from a symmetric Lorentzian line and, therefore, $|1/q|$ is a measure of the degree of BWF coupling. The $1/q$ value is also correlated with Γ [33]. The values of the fitting parameters are listed in Table 2.

Four Lorentzian lines located at 1546, 1565, 1594 and 1610 cm^{-1} can be assigned to the E_2^{LO}, unresolved (A^{TO}+E_1^{LO}), unresolved (A^{LO}+E_1^{TO}) and E_2^{TO} modes, respectively, in the G band associated with semiconducting SWNTs [20,28], while the Lorentzian and BWF lines located at 1580 and 1535 cm^{-1} can be assigned to the A^{TO} and A^{LO} modes, respectively, in the G band associated with metallic SWNTs [28], as presented in Table 2. Note that the relative ordering of A^{TO} and A^{LO} are opposite to those in Ref. [18]. These spectral features of the G band also suggest that both semiconducting and metallic SWNTs are resonant in the Raman spectrum of the as-grown bundles of SWNTs in Figure 4(a).

As shown in Figure 4(b), significant changes in the RBMs and the G band are observed for DNA-wrapped SWNTs in aqueous solution. For the RBMs, an up-shift of 2-5 cm^{-1} relative to as-grown bundles of SWNTs is observed, being similar to that in Figure 3. The G band for the DNA-wrapped SWNTs in aqueous solution is best fitted with two Lorentzian lines and one BWF line, as shown in Figure 4(b). One Lorentzian peak located at 1590 cm^{-1} can be assigned to the unresolved (A^{LO}+E_1^{TO}) mode, so-called the G^+ band associated with semiconducting SWNTs, as presented in Table 2. The other Lorentzian line and one BWF line located at 1580 and 1570 cm^{-1} can be assigned to the A^{TO} and A^{LO} modes, so-called the G^+ and G^- bands, respectively, associated with metallic SWNTs, also presented in Table 2. The change in G band for semiconducting SWNTs is consistent with that in Figure 3.

The width of G^+ band associated with metallic SWNTs is narrowed by DNA-wrapping, similar to that for semiconducting SWNTs. We note that a significant change in the BWF line (G^- band) associated with metallic SWNTs is observed for DNA-wrapped SWNTs in aqueous solution. As shown in Figure 4(b), the broad and asymmetric feature of the BWF line is almost vanished by DNA-wrapping in aqueous solution. The BWF feature strongly reappears after the DNA-wrapped SWNTs are dried in air, as shown in Figure 4(c). The corresponding changes in the values of BWF parameters are presented in Table 2. In addition, the BWF profile for dried samples is almost recovered to that in aqueous solution after they are dipped in water. The change in the BWF profile occurs reversibly by air-drying and hydrating of the DNA-wrapped SWNTs.

The observed change in the BWF line, or BWF coupling, due to air-drying and hydrating means that the electronic properties in DNA-wrapped SWNTs can be changed by the desorption and absorption of water around them. These changes in the electronic properties might be attributed to the bundling and debundling or isolation of DNA-wrapped SWNTs. On the other hand, the G band profile associated with semiconducting SWNTs, shown in Figures 5(b) and (c), is not significantly changed even by air-drying and hydrating of the DNA-wrapped SWNTs. These results show that the BWF line associated with metallic SWNTs is very sensitive to the surrounding environment, in comparison with other various Raman modes.

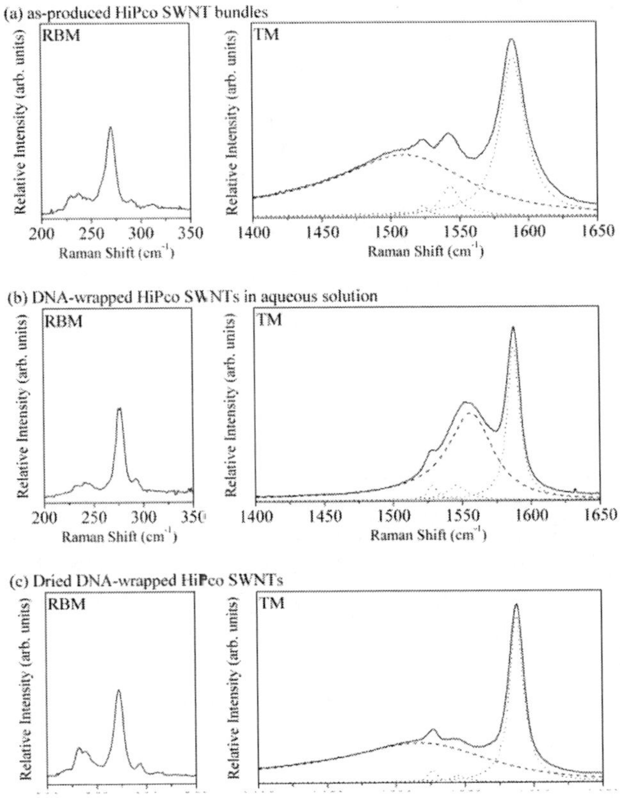

Figure 5. RBM and G bands in typical Raman spectra for (a) as-produced HiPco SWNT bundles, (b) DNAwrapped HiPco SWNTs in aqueous solution, and (c) dried DNA-wrapped HiPco SWNTs, taken with an excitation energy of 2.33 eV (532 nm) at 296 K. The G band is fitted with three Lorentzian (dotted) lines and one BWF (dashed) line. [11]

A similar change in the BWF line has been observed for DNA-wrapped SWNTs from HiPco and CoMoCAT samples with smaller mean diameters of ~ 0.9 and ~ 0.8 nm, respectively [10,11], as shown in Figure 5. Therefore, it seems that such a change in the BWF line is independent of the mean diameter of the SWNTs wrapped by DNA in the range of ~ 0.8 to 1.5 nm.

4. Humidity Dependence of DNA-Wrapped SWNTs

It is well known that the conformation and electrostatic property of DNA are sensitive to surrounding water [40]. It is expected that the properties of DNA-wrapped SWNTs are influenced by the change in the properties of DNA depending on surrounding water. Therefore, the properties of DNA-wrapped SWNTs associated with water vapor in air are of interest for further applications. In this section, humidity dependence of the Raman spectrum of DNA-SWNT hybrids in air is reviewed.

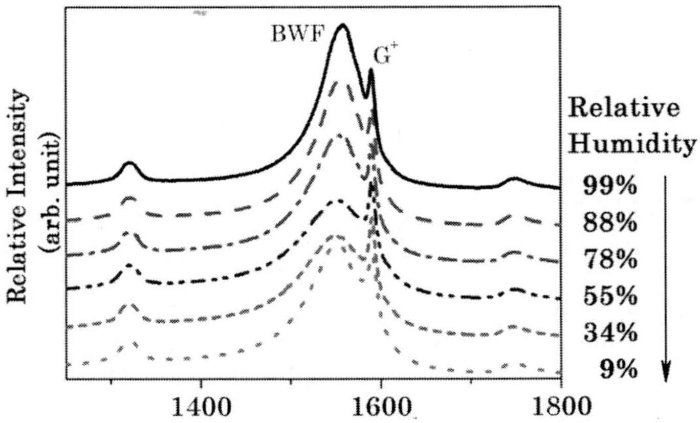

Figure 6. The relative humidity dependence of the G band in the Raman spectra for DNA-SWNT hybrids in air, obtained with an excitation energy of 1.96 eV. [13]

G band (1400 – 1600 cm^{-1}) in the Raman spectra of DNA-SWNT hybrids under various humidity conditions is shown in Figure 6. Note that the DNA-SWNT hybrids are obtained from SWNTs with a mean diameter of 1.5 nm synthesized by arc electric charge method. It should be noted that the intensity profile of the G band strongly depends on the relative humidity. In addition, the change in the intensity profile is reversible with the humidity. On the other hand, for pristine SWNTs, no humidity dependence is observed in the G band. Thus, the humidity dependence of the G band is one of interesting features for DNA-SWNT hybrids in air.

To clarify the humidity dependence for DNA-SWNT hybrids, the G band is analyzed in detail. The G band is fitted with a Lorentzian line (G^+) and a broad and asymmetric BWF line as mentioned above. The $1/q$ is correlated with Γ where both $1/q$ and Γ are fitting parameters for BWF line as mentioned above. The fitting results show that the BWF line strongly depends on the relative humidity, while the Lorentzian line exhibit no significant humidity dependence.

The humidity dependences of the $|1/q|$, $FWHM_{BWF}$ (2Γ) and I_{BWF}/I_G^+ (the relative peak intensity of the BWF line to the Lorentzian line) used to fit the BWF line are shown in Figure 7. It is found that these fitting parameters strongly depend on the humidity. As seen in Figure 7 (a) and (b), the values of $|1/q|$ and $FWHM_{BWF}$ increase with decreasing humidity. The values reach a maximum at the humidity of 55 %. It is of interest that further decrease of the humidity leads to the decrease of $|1/q|$ and $FWHM_{BWF}$. On the other hand, for I_{BWF}/I_G^+, its value decreases with decreasing humidity and reaches a minimum at the humidity of about 55 %, as seen in Figure 7 (c). Then its value increases with decreasing humidity. The behavior of I_{BWF}/I_G^+ is quite consistent with those of $|1/q|$ and $FWHM_{BWF}$. Similar behaviors were also observed for DNA-SWNT hybrids prepared from HiPco SWNTs with mean diameter smaller than the electric arc discharge SWNTs. These results show that the BWF coupling increases with decreasing humidity in the higher humidity range above 55 %, while it decreases with decreasing humidity in the lower humidity range below 55 %. The change of the BWF coupling can be attributed to the electronic property in DNA-SWNT hybrids depending on the humidity. According to previous reports [11,34,35], the BWF coupling can be also related to the bundling of SWNTs. Thus, the decrease of the humidity in the higher humidity range can occur the bundling of DNA-SWNT hybrids, while the decrease of the humidity in the lower humidity range might produce the debundling.

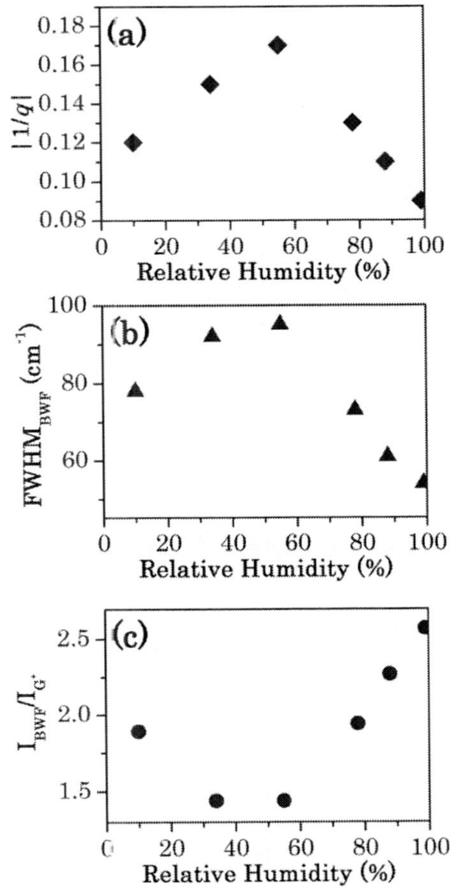

Figure 7. Relative humidity dependences of: (a) $|1/q|$, (b) $FWHM_{BWF}$ (2Γ), and (c) I_{BWF}/I_G^+ (the relative peak intensity of the BWF line to the Lorentzian line) used to fit the BWF line in the G band Raman line for DNA-SWNT hybrids shown in Figure 6.[13]

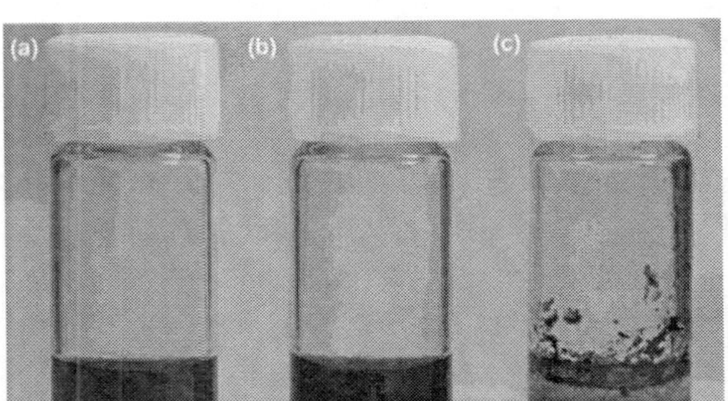

Figure 8. Photographs of DNA-SWNT hybrid solutions prepared from (a) HiPco-SWNTs and (b) Arc-SWNTs. (c) Arc-SWNTs poorly dispersed in aqueous solution without DNA is also shown as a reference. [14]

Let us consider the mechanism of such humidity dependence of the BWF line for DNA-SWNT hybrids. It is well known that the water is easily absorbed on DNA in air [40]. The absorption behavior of the water strongly depends on not only the humidity but also the absorption site in DNA. As main absorption sites of the water, there are phosphoric acid groups, the carbonyl groups, ring nitrogen atoms in DNA [40]. In the higher humidity range above 60 %, the desorption of water predominantly occur on carbonyl groups and/or ring nitrogen atoms in DNA. In the lower humidity range below 60 %, the desorption of water can occur on not only carbonyl groups and/or ring nitrogen atoms but also phosphoric acid groups of DNA backbone. Such desorption of water might change the structure of DNA in DNA-SWNT hybrids, as A or B form for double-stranded DNA [40]. The particular humidity of 60 % at which the desorption site of the water in DNA changes is in good agreement with 55 % relative humidity in the humidity dependence of the BWF line as seen in Figure 7. Thus, the change in the desorption site of water in DNA can vary the sign of the humidity dependence of the electronic property, or bundling effect, in DNA-SWNT hybrids. From these results, it is suggested that such DNA-SWNT hybrid is a promising material for applications, such as humidity sensors, even in air.

5. Morphologies of DNA-Wrapped SWNTs

Despite significant progress in the practical exploitation of the unique features of DNA-SWNT hybrids, very little is known how SWNTs are wrapped by DNA, which a key issue for the basic understanding of DNA-SWNT interaction as well as engineering applications of DNA-SWNT hybrids. In this section, the morphologies of DNA-CNT hybrids by atomic force microscopy (AFM) are discussed.

Figures 8(a) and (b) show photographs of DNA-SWNT hybrid solutions prepared from HiPco- and Arc-SWNTs with mean diamters of 0.9 and 1.4 nm, respectively. Note that these solutions are prepared without HEPES buffer for AFM observation of DNA-SWNT hybrids. As a reference, the photograph of Arc-SWNTs poorly dispersed in aqueous solution without DNA is also shown in Figure 8(c). From the comparison in Figure 8, it is obvious that DNA-SWNT hybrids are well-dispersed in both solutions. The Raman spectra exhibited the vanishing of E_2 mode and the reduction of BWF line in G band for both hybrid solutions. These Raman features, which are in good agreement with those with HEPES buffer [11-13] as shown in Figures 4 and 5, indicate the debundling or isolation of SWNTs. Thus, the debundling of SWNTs by DNA-wrapping for both hybrid solutions occurs even without HEPES.

The DNA-SWNT hybrids on the mica substrate, prepared from the solution, are observed by AFM. The DNA-HiPco-SWNT hybrids from original solution without dilution exhibit their high density and overlaps. These lead to the difficulty of measuring the exact diameter of each hybrid. To separate each hybrid, the solution is diluted with water. Figure 9(a) shows a typical AFM image of DNA-HiPco-SWNT hybrids from the solution diluted five times. The line contrasts correspond to the hybrids, while the white particles can be related to DNA aggregations. In addition, the surface undulation of each hybrid as seen in Figure 9(a) would be attributed to DNA wrapping SWNTs. The heights of the hybrids are measured from AFM images of various areas on the mica substrate as

Figure 9(a). Note that the top height for each hybrid with the undulation is selected as the diameter of the hybrid. Figure 10(a) shows the histogram of the diameters of DNA-HiPco-SWNT hybrids. The mean diameter of the hybrids is estimated to be 1.94 nm. This magnitude is larger than the mean diameter (~0.9 nm) of pristine HiPco-SWNTs without DNA by about 1.0 nm, which would be consistent with the thickness of a DNA [40]. Thus, it is suggested that, in DNA-HiPco-SWNT hybrid, one SWNT is wrapped by a layer of DNA.

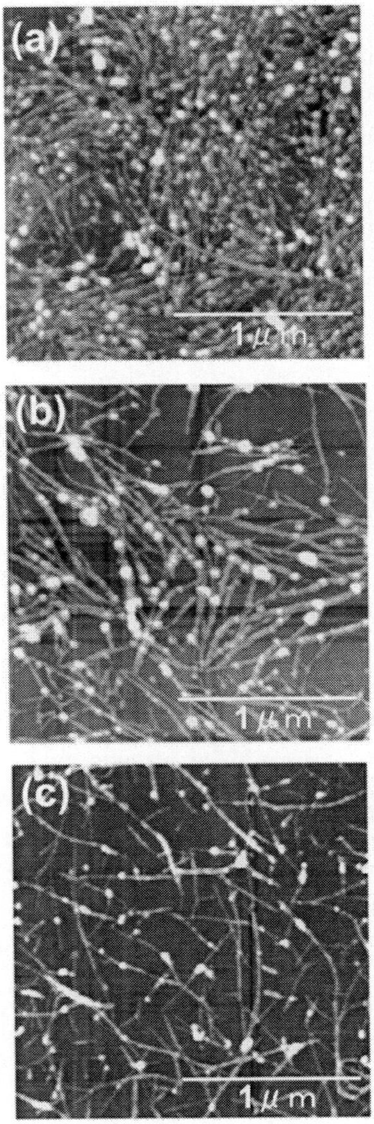

Figure 9. AFM images for DNA-HiPco-SWNT hybrids on mica substrates, prepared from aqueous solutions (a) diluted five times, (b) diluted ten times, and (c) diluted twenty times. [14]

AFM images are also observed for further diluted DNA-HiPco-SWNT hybrid solution. Figures 9(b) and (c) show AFM images of DNA-HiPco hybrids prepared from solutions diluted ten and twenty times, respectively. The histograms of diameters of the diluted hybrids corresponding to Figures 9(b) and (c) are shown in Figures 10(b) and (c), respectively. It is found that the density of the hybrids on the substrate is clearly reduced with dilution as seen in Figures 9(b) and (c). On the other hand, as shown in Figures 10(b) and (c), the mean diameters of the hybrids are estimated to be 1.87 and 1.81 nm for solutions diluted ten and twenty times, respectively. Thus, no significant change in the mean diameter for DNA-HiPco-SWNT hybrids is observed with dilution.

DNA-Arc-SWNT hybrids prepared by similar procedure as mentioned above are observed by AFM. Figure 10(a) show a typical AFM image of DNA-Arc-SWNT hybrids prepared from original solution without dilution. The DNA-Arc-SWNT hybrids are separately observed as seen in Fig, 10(a). Some DNA aggregations are also observed as particles on the substrate. The density of DNA-Arc-SWNT hybrids is low compared with that of DNA-HiPco-SWNT one without dilution. The diameters of DNA-Arc-SWNT hybrids are much larger than those of DNA-HiPco-SWNT ones. The mean diameter of DNA-Arc-SWNT hybrids is estimated to be 3.74 nm. This magnitude is more than twice as large as those of pristine Arc-SWNTs (~1.4 nm) without DNA. This large diameter can not be simply explained by the model in which one SWNT is wrapped by a layer of DNA (~1 nm) as suggested for DNA-HiPco-SWNT hybrids. The large diameter of 3.74 nm can corresponds to a bundle containing seven close-packed SWNTs (each diameter of 1.4 nm), or to one SWNT wrapped with several layers of DNA. These results show that the mechanism of DNA-wrapping for large diameter Arc-SWNTs is different from that for small diameter HiPco-SWNTs.

DNA-Arc-SWNT hybrid solution is also diluted with water. The AFM images and diameter distribution of the diluted DNA-Arc-SWNT hybrids are shown in Figures 11 and 12. It should be noted that the diameter of DNA-Arc-SWNT hybrids is drastically reduced with dilution. The mean diameter of the hybrids from the solution diluted ten times reaches 2.53 nm, which is close to one Arc-SWNT (~1.4 nm) wrapped by a layer of DNA (~1.0 nm). Namely, the diameter of the hybrids is reduced by 1.2 nm even when it is diluted only ten times. In contrast with such reduction of diameter, the number of DNA aggregations on the substrate increases with dilution. From these results, it is predicted that, in DNA-Arc-SWNT hybrid solution, the large amount of DNA is bound or wrapped around Arc-SWNTs. The DNA can be detached from the hybrids when the solution is diluted with water. Consequently, the detached DNA is precipitated and aggregated when the solution is air-dried on the substrate. Such behaviors of DNA in DNA-Arc-SWNT hybrid solution with dilution can explain AFM results as mentioned above. Thus, it is suggested that large diameter Arc-SWNTs are wrapped by several layers of DNA in aqueous solution, differing from DNA-HiPco SWNT hybrids. Such discrepancy in DNA-wrapping mechanism for different diameter SWNTs can reflect their chemical and physical properties, such as separation and photoluminescence.

DNA-SWNT hybrids can be prepared with not only single- but also double-stranded DNA solutions. Recently, the AFM study on DNA-HiPco-SWNT hybrids prepared with double-stranded DNA solution was reported [41]. The diameter of the hybrids was also reduced with dilution, as similar to the results with single-stranded DNA as mentioned above. In the report, it was suggested that the origin of the diameter reduction is due to the spontaneous debundling of SWNTs in the diluted solution. However, it is expected that the wrapping mechanism of single-stranded DNA for SWNTs is different from that of double-stranded one, since the structure and hydrophobicity of single-stranded DNA are quite different from those of double-stranded DNA. Therefore, it is difficult to simply compare DNA-wrapping mechanism between them.

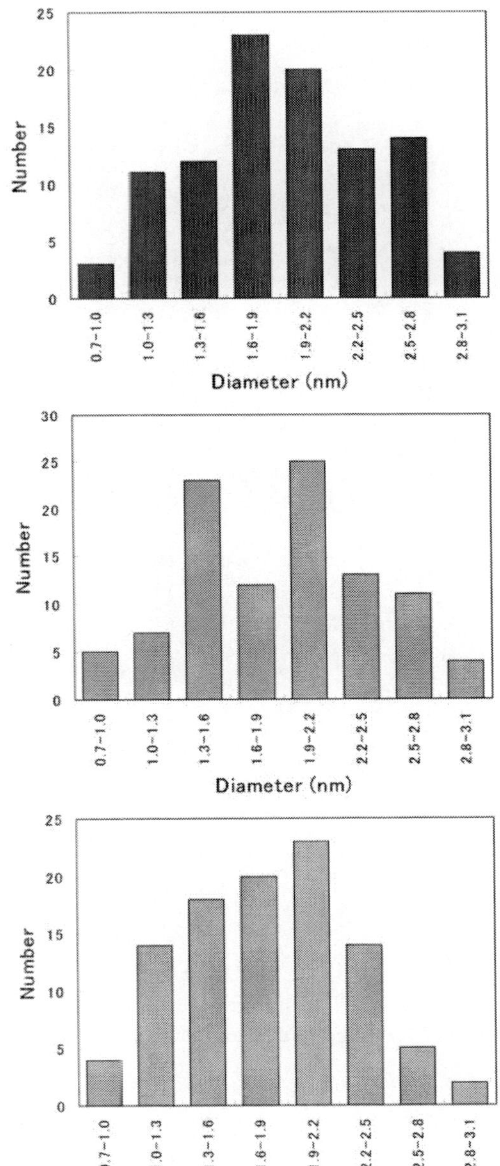

Figure 10. Histograms of diameters for DNA-HiPco-SWNT hybrids prepared from aqueous solutions (a) diluted five times, (b) diluted ten times, and (c) diluted twenty times, corresponding to Figure 9. [14]

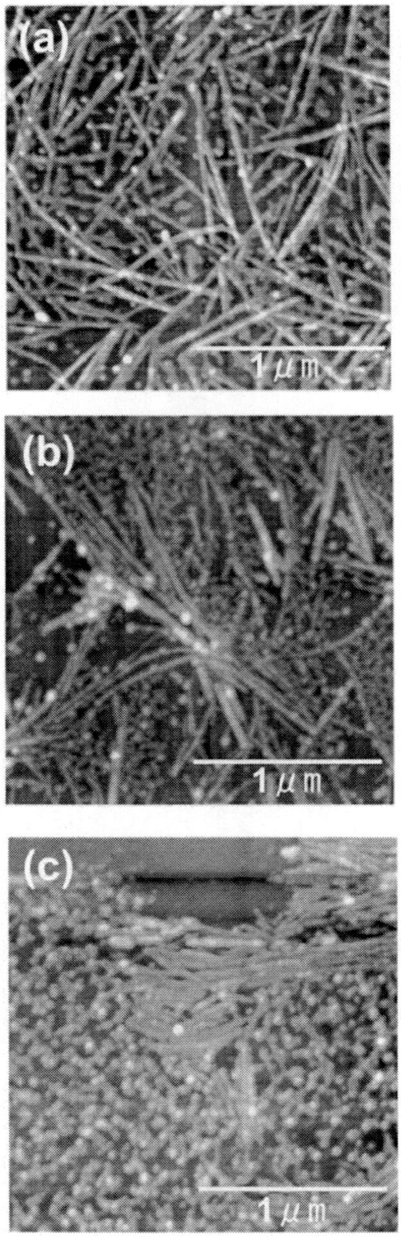

Figure 11. AFM images for DNA-Arc-SWNT hybrids on mica substrates, prepared from aqueous solution (a) without dilution and (b) diluted five times, and (c) diluted ten times. [14]

More recently, DNA-SWNT hybrids were observed by scanning tunneling microscopy (STM) [42]. The STM images showed strands of DNA wrapping around (6,5) nanotubes at ~63° angle with a coiling period of 3.3 nm, in agreement with the theoretical predication. These results provide more detailed insight into the self-assembling mechanism of DNA-SWNT hybrids.

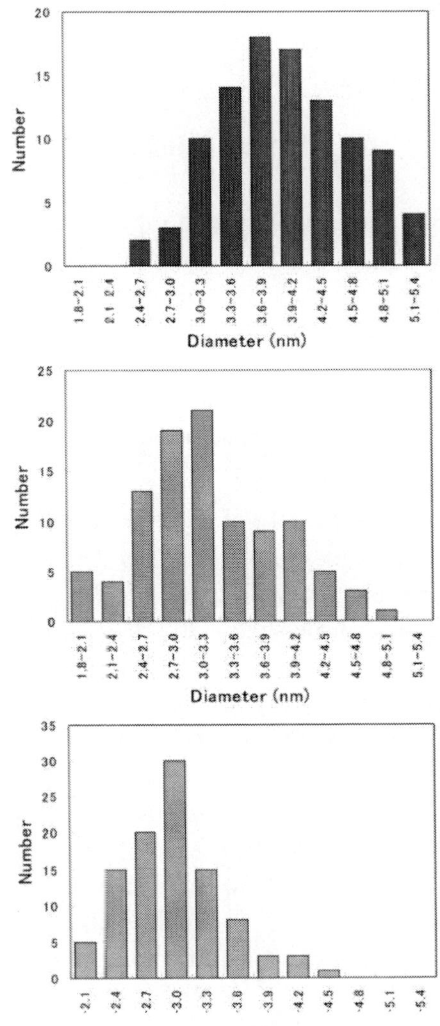

Figure 12. Histograms of diameters for DNA-Arc-SWNT hybrids prepared from aqueous solution (a) without dilution and (b) diluted five times, and (c) diluted ten times, corresponding to Figure 11. [14]

6. CONCLUSION

Raman spectra of SWNTs are significantly influenced by DNA wrapping. Especially, the broad and asymmetric features of G^- band associated with metallic SWNTs are vanished by DNA-wrapping in aqueous solution. The metallic G^- band appears again when the DNA-wrapped SWNTs are dried in air. The change in metallic G^- band of DNA-wrapped SWNTs can be explained by the interaction between SWNT and DNA whose conformation and electrostatic properties is easily changed by the absorption and desorption of surrounding water. In addition, the DNA-wrapping mechanism for smaller diameter SWNTs is different from that of larger ones. It seems that the difference affects the physical and chemical properties such as photoluminescence and separation. Finally , the combination of unique electronic and optical properties of SWNTs with the functionalization of DNA promises a broad range of applications, especially in biological systems including medicine, drug delivery, and cancer therapy.

ACKNOWLEDGMENTS

The author acknowledges the contributions to the work presented here by all past and present collaborators, especially Dr. Hironori Kawamoto, Dr. Takashi Uchida, Mr. Sho Toita, Mr. Dongchul Kang, Ms. Mari Hakamatsuka, and Emeritus Prof. Kenichi Kojima. This work was supported in part by a Strategic Research Project in Yokohama City University.

REFERENCES

[1] Nakashima, N; Okuzono, S; Murakami, H; Nakai, T; Yoshikawa, K. DNA Dissolves Single-walled Carbon Nanotubes in Water. *Chem. Lett.*, 2003, 32, 456-457.

[2] Zheng, M; Jagota, A; Semke, ED; Diner, BA; Mclean, RS; Lustig, SR; Richardson, RE; Tassi, NG. DNA-assisted dispersion and separation of carbon nanotubes. *Nature Mater.*, 2003, 2, 338-342.

[3] Zheng, M; Jagota, A; Strano, MS; Santos, AP; Barone, P; Chou, SG; Diner, BA; Dresselhaus, MS; Mclean, RS; Onoa, GB; Samsonidze, GG; Semke, ED; Usrey, M; Walls, DJ; Structure-Based Carbon Nanotube Sorting by Sequence-Dependent DNA Assembly. *Science,* 2003, 302, 1545-1548.

[4] Chou, SG; Plentz, F; Jiang, J; Saito, R; Nezich, D; Ribeiro, HB; Jorio, A; Pimenta, MA; Ge. Samsonidze, G; Santos, AP; Zheng, M; Onoa, GB; Semke, ED; Dresselhause, G; Dresselhaus, MS. Phonon-Assisted Excitonic Recombination Channels Observed in DNA-Wrapped Carbon Nanotubes Using Photoluminescence Spectroscopy. *Phys. Rev. Lett.,* 2005, 94, 127402.

[5] Barone, PW; Baik, S; Heller, DA; Strano, MS; Near-infrared optical sensors based on single-walled carbon nanotubes. *Nat. Mater.*, 2004, 4, 86-92.

[6] Staii, C; Johnson, AT. DNA-Decorated Carbon Nanotubes for Chemical Sensing. *Nano Lett.,* 2005, 5, 1774-1778.

[7] Heller, DA; Jeng, ES; Yeung, TK; Martinez, BM; Moll, AE; Gastala, JB; Strano, MS. Optical Detection of DNA Conformational Polymorphism on Single-Walled Carbon Nanotubes. *Science,* 2006, 311, 508-511.

[8] Gui, EL; Li, LJ; Lee, PS; Lohani, A; Mhaisalkar, SG; Cao, Q; Kang, SJ; Rogers, JA; Tansil, NC; Gao, Z. Electrical detection of hybridization and threading intercalation of deoxyribonucleic acid using carbon nanotube network field-effect transistors. *Appl. Phys. Lett.*, 2006, 89, 232104.

[9] Xu, Y; Pehrsson, PE; Chen, L; Zhang, R; Zhao, W. Double-Stranded DNA Single-Walled Carbon Nanotube Hybrids for Optical Hydrogen Peroxide and Glucose Sensing. *J. Phys. Chem. C,* 2007, 111, 8638-8643.

[10] Chou, SG; Ribeiro, HB; Barros, EB; Santos, A; Nezich, PD; Ge. Samsonidze, G; Fantini, C; Pimenta, MA; Jorio, A; Filho, FP; Dresselhaus, MS; Dresselhaus, G; Saito, R; Zheng, M; Onoa, GB; Semke, ED; Swan, AK; Ünlü, MS; Goldberg, BB. Optical characterization of DNA-wrapped carbon nanotube hybrids. *Chem. Phys. Lett.,* 2004, 397, 296-301.

[11] Kawamoto, H; Uchida, T; Kojima, K; Tachibana, M; The feature of the Breit-Wigner-Fano Raman line in DNA-wrapped single-wall carbon nanotubes. *J. Appl. Phys.*, 2006, 99, 094309.

[12] Kawamoto, H; Uchida, T; Kojima, K; Tachibana, M. G band Raman features of DNA-wrapped single-wall carbon nanotubes in aqueous solution and air. *Chem. Phys. Lett.*, 2006, 432, 172-176.

[13] Kawamoto, H; Uchida, T; Kojima, K; Tachibana, M; Raman study of DNA-wrapped single-wall carbon nanotube hybrids under various humidity conditions. *Chem. Phys. Lett.*, 2006, 431, 118-120.

[14] Toita, S; Kang, D; Kobayashi, K; Kawamoto, H; Kojima, K; Tachibana, M. Atomic force microscopic study on DNA-wrapping for different diameter single-wall carbon nanotubes. *Diamond & Related Materials*, 2008, 17, 1389-1393.

[15] Dresselhaus, MS; Dresselhaus, G; Saito, R; Jorio, A. Raman spectroscopy of carbon nanotubes. *Phys. Rep.*, 2005, 409, 47-99.

[16] Thomsen, C; Reich, S; in: Cardona, M; Merlin, R; (Eds.), Light Scattering in Solids IX, Springer-Verlag, Berlin Heidelberg, 2007, 115.

[17] Bandow, S; Asaka, S; Saito, Y; Rao, AM; Grigorian, L; Richter, E; Eklund, PC. Effect of the Growth Temperature on the Diameter Distribution and Chirality of Single-Wall Carbon Nanotubes. *Phys. Rev. Lett.*, 80, 1998, 3779-3782.

[18] Jorio, A; Dresselhaus, G; Dresselhaus, MS; Souza, M; Dantas, MSS; Pimenta, MA; Rao, AM; Saito, R; Liu, C; Cheng, HM. Polarized Raman Study of Single-Wall Semiconducting Carbon Nanotubes. *Phys. Rev. Lett.,* 2000, 85, 2617-2620.

[19] Kataura, H; Kumazawa, Y; Maniwa, Y; Umezu, I; Suzuki, S; Ohtsuka, Y; Achiba, Y. Optical Properties of Single-Wall Carbon Nanotubes. *Synth. Met.*, 1999, 103, 2555-2558.

[20] Brown, SDM; Jorio, A; Corio, P; Dresselhaus, MS; Dresselhaus, G; Saito, R; Kneipp, K. Origin of the Breit-Wigner-Fano lineshape of the tangential *G*-band feature of metallic carbon nanotubes. *Phys. Rev. B,* 63, 2001, 155414.

[21] Strano, MS; Huffman, CB; Moore, VC; O'Connell, MJ; Haroz, EH; Hubbard, J; Miller, M; Rialon, K; Kittrell, C; Ramesh, S; Hauge, RH; Smalley. RE. Reversible, Band-Gap-Selective Protonation of Single-Walled Carbon Nanotubes in Solution. *J. Phys. Chem. B,* 2003, 107, 6979-6985.

[22] Fantini, C; Jorio, A; Souza, M; Strano, MS; Dresselhaus, MS; Pimenta, MA. Optical Transition Energies for Carbon Nanotubes from Resonant Raman Spectroscopy: Environment and Temperature Effects. *Phys. Rev. Lett.* 2004, 93, 147406.

[23] Ohno, Y; Iwasaki, S; Murakami, Y; Kishimoto, S; Maruyama, S; Mizutani, T. Chirality-dependent environmental effects in photoluminescence of single-walled carbon nanotubes. *Phys. Rev. B*, 2006, 73, 235427.

[24] Duque, JG; Cognet, L; Nicholas, A; Vasquez, GP; Nicholas, N; Schmidt, HK; Pasquali, M. Stable Luminescence from Individual Carbon Nanotubes in Acidic, Basic, and Biological Environments. *J. Am. Chem. Soc.* 2008, 130, 2626-2633.

[25] Kang, D; Kato, K; Kojima, K; Uchida, T; Tachibana, M; Phonon control in metallic carbon nanotubes due to laser-induced defects. *Appl. Phys. Lett.,* 2008, 93, 133102.

[26] Milnera, M; Kürti, J; Hulman, M; Kuzmany, H. Periodic Resonance Excitation and Intertube Interaction from Quasicontinuous Distrbuted Helicities in Single-Wall Carbon Nanotubes. *Phys. Rev. Lett.,* 2000, 84, 1324-1327.

[27] Jorio, A; Fantini, C; Pimenta, MA; Capaz, RB; Ge. Samsonidze, G; Dresselhaus, G; Dresselhaus, MS; Jiang, J; Kobayashi, N; Grüneis, A; Saito, R. Resonance Raman spedtroscopy (n, m)-dependent effects in small-diameter single-wall carbon nanotubes. *Phys. Rev. B,* 2005, 71, 075401.

[28] Popov, VN; Lambin, P. Radius and chirality dependence of the radial breathing mode and the *G*-band phonon modes of single-walled carbon nanotubes. *Phys. Rev. B* 2006, 73, 085407.

[29] Jorio, A; Souza Filho, AG; Dresselhaus, G; Dresselhaus, MS; Swan, AK; Ünlü, MS; Goldberg, BB; Pimenta, MA; Hafner, JH; Lieber, CM; Saito, R. *G*-band resonant Raman study of 62 isolated single-wall carbon nanotubes. *Phys. Rev. B,* 2002, 65, 155412.

[30] Rao, AM; Eklund, PC; Bandow, S; Thess, A; Smalley, RE. Evidence for charge transfer in doped carbon nanotube bundles from Raman scattering. *Nature,* 1997, 388, 257-259.

[31] Dresselhaus, MS; Dresselhaus, G. Intercalation compounds of graphite. *Adv. Phys.* 1981, 30, 139-326.

[32] Enoki, T; Endo, M; Suzuki, M. Graphite Intercalation Compounds and Applications, Oxford University Press, New York, 2003.

[33] Klein, MV; in: Cardona M; (Ed.), Light Scattering in Solids I, Springer-Verlag, Berlin, 1983, 169.

[34] Uchida, T; Tachibana, M; Kurita, S; Kojima, K. Temperature dependence of the Breit-Wigner-Fano Raman line in single-wall carbon nanotube bundles. *Chem. Phys. Lett.* 2004, 400, 341-346.

[35] Jiang, C; Kempa, K; Zhao, J; Schlecht, U; Kolb, U; Basché, T; Burghard, M; Mews, A. Strong enhancement of the Breit-Wigner-Fano Raman line in carbon nanotube bundles caused by plasmon band formation. *Phys. Rev. B,* 2002, 66, 161404.

[36] Paillet, M; Poncharal, Ph; Zahab, A; Sauvajol, JL; Meyer, JC; Roth, S. Vanishing of the Breit-Wigner-Fano Component in Individual Single-Wall Carbon Nanotubes. *Phys. Rev. Lett.* 2005, 94, 237401.

[37] Lazzeri, M; Piscanec, S; Mauri, F; Ferrari, AC; Robertson, J. Phonon linewidths and electron-phonon coupling in graphite and nanotubes. *Phys. Rev. B,* 73, 2006, 155426.

[38] Piscanec, S; Lazzeri, M; Robertson, J; Ferrari, AC; Mauri, F. Optical Phonons in carbon nanotubes: Kohn anomalies, Peierls distortions, and dynamic effects. *Phys. Rev. B,* 2007, 75, 035427.

[39] Farhat, H; Son, H; Ge. Samsonidze, G; Reich, S; Dresselhaus, MS; Kong, J. Phonon Softening in Individual Metallic Carbon Nanotubes due to the Kohn Anomaly. *Phys. Rev. Lett.,* 2007, 99, 145506.

[40] Saenger, W; in: Cantor, CR; Principles of Nucleic Acid Structure, Springer, New York, 1984.

[41] Cathcart, H; Quinn, S; Nicolosi, V; Kelly, JM; Blau, WJ; Coleman, JN; Spontaneous Debundling of Single-Walled Carbon Nanotubes in DNA-Based Dispersions. *J. Phys. Chem. C,* 2007, 111, 66-74.

[42] Yarotski, DA; Kilina, SV; Talin, AA; Tretiak, S; Prezhdo, OV; Balatsky, AV; Taylor, AJ. Scanning Tunneling Microscopy of DNA-Wrapped Carbon Nanotubes. *Nano Lett.*, 2009, 9, 12-17.

In: Horizons in DNA Research, Volume 1
Editor: Jason R. Chesterton, pp. 225-234
ISBN: 978-1-60876-968-1

Chapter 10

ANALYSIS OF MITOCHONDRIAL DNA SEQUENCE VARIATIONS IN TYPE 2 DIABETES PATIENTS CHARACTERIZED FOR HAPLOGROUP AFFILIATION

Antonella Cormio[1], Francesco Milella[1], Maurizio Marra[2], Maria Pala[3], Angela Maria Serena Lezza[1], Anna Rita Bonfigli[2], Claudio Franceschi[4,5], Palmiro Cantatore[1,6] and Maria Nicola Gadaleta[1,6*]

[1]Dipartimento di Biochimica e Biologia Molecolare "E. Quagliariello", Università di Bari, Bari, Italy.
[2]Unità di Diabetologia, INRCA Ancona, Italy.
[3]Dipartimento di Genetica e Microbiologia, Università di Pavia, Pavia, Italy.
[4]Centro Interdipartimentale "L. Galvani" (CIG), Università di Bologna, Bologna, Italy.
[5]Centro di Biologia Molecolare e Genetica, INRCA, Ancona, Italy.
[6]Istituto di Biomembrane e Bioenergetica CNR, Bari, Italy.

ABSTRACT

The knowledge of the haplogroups of the analyzed subjects allows to prevent errors in the analysis of the association between mtDNA variations and some pathologies like diabetes. In this paper a cohort of type 2 diabetic Italian patients with known mitochondrial genetic background (haplogroup) were investigated for the presence in the blood of the 3243A>G mutation and other variants of close to $tRNA^{Leu(UUR)}$ gene (16S/ND1 region) of mitochondrial DNA (mtDNA). The 3243A>G mutation was present in 0.4% of analyzed patients in agreement with the reported frequencies of such mutation in other European groups. This result suggests a limited role for the 3243A>G mutation in the pathogenesis of diabetes also in the Italian population. In the 16S/ND1 region 15 base changes were found. The knowledge of the haplogroup of the analyzed individuals

* Corresponding author: Tel. +39-080-5443309, Fax: +39-080-5443403, E.mail: m.n.gadaleta@biologia.uniba.it

allowed to distinguish between haplogroup-specific polymorphisms and variants potentially associated to diabetes.

INTRODUCTION

Type 2 (non insulin dependent) diabetes mellitus (T2DM) is characterized by chronic hyperglycaemia and by the type-specific impaired insulin release and/or sensitivity (DeFronzo 1988, Beck-Nielsen*et al.* 1994). Many epidemiologic studies have noted a maternal excess in the transmission of type 2 diabetes (Martin *et al.* 1985, Young *et al.* 1995, Simmons *et al.* 1995). As mitochondrial DNA (mtDNA) is passed down along the maternal line, it was hypothesized that mtDNA defects might contribute to the excess maternal transmission (Alcolado *et al.* 2002). Therefore, attention has focused on the eventual role of mtDNA variants in producing mitochondrial dysfunction and diabetes. The best defined diabetogenic mtDNA defect is the A to G substitution at 3243 nt in the $tRNA^{Leu(UUR)}$ gene (van den Ouweland *et al.* 1994, Maassen and Kadowaki, 1996). Following these observations many other base changes in the proximity of the $tRNA^{Leu(UUR)}$ gene (Nakagawa *et al.* 1995; Hirai *et al.* 1996; McCarthy *et al.* 1996; Odawara *et al.* 1996; Yu *et al.* 2004; Tang *et al.* 2006, Liao *et al.* 2008). and also in other parts of the mtDNA genome, in particular in the D-loop region, were identified in diabetic patients (Poulton *et al.* 2002, Kamiya and Aoki 2003, Saxena *et al.* 2006, Park *et al.* 2008, Liao *et al.* 2008, Cormio *et al.* 2009). The association of these base changes to diabetes is controversial (Lam *et al.* 2001, Ohkubo *et al.* 2001, Chinnery *et al.* 2005, Mohlke *et al.* 2005). In fact the evaluation of the pathogenicity of a particular mtDNA base change found in patients must take into account also the mitochondrial background (haplogroup) in which that base change arises. The sequential accumulation of new mtDNA mutations along radiating maternal lineages has given rise, in fact, to haplogroups (groups of haplotypes) which tend to be restricted to specific geographic areas and /or ethnic groups. Each haplogroup is characterized by a distinguishing combination of mutations (Torroni *et al.* 2006). Therefore, the preferential association of a clinical disorder with a specific haplogroup is suggestive that one or more of the haplogroup-specific sequence changes might play a role in the disease expression (Carelli *et al.* 2006). As far as T2DM, no statistically significant association between any European mtDNA haplogroup and diabetes was found until now (Mohlke *et al.* 2005, Chinnery *et al.* 2007). However, other base changes, not haplogroup-specific, might have a role in the pathogenesis of the disease.

To provide a contribution to the understanding of the role of mtDNA sequence variations in the pathogenesis of T2DM we analyzed the mtDNA extracted from blood of 502 T2DM Italian patients and 503 Italian controls with known mitochondrial haplogroup. In particular, we analyzed the $tRNA^{Leu(UUR)}$ gene region (16S/ND1 region). We found that the 3243A>G mutation was present only in diabetic patients at a very low frequency (0.4%). We detected also other base changes in the 16S/ND1 region; in some cases the knowledge of the haplogroups of the investigated subjects allowed to discriminate between haplogroup-specific polymorphisms and variants potentially associated to diabetes.

Methods

Subjects

A total of 502 Italian patients with type 2 diabetes (the diagnosis was made according to the American Diabetes Association criteria), 260 men and 242 women, 34-90 years old, mean ± SD 66± 8 years, were recruited by the Diabetology Unit, INRCA Ancona, Italy. 503 Italian subjects in good health (without diabetic familiarity), 190 men and 313 women, 34-90 years old, mean ± SD 57± 11.6 years, were enrolled as a control group. Detailed medical history was recorded for each analyzed subjects. All patients gave their written informed consent before they enrolled in the study, which was approved by the local Ethics Committees.

Isolation and amplification of DNA

Total DNA was prepared from peripheral blood using Quiamp DNA Blood Maxi Kit (Qiagen) and stored at -20°C. The 16S/ND1 region was amplified by PCR from total DNA with primers 3.1 For (3162-3182) - 3.4 Rev (3420-3400) to obtain a product of 259 bp and primers 2.3 For (2364-2386) e 4.2 Rev (4249-4228) to obtain a product of 1.9 Kb. Nucleotide numbering is according to the Cambridge Reference Sequence (CRS) (Anderson *et al.*1981). The reaction was performed in a 50 µl volume containing 1X PCR buffer (10 mM Tris-HCl pH 8.3, 50 mM KCl, 1.5 mM $MgCl_2$, 0.001 % (w/v) gelatin), 0.2 mM dNTPs, 0.6 µM of the chosen pair primers, 2.5 U JumpStart-REDTaq DNA Polymerase (Sigma Aldrich) and 100 ng of total DNA. Reaction conditions were: 94°C for 2 min and 35 cycles of 94°C for 1 min, 58°C for 1 min and 72°C for 1 min. The final extension proceeded for 8 min. The capability of the primer pairs to amplify nucleus-embedded mtDNA pseudogenes was investigated by performing the PCR on 400 ng of total DNA extracted from a human osteosarcoma-derived mtDNA-deficient cell line (143B, ρ^0). Control experiments on DNA extracted from blood and 143 B ρ^+ cells were also performed.

RFLP analysis of the 3243A>G mutation

The 1.9 Kb PCR product from the 16S/ND1 region produced with primers 2.3 For e 4.2 Rev was purified by using a Quiaquick PCR-purification Kit (QUIAGEN) and digested with Apa I restriction enzyme at 37°C for 10 h. The 3243A>G mutation creates a new site for Apa I. Subsequently an electrophoretic separation was carried out in a 5% polyacrylamide gel in TBE 1X. The undigested and digested fragments were revealed by silver staining.

Sequencing of 16S/ND1 regions

To facilitate the direct sequencing of PCR-amplified products, 3.1 For and 3.4 Rev primers were tagged with M13 sequence as reported in (Cormio *et al.* 2005). PCR products were checked by agarose gel electrophoresis, purified by using a Quiaquick PCR-purification

Kit (QUIAGEN) and sequenced by MWG Biotech facilities. All sequences were analyzed with Chromas program and compared with the CRS with BLASTN program.

Statistics

Statistical analysis was carried out by using SPSS Base 11.5 software. Chi square test was used. Statistical significance was set at $p<0.05$.

Results and Discussion

Detection of the 3243A>G mutation and identification of other sequence variants in the 16S/ND1 region.

It is known that the 3243A>G mutation occurs in maternal diabetic patients at different levels of heteroplasmy, ranging from 10% to 30% in the blood (Procaccio *et al.* 2006). Before analyzing the sampled population for the presence of such mutation we determined the lowest level of experimentally detectable heteroplasmy, comparing the sensitivity of two types of analyses, namely RFLP and direct sequencing. For this purpose we carried out preliminary experiments in which DNA with different levels of heteroplasmy was amplified by PCR and subjected to RFLP and sequencing analysis; we found (data not shown) that the two methods had a comparable sensitivity and that we were able to detect a heteroplasmy level as low as 5-10% of the initial template. Since the analysis by direct sequencing allows to gather information not only on the 3243A>G mutation, but also on other variations eventually present in the $tRNA^{Leu(UUR)}$ region, we decided to use this method to analyze all the samples. When the 3243A>G mutation was found or in case of ambiguity the RFLP analysis was also performed.

To analyze the sampled population for the presence of the 3243A>G and other mutations in the 16S/ND1 region we extracted the DNA from the peripheral blood of 502 T2DM patients and 503 control subjects and amplified and then sequenced a 259 bp fragment containing a short part of such region (3162-3420). We detected the 3243A>G mutation only in two diabetic patients (0.4% of the analyzed patients). Fig 1 (Panels A1 and A2) shows the electropherogram of part of the sequence of the 16S/ND1 region from these two diabetic patients. The RFLP analysis of the same two patients was also performed by amplifying a 1.9 kb region, encompassing the 3'-end of the 16S rRNA gene, the $tRNA^{Leu(UUR)}$ gene and the 5'-end of the ND1 gene. The 1.9 Kb product was purified and digested with Apa I, which cleaves the 3243A>G mutated mtDNA at one site. Fig. 1 B shows the presence in lanes 1 and 2 of two bands of 0.9 Kb and 1 Kb in the DNA from the two patients, which correspond to those exhibited by a positive control (PC), a MELAS patient with 90% heteroplasmy for the 3243A>G mutation. Assuming that mutated and not mutated mtDNAs were amplified with the same efficiency we could estimate, from both methods, a heteroplasmy level around 5-10% in the blood of these two individuals. The affected patients are two females (64 years and 42 years) belonging to the H haplogroup in which diabetes was diagnosed precociously, respectively, at 39 and 41 years. The NC lane is a negative control showing only the wild-type non-mutated sequence. The 64 years old patient is affected also by retinopathy.

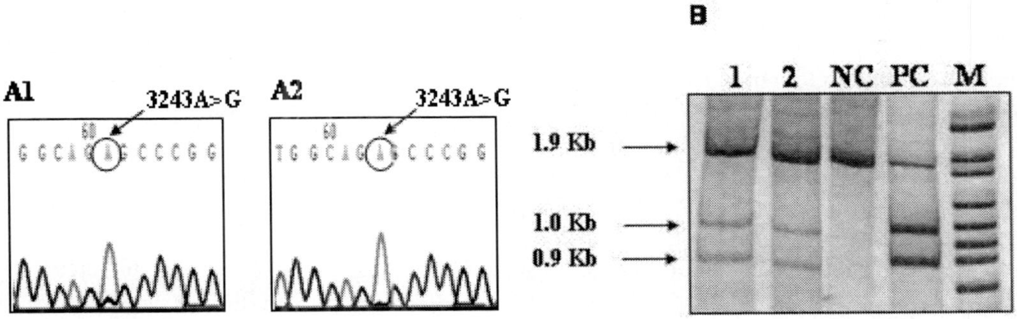

Figure 1. Sequence and RFLP analysis for detection of the 3243A>G mutation. Panels A1 and A2 show electropherogram of part of the sequence of the 16S/ND1 region of two diabetic patients (patient 1, 64 years old and patient 2, 42 years old). Panel B shows the undigested (1.9 Kb) and digested (about 1Kb and 0.9 Kb) PCR products obtained after the treatment with Apa I and revealed by silver staining. Patients 1, 2 and positive control PC (a heteroplasmic MELAS patient with 90% mutant DNA) show the digested products, negative control NC (a subject without the 3243A>G mutation) only the undigested product. M, DNA size standard (3000, 2000, 1500, 1200, 1031, 900, 800, 700 bp).

The sequencing of the 259 bp PCR product from the 16SrRNA/ND1 regtion of 1005 subjects revealed also 15 different types of sequence variants with respect to the Cambridge Reference Sequence (CRS). The total base changes found, their localization, the eventual aminoacidic changes produced and their frequencies in diabetics and controls are reported in table 1 together with the number of diabetic patients and controls bearing each base change, the haplogroup affiliation and the total number of patients and controls belonging to that haplogroup. As shown in Table 1 we detected four changes in the 16S rRNA gene (3197T>C, 3204C>T, 3213A>G, 3221A>G), one in the tRNA$^{Leu(UUR)}$ gene (3290T>C) and 10 changes in the ND1 gene (3308T>C, 3316G>A, 3333C>T, 3337G>A, 3348A>G, 3357G>A, 3372T>C, 3394T>C, 3396T>C and 3397A>G). All changes are already reported in mitochondrial databases (MITOMAP: http://www.mitomap.org; mtDB: http://www.genpat.uu.se/mtDB); they were mostly homoplasmic. We did not find statistically significant differences in sequence variants frequencies between controls and patients; however, the analysis of the sequence variants allows to draw some considerations. Sequence variants found prevalently in diabetic patients include the 3204C>T and 3221A>G changes in the 16S rRNA gene and the 3333C>T change in ND1. The frequency of the 3204C>T variation was very low in the entire analyzed population (about 0.4%); however, since the variation was found only in subjects belonging to U4 haplogroup, its frequency rose to about 30% and 12% when referred to U4 diabetics and controls, respectively. The most frequent change in the analyzed population is the 3197T>C (about 8% both in controls and patients). This is a known U5 haplogroup-specific polymorphism (Achilli *et al.* 2005). It was found not only in all the patients and controls belonging to this haplogroup, but also in four controls and in one patient belonging to different haplogroups. The 3316G>A, 3348A>G and 3394T>C changes, known to be D4e1, U6a and J1c1 haplogroup-specific polymorphisms, respectively (Palanichamy *et al.* 2004; Achilli *et al.* 2005), were found in all the diabetic patients and controls belonging to these haplogroups. Interestingly, we detected the 3316G>A base change, that causes an Ala-Thr change in ND1, also in two diabetic individuals belonging to other haplogroups (U5 and T2). A similar situation concerns the 3394T>C change. This mutation, that was associated with

Leber Hereditary Optic Neuropathy (LHON) (Johns *et al.* 1992), causes a Tyr-His change in a conserved ND1 position. In addition to the known association with a specific subset of J1 (J1c1) (Palanichamy *et al.* 2004; Carelli *et al.* 2006), we found this change also in two patients belonging to haplogroup H.

CONCLUSION

The role of mtDNA sequence variants in the pathogenesis of type 2 diabetes is still controversial. Although the association of many diseases with mutations in mitochondrial genes has been clearly established, there is now a growing body of evidence arguing for a more subtle role of mtDNA variants in the disease process. Since mtDNA can be assigned to different haplogroups associated with different ethnic groups, to assess whether a mutation is pathogenetic or not, it is necessary to know the mitochondrial background in which such mutation arises. Recently a detailed study of mtDNA variation in the East Asian population has highlighted the need for a deep analysis based on the haplogroup phylogenesis to assess the association of mtDNA base changes with a disease (Kong *et al.* 2006).

In this paper a careful study about the role of mtDNA sequence variations in the pathogenesis of type 2 diabetes in Italian population from a restricted Italian region was carried out for the first time in subjects with known mitochondrial background. In our samples, no association between T2DM and mtDNA haplogroups was found (A.Torroni, personal communication). We found the 3243A>G $tRNA^{Leu\ (UUR)}$ in two female patients belonging to the haplogroup H. The frequency of the 3243A>G mutation in the analyzed diabetic patients was 0.4%, in agreement with the reported frequencies of such mutation in other European groups (Newkirk *et al.* 1997; Saker *et al.* 1997). This result suggests a limited role of the 3243A>G mutation in the pathogenesis of diabetes also in the Italian population. As far as the other mtDNA variants here found in the 16S/ND1 region, no one reaches a statistically significant difference in patients relative to controls to be considered pathogenetic for diabetes. Some of them, such as the 3316G>A and 3394T>C base changes which have been reported to be associated with the disease (Nakagawa *et al.* 1995; Hirai *et al.* 1996, Odawara *et al.* 1996; Yu *et al.* 2004; Tang *et al.* 2006), are haplogroup-specific polymorphisms. However, in our analyzed subjects, they occur also in diabetic patients with different mtDNA backgrounds. The presence of the mutations 3316G>A and 3394T>C only in diabetic patients belonging to additional haplogroups might indicate a potential risk factor for diabetes. However, the number of individuals found in such conditions is too low to allow a statistical evaluation.

These results confirm a limited role of mtDNA variations in the mtDNA $tRNA^{Leu(UUR)}$ gene region in the pathogenesis of diabetes in the Italian population and show that previous studies on the association between some mtDNA base changes and diabetes, reported without haplogroup affiliation, may be misleading.

Table 1. MtDNA variations in the 16S/ND1 region of diabetic patients and controls.

Change	3197T>C	3204C>T	3213A>G	3221A>G	3243A>G	3290T>C	3308T>C	3316G>A	3333C>T	3337G>A	3348A>G	3357G>A	3372T>C	3394T>C	3396T>C	3397A>G
Locus	16SrRNA	16SrRNA	16SrRNA	16SrRNA	$tRNA^{Leu}$	$tRNA^{Leu}$	ND1	ND1	ND1	ND1	ND1	ND1	ND1	ND1	ND1	ND1
Aa change							Met-Thr	Ala-Thr		Val-Met				Tyr-His		Met-Val
N	78	4	1	3	2	1	1	11	1	1	3	2	1	6	2	3
Diabetics	8.0%	0.6%	0.0%	0.6%	0.4%	0.0%	0.0%	1.2%	0.2%	0.0%	0.4%	0.2%	0.0%	0.8%	0.0%	0.0%
Controls	7.6%	0.2%	0.2%	0.0%	0.0%	0.2%	0.2%	1.0%	0.0%	0.2%	0.2%	0.2%	0.2%	0.4%	0.4%	0.6%
Haplogroup																
U5 (39D, 34C)	39D, 34C							1D								
H (92D, 95C)	4C				2D	1C			1D			1C, 1D	1C	2D	2C	1C
H1 (48D, 53C)				2D												
U4 (11D, 8C)		3D, 1C														
K1 (28D, 34C)			1C							1C						
D4e1 (4D,5C)								4D, 5C								
X2 (15D, 14C)																1C
HV (5D, 9C)	1D															
U3a (9D, 4C)				1D												
U6a (2D, 1C)											2D, 1C					
L1b (1C)							1C									
T2 (58D, 47C)								1D								
J1c1 (2D, 2C)														2D, 2C		1C

In the top part of the table the type of change, its localization, the eventual aminoacidic change, the total number of subjects having that change (N), the frequency of the change in diabetics and controls are reported. In the bottom part of the table, for each haplogroup, the analyzed number of diabetics (D) and controls (C) is reported in brackets. In the corresponding row, the numbers of diabetics and controls having the above reported changes are reported. H, subjects that cannot be classified into specific subhaplogroups.

ACKNOWLEDGMENTS

We are very grateful to Antonio Torroni for haplogroups analysis and for helpful comments, to Elvira Greco for technical assistance and to T. Rabillaud for the gift of osteosarcoma 143B ρ^{o} and ρ^{+} cells. This work was supported by MIUR- FIRB 2001, University of Bari, Fondi Ateneo 2007 and MIUR-CNR N157 - WP1.

REFERENCES

[1] Achilli, A; Rengo, C; Battaglia, V; Pala, M; Olivieri, A; Fornarino, S; Magri; C; Scozzari, R; Babudri, N; Santachiara-Benerecetti, AS; Bandelt, HJ; Semino, O; Torroni, A. Saami and Berbers - an unexpected mitochondrial DNA link. *Am. J. Hum. Genet.*, 2005, 76, 883-886.

[2] Alcolado, JC; Laji, K; Gill-Randall, R. Maternal transmission of diabetes. *Diabet. Med.*, 2002, 19, 89-98.

[3] Anderson, S; Bankier, AT; Barrell, BG; de Bruijn, MH; Coulson, AR; Drouin, J; Eperon, IC; Nierlich, DP; Roe, BA; Sanger, F; Schreier, PH; Smith, AJ; Staden, R; Young, IG. Sequence and organisation of the human mitochondrial genome. *Nature,* 1981, 290, 457-465.

[4] Beck-Nielsen; Hother-Nielsen, O; Vaag, A; Alford, F. Pathogenesis of type 2 (non-insulin *Diabetologia,* 1994, 37, 217-221.

[5] Carelli; V; Achilli, A; Valentino, ML; Rengo, C; Semino, O; Pala, M; Olivieri, A; Mattiazzi, M; Pallotti, F; Carrara, F; Zeviani, M; Leuzzi, V; Carducci, C; Valle, G; Simionati, B; Mendieta, L; Salomao, S; Belfort, R; Sadun, AA; Torroni, A. Haplogroup effects and recombination of mitochondrial DNA: novel clues from the analysis of Leber hereditary optic neuropathy pedigrees. *Am. J. Hum. Genet.*, 2006, 78, 564-574.

[6] Chinnery, PF; Elliott, HR; Patel, S; Lambert, C; Keers, SM; Durham, SE; McCarthy, MI; Hitman, GA; Hattersley, AT; Walker, M. Role of the mitochondrial DNA 16184-16193 poly-C tract in type 2 diabetes. *Lancet*, 2005, 366, 1650-1651.

[7] Chinnery, PF; Mowbray, C; Patel, SK; Elson, JL; Sampson, M; Hitman, GA; McCarthy, MI; Hattersley, AT; Walker, M. Mitochondrial DNA haplogroups and type 2 diabetes: a study of 897 cases and 1010 controls. *J. Med. Genet.*, 2007, 44, e80.

[8] Cormio, A; Milella, F; Vecchiet, J; Felzani, G; Gadaleta, MN; Cantatore, P. Mitochondrial DNA mutations in RRF of healthy subjects of different age. *Neurobiol. Aging.,* 2005, 26, 655-664.

[9] Cormio, A; Micella, F; Marra, M; Pala, M; Lezza, AMS; Bonfigli, B; Franceschi, C; Cantatore, P; Gadaleta, MN. Variations at the H-strand replication origins of mitochondrial DNA and mitochondrial DNA content in the blood of type 2 diabetes patients. *BBA,* in press (2009).

[10] De Fronzo, RA. Lilly Lecture 1987. The triumvirate: beta-cell, muscle, liver. A collusion responsible for NIDDM. *Diabetes,* 1988, 37, 667-687.

[11] Hirai, M; Suzuki, S; Onoda, M; Hinokio, Y; Ai, L; Hirai, A; Ohtomo, M; Komatsu, K; Kasuga, S; Satoh, Y; Akai, H; Toyota, T. Mitochondrial DNA 3394 mutation in the

NADH dehydrogenase subunit 1 associated with non-insulin-dependent diabetes mellitus. *Biochem. Biophys. Res. Commun.,* 1996, 219, 951-955.

[12] Johns, DR; Neufeld, MJ; Park, RD. An ND-6 mitochondrial DNA mutation associated with Leber hereditary optic neuropathy. *Biochem. Biophys. Res. Commun.*, 1992, 87, 1551-1557.

[13] Kamiya, J; Aoki, Y. Associations between hyperglycaemia and somatic transversion mutations in mitochondrial DNA of people with diabetes mellitus. *Diabetologia,* 2003, 46, 1559-1566.

[14] Kong, QP; Bandelt, HJ; Sun, C; Yao, YG; Salas, A; Achilli, A; Wang, CY; Zhong, L; Zhu, CL; Wu, SF; Torroni, A; Zhang, YP. Updating the East Asian mtDNA phylogeny: a prerequisite for the identification of pathogenic mutations. *Hum. Mol. Genet.*, 2006, 15, 2076-2086.

[15] Lam, C-W; Yang, T; Tsang, M-W; Pang, C-P. Homoplasmic 3316G/A in the ND1 gene of the mitochondrial genome: a pathogenic mutation or a neutral polymorphism? *J. Med. Genet.*, 2001, 38.

[16] Liao, WQ; Pang, Y; Yu, CA; Wen, JY; Zhang, YG; Li, XH. Novel mutations of mitochondrial DNA associated with type 2 diabetes in Chinese Han population. *Tohoku J. Exp Med.,* 2008, 215, 377-384.

[17] Maassen, JA and Kadowaki T. Maternally inherited diabetes and deafness: a new diabetes subtype. *Diabetologia,* 1996, 39, 375-382.

[18] Martin, AO; Simpson, JL; Ober, C; Freinkel, N. Frequency of diabetes mellitus in mothers of probands with gestational diabetes: possible maternal influence on the predisposition to gestational diabetes. *Am J. Obstet. Gynecol.,* 1985, 151, 471–475.

[19] McCarthy, M; Cassell, P; Tran, T; Mathias, L; 't Hart, LM; Maassen, JA; Snehalatha, C; Ramachandran, A; Viswanathan, M; Hitman, GA. Evaluation of the importance of maternal history of diabetes and of mitochondrial variation in the development of NIDDM. *Diabet. Med.*, 1996, 13, 420-428.

[20] Mohlke, KL; Jackson, AU; Scott, LJ; Peck, EC; Suh, YD; Chines, PS; Watanabe, RM; Buchanan, TA; Conneely, KN; Erdos, MR; Narisu, N; Enloe, S; Valle, TT; Tuomilehto, J; Bergman, RN; Boehnke, M; Collins, FS. Mitochondrial polymorphisms and susceptibility to type 2 diabetes-related traits in Finns. *Hum. Genet.,* 2005, 118, 245-254.

[21] Nakagawa, Y; Ikegami, H; Yamato, E; Takekawa, K; Fujisawa, T; Hamada, Y; Ueda, H; Uchigata, Y; Miki, T; Kumahara, Y; Ogihara, T. A New Mitochondrial DNA Mutation Associated with Non-Insulin-Dependent Diabetes Mellitus. *Biochem. Biophys. Res. Commun.,* 1995, 209, 664-668.

[22] Newkirk, JE; Taylor, RW; Howell, N; Bindoff, LA; Chinnery, PF; Alberti, KG; Turnbull, DM; Walzer, M. Maternally inherited diabetes and deafness: prevalence in a hospital diabetic population. *Diabet. Med.,* 1997, 14, 457-460.

[23] Odawara, M; Sasaki, K; Yamashita, K. A G-to-A substitution at nucleotide position 3316 in mitochondrial DNA is associated with Japanese non-insulin-dependent diabetes mellitus. *Biochem. Biophys. Res. Commun.,* 1996, 227, 147-151.

[24] Ohkubo, K; Yamano, A; Nagashima, M; Mory, Y; Anzai, K; Akeki, Y. *et al.* Mitochondrial gene mutations in the $tRNA^{Leu(UUR)}$ region and diabetes: prevalence and clinical phenotypes in Japan. *Clin Chem.,* 2001, 47, 1641-1648.

[25] Palanichamy, MG; Sun, C; Agrawal, S; Bandelt, HJ; Kong, QP; Khan, F; Wang, CY; Chaudhuri, TK; Palla, V; Zhang, YP. Phylogeny of mitochondrial DNA macrohaplogroup N in India, based on complete sequencing: implications for the peopling of South Asia. *Am. J. Hum. Genet.*, 2004, 75, 966-978.

[26] Park, KS; Chan, JC; Chuang, LM; Suzuki, S; Araki, E; Nanjo, K; Ji, L; Ng, M; Nishi, M; Furuta, H; Shirotani, T; Ahn, BY; Chung, SS; Min, HK; Lee, SW; Kim, JH; Cho, YM; Lee, HK. Study Group of Molecular Diabetology in Asia. A mitochondrial DNA variant at position 16189 is associated with type 2 diabetes mellitus in Asians. *Diabetologia*, 2008, 51, 602-608.

[27] Poulton, J; Luan, J; Macaulay, V; Hennings, S; Mitchell, J; Wareham, NJ. Type 2 diabetes is associated with a common mitochondrial variant: evidence from a population-based case-control study. *Hum. Mol. Genet.,* 2002, 11, 1581-1583.

[28] Procaccio, V; Neckelmann, N; Paquis-Flucklinger, V; Bannwarth, S; Jimenez, R; Davila, A; Poole, JC; Wallace, DC. Detection of low levels of the mitochondrial tRNALeu(UUR) 3243A>G mutation in blood derived from patients with diabetes. *Mol. Diagn. Ther.,* 2006, 10, 381-389.

[29] Saker, PJ; Hattersley, AT; Barrow, B; Hammersley, MS; Horton, V; Gillmer, MD; Turner, RC. UKPDS 21: low prevalence of the mitochondrial transfer RNA gene (tRNA(Leu(UUR)) mutation at position 3243bp in UK Caucasian type 2 diabetic patients. *Diabet. Med.,* 1997, 14, 42-45.

[30] Saxena, R; de Bakker, PI; Singer, K; Mootha, V; Burtt, N; Hirschhorn, JN; Gaudet, D; Isomaa, B; Daly, MJ; Groop, L; Ardlie, KG; Altshuler, D. Comprehensive association testing of common mitochondrial DNA variation in metabolic disease. *Am. J. Hum. Genet.,* 2006, 79, 54-61.

[31] Simmons, D; Galtand, BA; Leakehe, L; Fleming, C. Frequency of diabetes in family members of probands with non-insulin dependent diabetes mellitus. *J. Intern. Med.,* 1995,237, 315–321.

[32] Tang, DL; Zhou, X; Li, X; Zhao, L; Liu, F. Variation of mitochondrial gene and the association with type 2 diabetes mellitus in a Chinese population. *Diabetes. Res. Clin. Pract.,* 2006, 73, 77-82.

[33] Torroni, A; Achilli, A; Macaulay, V; Richards, M; Bandelt, HJ. Harvesting the fruit of the human mtDNA tree. *Trends Genet.,* 2006, 22, 339-345.

[34] Van den Ouweland, JM; Lemkes, HH; Trembath, RC; Ross, R; Velho, G; Cohen, D; Froguel, P; Maassen, JA. Maternally inherited diabetes and deafness is a distinct subtype of diabetes and associates with a single point mutation in the mitochondrial tRNA(Leu(UUR)) gene. *Diabetes,* 1994, 43, 746-751.

[35] Young, CA; Kumar, S; Young, MJ; Boulton, AJM. Excess maternal history of diabetes in Caucasian and Afro-origin non-insulin dependent diabetic patients suggests dominant maternal factors in disease transmission. *Diabetes Res. Clin. Pract.,* 1995, 28, 47–49.

[36] Yu, P; Yu, DM; Liu, DM; Wang, K; Tang, XZ. Relationship between mutations of mitochondrial DNA ND1 gene and type 2 Diabetes. *Chin. Med. J.,* 2004, 117, 985-989.

In: Horizons in DNA Research, Volume 1
Editor: Jason R. Chesterton, pp. 235-240

ISBN: 978-1-60876-968-1

Chapter 11

REGULATIONAL ROLE OF MITOCHONDRIA ON AGEING AND SENESCENCE

Hiroshi Kondoh**[**1], Haruhisa Inoue [2], Akifumi Takaori-Kondo[3], Takumi Mikawa[1] and Manabu Minami [4]

[1]Department of Geriatric Medicine Graduate School of Medicine, Kyoto University, 54 Kawahara-cho, Shogoin, Sakyo-ku, Kyoto, 606-8507, Japan.

[2]Center for iPS Cell Research and Application (CiRA) Institute for integrated Cell-Material Sciences (iCeMS) Kyoto University, JST YAMANAKA iPS Cell Project Office, Kyoto Research Park Building 2, Third floor, 134 Chudoji Minami-machi, Shimogyo-ku, Kyoto 600-8813, Japan.

[3]Department of Hematology & Oncology Graduate School of Medicine, Kyoto University.

[4]Department of Clinical Innovative Medicine Translational Research Center, Kyoto University.

ABSTRACT

Mitochondria is the major supplier of intracellular ROS (Radical oxygen species). Since the proposal of free radical theory by Harman in 1956, there has been a long dispute about the impact of mitochondria on longevity. Recently it was reported that several mouse models of mitochondria mutant display the phenotype of premature ageing or age-related disease. Moreover ectopic expression of mitochondria-targeted radical scavenger extends organismal life span in mice

These suggest that mitochondria dysfunction intriguingly correlates with ageing and senescence. However, opposed or different ideas on the role of mitochondria on longevity are also proposed in the research of yeast genetics. Here we review the recent advance on molecular link between mitochondria and senescence. We propose that high efficient mitochondria model might explain clearly the importance of mitochondria on ageing and senescence.

[*] Corresponding author: E-mail: hkondoh@kuhp.kyoto-u.ac.jp

INTRODUCATION

In the free radical theory of ageing proposed by Harman in 1956, he described that progressive accumulation of oxidative damage would lead to age-related loss of physiological function (Harman, 1956) . Mitochondria is known as major energy source inside the cell, while it also produces about 90% of intrinsic reactive oxygen species (ROS). It has been suspected that the mitochondria is a crucial determinant of organismal lifespan. Later Harman modified his initial theory, as he renamed it the mitochondrial theory of ageing (Harman 1972).

Current model on mitochondrial theory of ageing suggests that ROS derived from mitochondrial respiratory chain damages mitochondrial DNA (mtDNA), leading to further mitochondrial dysfunction, which could possibly explain ageing-related dysfunction of organs or tissues. Indeed it was observed by several groups that an increase of of ROS generation with age was accompanied with downregulated mitochondrial function and increased oxidative damage to mtDNA (Van Remmen, 2001) . Thus a close correlation between ageing and mitochondrial dysfunction has been well established.

However, the question is whether mitochondrial damage or dysfunction would have a causal effect on ageing or lifespan. Here we overview the recent data on genetical analysis in experimental models, suggesting both the clues and questions for understanding the mitochondrial theory of ageing.

1. MITOCHONDRIAL RADICAL SCAVENGERS ARE ESSENTIAL FOR VIABILITY IN MICE

The balance between ROS generation and antioxidant defense might have great impact on organismal lifespan. Mitochondria itself contains an intricate antioxidant defense system to scavenge ROS immediately, which would be crucial to restore mitochondrial function. They are manganese superoxide dismutase MnSOD (Sod2), the mitochondrial form of thioredoxin, thioredoxin 2 (Trx2) and others.

Knockout mice against these mitochondrial radical scanengers are generated recently. First, the data on Sod2$^{-/-}$ mice provided strong evidence that mitochondrial dysfunction could trigger age-related diseases (Li, 1995) . Sod2 is the crucial enzyme that scavenges superoxide in the mitochondrial matrix, whose ablation causes early postnatal death (average lifespan is 0-4 weeks). This is due to dilated cardiomyopathy and neurodegeneration (Lebovitz, 1996) , quite similar to the symptoms of human genetical mitochondria disease. It was also observed that mitochondria was heavily damaged in these mice. But, it was quite difficult to assess the other ageing-assoiated phenotype, as Sod2$^{-/-}$ mice die early. Moreover the average lifespan of Sod2$^{+/-}$ mice was normal, although they suffer partial increase of oxidative stress.

Second, Trx2 knockout mice were generated (Perez, 2008) , as Trx2 was expected as another important antioxidant enzyme in mitochondria. But Trx2$^{-/-}$ mice displayed embryonic lethality, which made it quite difficult to analyze their ageing phenotype, again. Heterozygous knockout mice showed decreased mitochondrial function with increased oxidative damage, while their lifespan were not affected. Thus the data on Sod2 or Trx2 knockout mice could not completely explain the mitochondrial theory of ageing, although they are quite essential antioxidant enzymes for viability of mice.

2. Mitochondria Dysfunction Provokes Premature Ageing and Ageing-Related Disease

Then a different approach was performed to generate mitochondrial dysfunctional models by two groups. They generated knock-in mice of point mutated mitochondrial DNA polumerase (POLG) whose proofreading activity is known to be required to maintain integrity of mitochondrial DNA. Homozygous mic for this mutation (Polγ$^{D257A/D257A}$) showed 70% reduction of lifespan compared to wild type, associated with accumulated mitochondrial DNA mutations (Trifunovic, 2004). Surprisingly, the mice suffered premature ageing phenotype; body weight loss, gray hair, kyphosis, osteoporosis, muscle loss and others. Thus these mice provided strong evidence for the mitochondrial theory of ageing.

Although Polγ$^{D257A/D257A}$ mice clearly disclosed the causal effect of mtDNA mutations on ageing, no significant difference on oxidative damage or mitochondrial ROS production was observed between these mice and wild type. Thus it has been still a big question whether mitochondrial damage caused by oxidative stress is crucial for lifespan or not.

One breakthrough was the observation on transgenic mice overexpressing catalase targeted to mitochondria (mCAT mice) (Schriner, 2005) . In these mice, mitochondrial ROS production was reduced and the lifespan was increased by 20%. Moreover these mice showed less mitochondrial DNA mutations, less ageing symptoms. Taken together with the data on Polγ$^{D257A/D257A}$ mice, mitochondrial dysfunction is a crucial determinant of murine life span (Figure 1).

3. Mitochondrial Activity and Cellular Ageing

Next question could be whether mitochondrial activity itself could affect ageing process or not. It was easily assumed that reduction of mitochondrial activity would result in downregulation of mitochondrial ROS production. But the problem is that the impact of respiratory inhibition on organismal lifespan is quite difficult to evaluate. So far, the experiments on cultured cells would be applicable for this purpose. Indeed recent reports by two different groups suggested that enhanced glycolysis is accompanied by 30% reduction of mitochondrial respiration of cultured cells (Kondoh et al 2007, Matoba et al. 2006). In these setting, primary cells bypass senescence with less oxidative damage. Enhanced glycolysis is well known property of cancerous cells and tissues, called the Warburg effect (Warburg 1930). Thus it is possible that enhanced glycolysis with reduced mitochondrial function could be a booster for cell proliferation not only as energy supply but also as radical scavenger (Kondoh et al. 2005).

However, the fact that we should notice very carefully on discussion about longevity and mitochondrial activity is the results observed in yeast genetics. The hypothesis that enhanced glycolysis can promote proliferation would not be applied in every species. It seems quite opposite in case of budding yeast. In yeast, Sir2 gene was isolated as longevity gene, which can be activated in calorie-restricted condition(Sinclair et al., 2006). Calorie restriction is also well known to extend the life sapn of mouse and yeast. When calorie restricted, glycolysis declines in yeast, accompanied by the elevation of mitochondrial respiratory rate.

Sir2 works as NAD-dependent histone deacetylase (Imai et al. 2000). Activated respiration during calorie restriction is a mode of energy production that creates NAD as a by-product and lowers NADH. In consequence, NAD-dependent Sir2 activity is induced by calorie restriction. One of the main targets of Sir2 is transcriptional factor PGC-1 (nemoto et al. 2005). PGC-1 is quite essential for mitochondria genesis.Indeed Sir2 activation could facilitate mitochondrial respiration. Thus the longevity caused by Sir2 or calorie restriction in yeast is accompanied by increased mitochondrial respiration.

4. CONCLUSION AND ANOTHER RISING QUESTION

It seems that mitochondrial activity has opposing effect on longevity among different experimental models. MtDNA mutations acceralates murine ageing (Pol$\gamma^{D257A/D257A}$ mice), and radical scavenger in mitochondria could extend organismal lifespan (mCAT mice). On the other hand, sir2 activation extend yeast lifespan with enhanced mitochondrial respiration.

So far, it is quite difficult to understamd these difference in mitochondrial profile between long-lived yeast and long-lived mouse model. It is possible that ageing process is differently regulated among all speiceis. But is there any unifying hypothesis for these observations?

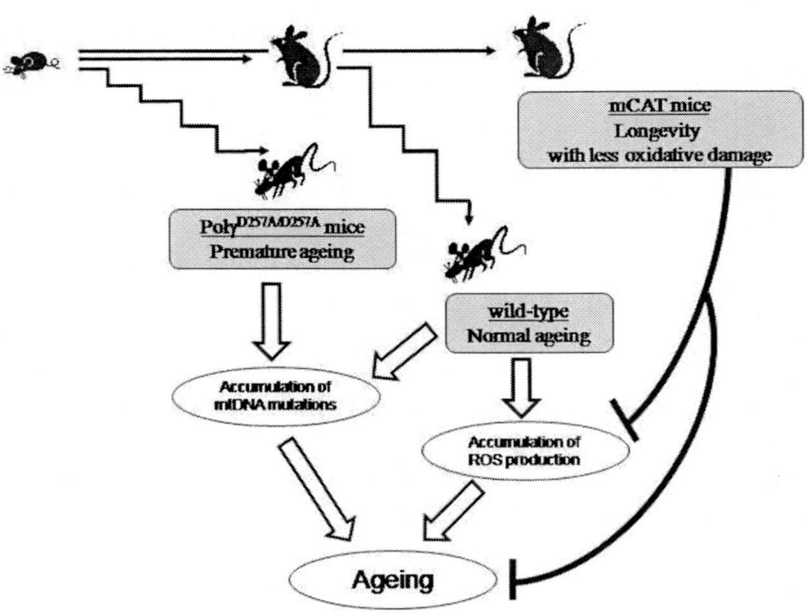

Figure 1 The impact of mitochondria on mice longevity. Poly$\gamma^{D257A/D257A}$ mice displayed premature senescence with mitochondria DNA mutations, while mCAT mice live longer with less oxidative damage. See details in the text.

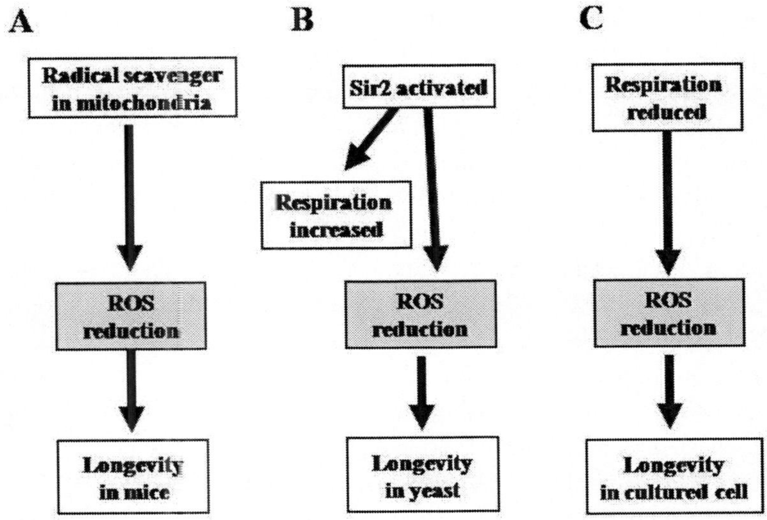

Figure 2 Longevity in different species. Pathway to longevity is shown in mice (A), in yeast (B), and in cultured cells (C). Please note that ROS reduction is a common feature, although mitochondrial profile is different among three.

As one possibility, we propose " high efficient mitochondria" is a key for ageing process. High efficient mitochondrial is supposed to be highly productive for energy with less ROS generation than others (Figure 2). It is known that sir2-activatation confers cells resistant to oxidative stress (Luo et al. 2001). Cells in mCAT mice also displayed normal mitochondrial activity with less ROS accumulation comapered to wildtype mice. Such a " high efficient mitochondria" could be favorable both for yeast and organismal longevity, which remains to be clarified in details in the future.

REFERENCE

Harman D. (1956) Aging: a theory based on free radical and radiation chemistry. *J Gerontol, 11*, 298-300

Harman, D. (1972) The biologic clock: the mitochondria? *J Am Geriatr Soc,* 20, 145-7

Imai, S., Armstrong, C. M., Kaeberlein, M., & Guarente, L. (2000). Transcriptional silencing and longevity protein Sir2 is an NAD-dependent histone deacetylase. *Nature, 403*, 795-800.

Kondoh, H., Lleonart, M. E., Gil, J., Beach, D., & Peters, G. (2005). Glycolysis and cellular immortalization. *Drug Discovery Today, 2*, 263-267.

Kondoh, H., Lleonart, M. E., Nakashima, Y., Yokode, M., Tanaka, M., Bernard, D., Gil, J., & Beach, D. (2007). A high glycolytic flux supports the proliferative potential of murine embryonic stem cells. *Antioxidants & redox signaling, 9*, 293-299.

Lebovitz, R. M., Zhang, H., Vogel, H., Cartwright, J., Jr., Dionne, L., Lu, N., Huang, S. & Matzuk, M. M. (1996) Neurodegeneration, myocardial injury, and perinatal death in mitochondrial superoxide dismutase-deficient mice. *Proc Natl Acad Sci U S A.*, *93*, 9782-7.

Li, Y., Huang, T. T., Carlson, E. J., Melov, S., Ursell, P. C., Olson, J. L., Noble, L. J., Yoshimura, M. P., Berger, C., Chan, P. H., Wallace, D. C. & Epstein, C. J. (1995). Dilated cardiomyopathy and neonatal lethality in mutant mice lacking manganese superoxide dismutase. *Nat Genet.*, *11*, 376-81

Luo, J., Nikolaev, A. Y., Imai, S., Chen, D., Su, F., Shiloh, A., Guarente, L., & Gu, W. (2001). Negative control of p53 by Sir2alpha promotes cell survival under stress. *Cell, 107*, 137-148.

Matoba, S., Kang, J. G., Patino, W.D., Wragg, A., Boehm, M., Gavrilova, O., Hurley, P. J., Bunz, F., & Hwang, P. M. (2006). p53 regulates mitochondrial respiration. *Science,* (New York, NY *312*, 1650-1653.

Nemoto, S., Fergusson, M. M., and Finkel, T. (2005). SIRT1 functionally interacts with the metabolic regulator and transcriptional coactivator PGC-1{alpha}. *The Journal of biological chemistry, 280*, 16456-16460.

Perez, V. I., Lew, C. M., Cortez, L. A., Webb, C. R., Rodriguez, M., Liu, Y., Qi, W., Li, Y., Chaudhuri, A., Van Remmen, H., Richardson, A., & Ikeno, Y. (2008). Thioredoxin 2 haploinsufficiency in mice results in impaired mitochondrial function and increased oxidative stress. *Free Radic Biol Med.*, *44*, 882-92

Schriner, S. E., Linford, N. J., Martin, G. M., Treuting, P., Ogburn, C. E., Emond, M., Coskun, P. E., Ladiges, W., Wolf, N., Van Remmen, H., et al. (2005). Extension of murine life span by overexpression of catalase targeted to mitochondria. *Science,* (New York, NY *308*, 1909-1911.

Sinclair, D. A., & Guarente, L. (2006). Unlocking the secrets of longevity genes. *Scientific American, 294*, 48-51, 54-47.

Trifunovic, A., Wredenberg, A., Falkenberg, M., Spelbrink, J. N., Rovio, A. T., Bruder, C. E., Bohlooly, Y. M., Gidlof, S., Oldfors, A., Wibom, R., et al. (2004). Premature ageing in mice expressing defective mitochondrial DNA polymerase. *Nature, 429*, 417-423.

Van Remmen, H. & Richardson, A. (2001) Oxidative damage to mitochondria and aging. *Exp Gerontol.*, *36* p957-68

Warburg, O. (1930). The metabolism of tumors. *Constable*, London.

INDEX

A

B

C

D

E

F

G

H

I

J

K

N

O

P

S

T

U

V

W

X

Y

Z